ADEQUATE FOOD FOR ALL

Culture, Science, and Technology of Food in the 21st Century

ADEQUATE FOOD FOR ALL

Culture, Science, and Technology of Food in the 21st Century

Edited by

Wilson G. Pond
Cornell University
Ithaca, New York, U.S.A.

Buford L. Nichols
Baylor College of Medicine
Houston, Texas, U.S.A.

Dan L. Brown
Cornell University
Ithaca, New York, U.S.A.

CRC Press
Taylor & Francis Group
Boca Raton London New York

CRC Press is an imprint of the
Taylor & Francis Group, an **informa** business

CRC Press
Taylor & Francis Group
6000 Broken Sound Parkway NW, Suite 300
Boca Raton, FL 33487-2742

© 2009 by Taylor & Francis Group, LLC
CRC Press is an imprint of Taylor & Francis Group, an Informa business

No claim to original U.S. Government works
Printed in the United States of America on acid-free paper
10 9 8 7 6 5 4 3 2 1

International Standard Book Number-13: 978-1-4200-7753-7 (Hardcover)

Library of Congress Cataloging-in-Publication Data

Adequate food for all : culture, science, and technology of food in the 21st
   century / editors, Wilson G. Pond, Buford L. Nichols, and Dan L. Brown.
      p. cm.
   Includes bibliographical references and index.
   ISBN 978-1-4200-7753-7 (alk. paper)
   1. Food supply. 2. Agriculture. 3. Nutrition. I. Pond, Wilson G., 1930- II.
Nichols, Buford Lee, 1931- III. Brown, Dan L. IV. Title.

HD9000.5.A323 2009
338.1'9--dc22   2008040754

Visit the Taylor & Francis Web site at
http://www.taylorandfrancis.com

and the CRC Press Web site at
http://www.crcpress.com

We dedicate this book to the world's farmers as stewards of the land and soil and of the plants and animals that provide food for the rest of us.

Contents

SECTION I Food and Agriculture in Human History

SECTION II Foods by Choice

SECTION III The Required Nutrients

SECTION IV   Foods and Health

SECTION V   Food Production: Synergy of Science, Technology, and Human Ingenuity

SECTION VI   Global Food Security

Foreword

The timeliness of this book is obvious: Rice prices tripled between January and May 2008, and overall grain prices increased by 60% between 2006 and 2008 (World Bank, 2008). The rapid rise in food prices is linked to increased energy costs; rapid economic growth in Asia, especially in China and India, which has shifted food consumption patterns; urban migration; demand for grain for biofuel production; and decreased cereal supplies due to reduced planting in the world's bread baskets and adverse weather (Von Braun, 2007). The World Bank predicts that maize prices will remain more than 50% higher than 2004 levels through 2015, with similar projections for other grains. Food riots by people desperate because of their inability to feed their families have occurred in several countries in Africa and Asia. Providing *food security*, defined as "physical and economic access to sufficient, safe and nutritious food to meet dietary needs and food preferences for an active and healthy life" (World Food Summit, 1996) remains one of the greatest challenges of our time. This volume on nutrition and the food system covers many of the issues that we must address if we are to meet this goal.

Too often, we look within narrow academic disciplines in nutrition and agriculture for solutions to hunger and malnutrition without looking at the history of food and cultural beliefs, the social dimensions of food. In the opening chapters of this book, our dietary evolution from hunters and gatherers to agriculturalists is described with consideration of both the benefits and disadvantages of the resulting dietary changes. Although nutritionists usually emphasize dietary adequacy and the effects of what we eat on health and productivity, food plays many other important roles in our lives. All cultures have foods associated with holidays and feasts. Society must consider the cultural and religious aspects of food as we address malnutrition and embrace the goal of providing adequate food for all.

In many parts of the developing world, undernourished children live in households with their obese parents, who are at risk for hypertension and type 2 diabetes. Rapid increases in consumption of white sugar, refined flour, and oil contribute to this apparent paradox. While undernutrition threatens the physical and mental development of millions of children in the developing world, physicians in the United States are recommending treatment of American children as young as 8 years of age for high plasma cholesterol. Both under- and overnutrition have serious effects on children's development, which are discussed in the section on foods and health. It is becoming increasingly clear that the distinction between under- and overnutrition is narrow, and much attention must be devoted to alleviating the problems associated with the "nutrition transition" (Popkin, 2006).

In Part III of this book, what our bodies actually need is explored. This has been the focus of nutritionists for generations, leading to the identification of essential vitamins, minerals, and amino acids and for ways to avoid scurvy, beriberi, pellagra, kwashiorkor, and marasmus, conditions that were common and often fatal during the nineteenth and early twentieth centuries. Approximately 41% of women and

60% of children (Stoltzfus, 2003) in sub-Saharan Africa still are battling anemia, a preventable disease caused by insufficient iron due to malnutrition, but complicated by malaria, diarrhea, parasites, and other infections. For iron and other minerals, our understanding of bioavailability has increased during the last few decades, but our prediction of how much iron will be used by tissues still is poor. Often, this is because we have overlooked the effects of processing, cooking, and disease. It is only recently, largely due to HIV/AIDS that we have come to appreciate the effect of immune system function on nutrient requirements. This section of the book provides lay nutritionists and others in related disciplines with an overview of what is entailed in estimating nutrient requirements and supply, the primary task of many nutritionists, and considers how disease, activity, and physiological status modulate these requirements.

At Cornell, for the last 2 years we have had a group of students and faculty in nutrition and agriculture called FANG, the Food, Agriculture, and Nutrition Group. Part of the impetus for the formation of this group was a conference in which there was a heated debate about whether food-based nutritional solutions were effective. At face value, this is a peculiar question: If food is not the primary solution to malnutrition, what will we be eating? The basic issue is how best to supply micronutrients to populations at risk of deficiencies. Is supplementation the answer, or should we rely on fortification or dietary diversity? The answer to these questions depends on the situation at hand. In cases of acute deficiency, supplementation likely is most appropriate, but planning to ensure that diets routinely provide needed nutrients also is necessary. In rural areas of Africa, Asia, and Latin America, people often have neither economic nor physical access to vitamin and micronutrient supplements or to foods fortified after harvest. Biofortification through plant breeding also can remedy deficiencies, but this approach can accompany dietary diversity, which includes legumes, fruits, vegetables, and modest quantities of animal source foods. We are just beginning to understand the health benefits of dietary antioxidants and other nutriceuticals likely to be present in intact foods, but lacking in most supplements.

The political decision to support maize production, but not that of green leafy vegetables, has both dietary and environmental implications. High-fructose corn syrup, which in 1970 comprised less than 1% of the sweetener consumed in the United States and now is 42% of the sweetener market, has been linked to obesity (Bray et al., 2004). The economic advantage of high-fructose corn syrup over cane or beet sugar may be due in part to government agricultural subsidies.

As energy prices increase and supplies dwindle, we will need to consider food miles, how far food travels from the field to where it is consumed. Importation of out-of-season fruits likely will become prohibitively expensive. What will become of the farmers in Central America, Ghana, and Kenya whose livelihoods depend on sending their produce to the supermarkets of the north if these markets disappear?

Climate change will affect crop production with regional effects on food security (Brown and Funk, 2008; Lobell et al., 2008). While our predictive models are imperfect, there is little question that we must reduce our greenhouse gas production, improve the efficiency with which we use water, and develop proactive solutions for food production in areas of the world that will be adversely affected. Likewise, in the parts of the world where soil degradation is common and serious, especially in Africa, plant production will be maintained only if soil fertility is restored. The section of this volume devoted to the environment does an excellent job of outlining both the problems and possible solutions.

The most notable contribution of this book is the single-volume breadth of the coverage of issues facing our food system. The result is a volume that will benefit farmers, nutritionists, environmentalists, policy makers, and all of us who eat. My hope is that all of these groups will read and discuss this book for it will stimulate the type of broad-ranging solutions that we will need if we are to provide food security for all.

REFERENCES

Bray, G.A., S.J. Nielsen, and B.M. Popkin. 2004. Consumption of high-fructose corn syrup in beverages may play a role in the epidemic of obesity. *Am. J. Clin. Nutr.* 79:537–543.

Brown, M.E. and C.C. Funk. 2008. Food security under climate change. *Science* 319:580–581.

Lobell, D.B., M.B. Burke, C. Tebaldi, M.D. Mastrandrea, W.P. Falcon, and R.L. Naylor. 2008. Prioritizing climate change adaptation needs for food security in 2030. *Science* 319:607–610.

Popkin, B.M. 2006. Global nutrition dynamics: The world is shifting rapidly toward a diet linked with noncommunicable diseases. *Am. J. Clin. Nutr.* 84:289–298.

Stoltzfus, R.J. 2003. Iron deficiency: global prevalence and consequences. *Food Nutr. Bull.* 24:S99–S103.

Von Braun, J. 2007. The world food situation: new driving forces and required actions. *Food Policy Report* 18:1–27.

World Bank. 2008. *G8 Hokkaido-Tokyo Summit: Double Jeopardy: Responding to High Food and Fuel Prices.* Washington, DC: World Bank.

World Food Summit. 1996. The Rome declaration on world food security. *Pop. Dev. Rev.* 22:807–809.

Alice N. Pell[*]

[*] Alice N. Pell is vice provost for international relations at Cornell University, Ithaca, New York. She received the A.B. degree at Radcliffe College and, after service in the Peace Corps, earned an Ed.M. degree at Harvard University affiliated with the Center for Studies in Education and Development and M.S. and Ph.D. degrees in animal science at the University of Vermont. She has been on the faculty at Cornell University since 1990, serving in the Department of Animal Science and, since 2005, in the College of Agriculture and Life Sciences as Director of the Cornell International Institute for Food, Agriculture, and Development, prior to her appointment as vice provost for international relations in 2008.

Preface

Food is the sustenance of life. Throughout history, its availability has been affected by many factors, including population growth, technology, cultures, religion, economics, politics, education, and weather. This book looks at these factors and examines the myriad influences on the food supply today and the human responses to them. The expectation is that, through lessons learned, adequate food will be made available to the whole world in the future.

Anthropologists and other scholars have chronicled human evolutionary background, including the early history of dietary changes during our evolution, and written evidence of foods common in the early human diet is recorded in the Bible and other religious works. In this context, the continuum of preagricultural hunting and gathering and the progression of early agriculture and food production to current science and technology provide a rich background of information about the forces that have operated through the centuries.

Access to affordable food is the norm for some, an ongoing aspiration for others, but not a reality for all. We hope our interdisciplinary approach involving the joint efforts of chapter authors representing diverse expertise will challenge the reader to work toward finding solutions to these inequities. Recognition of food as a universal need and a common interest of people everywhere may be a point of union for the human spirit. The future holds opportunities and imperatives in the twenty-first century. This is our central theme and the reason for this book.

Six areas are addressed as the following sections: Food and Agriculture in Human History; Foods by Choice; The Required Nutrients; Foods and Health; Food Production: Synergy of Science, Technology, and Human Ingenuity; and Global Food Security. Experts in specific subject matter fields have been selected to author chapters covering topics in each of the six areas.

Most chapters end with a Conclusions section and perhaps a Study Topics section, the latter intended for use by college undergraduate students using the book as a course text.

"Feed me with the food that is needed to me."

Proverbs 30:8

Acknowledgments

We thank John Sulzycki, Senior Editor, Pat Roberson, Project Coordinator, and Gail Renard, Project Editor, at Taylor & Francis for their support, chapter authors for diligence and cooperation during the preparation of the book, Joanne Parsons for technical support throughout the process, and Marsha Pond for her insight and ideas.

Wilson Pond

Buford Nichols

Dan Brown

About the Editors

Wilson G. Pond was born in Minneapolis, Minnesota, February 16, 1930. He received his B.S. degree (1952) from the University of Minnesota and his M.S. degree (1954) and Ph.D. (1957) from Oklahoma State University. He served in the U.S. Army from 1954 to 1956. He began his career in 1957 at Cornell University, where he served successively as assistant professor, associate professor, and professor in the Department of Animal Science with a joint appointment on the faculty of the Graduate School of Nutrition. In 1978, he was named nutrition research leader at the Roman L. Hruska U.S. Meat Animal Research Center, Clay Center, Nebraska, and adjunct professor, Department of Animal Science, University of Nebraska, Lincoln. In 1990, he became a research animal scientist at the USDA Children's Nutrition Research Center, and adjunct professor, Department of Pediatrics, Baylor College of Medicine, Houston, Texas. In 1997, he returned to Cornell as a visiting professor, Department of Animal Science.

During his career, he has served as faculty committee chair or member of many M.S. and Ph.D. degree candidates in animal nutrition from a total of nine countries on six continents. Interdisciplinary nutrition research completed with his students and collaborators has been published in 325 refereed journal articles. He has authored, coauthored, or edited 18 books and 23 book chapters and has served as editor-in-chief of the *Journal of Animal Science* (1975–1978) and as co-editor with Alan W. Bell of the *Encyclopedia of Animal Science* (2005–2008). He is a past president of the American Society of Animal Science (ASAS) and has been the recipient of the American Feed Industry Association Research Award, ASAS (1969); the Gustav Bohstedt Award in Mineral Nutrition Research, ASAS (1979); the F.B. Morrison Award, ASAS, 1985; the Oklahoma State University Outstanding Graduate Student Alumni Award (1987); the USDA-ARS Outstanding Research Scientist of the Year Award (1994); and the ASAS Foundation Appreciation Club Award (2008). He is a fellow of ASAS, the American Association for the Advancement of Science (AAAS), and the American Society of Nutrition (ASN). Pond has one son, Kevin, who is professor and chair of the Department of Animal and Food Sciences at Texas Tech University, and has three grandchildren.

Buford Nichols M.S., M.D., is professor of pediatrics at Baylor College of Medicine (BCM). His interest in food and nutrition began in childhood when, as the child of American missionaries, he witnessed famine conditions in China. After completing his pediatric residency at Yale, he joined the pediatrics department at Baylor College of Medicine (BCM) in 1964 with the understanding that as associate director of the National Institutes of Health (NIH) General Clinical Research Center, he could do research in nutrition and gastroenterology. His primary BCM research was with malnourished infants with persistent diarrhea and clinical carbohydrate intolerance. The development of total parenteral nutrition (TPN) in 1989 proved to be life saving for these infants suffering from secondary malnutrition due to severe mucosal

damage and persistent diarrhea. This clinical success with TPN led to the establishment of the Section of Pediatric Gastroenterology and Nutrition on January 1, 1970,
at Texas Children's Hospital (TCH). Nichols was the founding chief of the gastrointestinal service. Between 1968 and 1971, clinical research was carried out on altered
energy metabolism and body potassium in infants with primary malnutrition living
in Houston, Jamaica, Mexico, and Guatemala. In 2008, the Gastroenterology and
Nutrition Service of TCH was recognized as among the top four in the United States.

Creation of the U.S. Department of Agriculture Agricultural Research Service,
Children's Nutrition Research Center (USDA/ARS CNRC) at BCM and TCH was
announced on November 2, 1978, and Nichols became the founding director. On
July 27, 1984, the House Agriculture Rural Development and Related Agencies
Subcommittee marked up funds for the construction of a facility for the CNRC "based
upon proximity to Baylor College of Medicine and Texas Children's Hospital," not-
ing that, "These institutions had conducted nutrition research for the past 20 years
and will provide ready access to newborn and maternity care and to pregnant and
lactating women and their unborn and newly born children." The early programs of
the CNRC were focused on growth and body composition; human lactation biology;
weaning transition; energy requirements; and protein requirements of normal lactating women and their infants. There have been more than 3000 peer-reviewed publi-
cations or invited reviews published by the CNRC scientists. Data from the CNRC
have been used to revise national and international Recommended Food Intakes for
infants and children. On July 1, 1992, Nichols became CNRC director emeritus and
returned to the laboratory to study the molecular aspects of human carbohydrate
digestion. The gene for rapid mucosal starch digestion, maltase-glucoamylase, was
cloned and expressed in recombinant form. The gene was ablated in mice to determine the interaction with other enzymes involved in starch and sugar digestion.

While at BCM and TCH for 45 years, Nichols had several recognitions. He served
as the president of the American College of Nutrition. He was a founding member
of the North American Society of Pediatric Gastroenterology and Nutrition and in
2002 received the Swachman Award from this society. He served on the Nutrition
Committee of the American Academy of Pediatrics for 10 years and in 1998 received
the academy's Nutrition Award. Nichols is a Fellow of the American Society of
Nutrition, for which he served as chief of the History of Nutrition Committee for 8
years. Nichols has published more than 300 peer-reviewed articles, invited reviews,
and book chapters. Nichols has coauthored eight books on child nutrition.

Dan Brown received his B.S. in animal science from the University of California
at Davis (UCD) in 1976 and was trained and certified by the UCD Agricultural
Education program to teach secondary school biology, agriculture, chemistry, and
mathematics in 1977. From 1977 until 1981, Brown studied the nutritional biochemistry of lactation in sheep and goats at Cornell University and received his Ph.D. in the
field of nutrition with minors in biochemistry and food science in 1981. Immediately
following graduate school, Dan Brown worked for Winrock International as a production systems specialist in Maseno, Kenya, during the early days of the Small
Ruminant Collaborative Research Support Program (CSRP). In Maseno, Brown
helped conduct intensive surveys of small farms in western Kenya and completed

several critical experiments concerning the nutritional constraints and their solutions as the CRSP team created the dual-purpose goat production systems for which it became known.

In 1983, Brown accepted a faculty position in the Animal Science Department at UCD, where he spent the next 11 years conducting research concerning the control of body nutrient reserves, limited resource animal agriculture, and nutritional toxicology. He also taught introductory animal science, upper division animal nutrition, sustainable animal agriculture, and smaller graduate courses in nutritional toxicology and international agriculture. During this time, he also conducted collaborative work in Kenya and with a series of foreign students and international scholars from Kenya, Spain, Ivory Coast, Hungary, Palestine, and Pakistan.

In 1994, Brown joined the Cornell Animal Science faculty. From 1994 to 2000, he spent half his time with nutritional toxicology research and teaching and half with youth extension. Then, from 2001 through 2003 Brown traded in the half-time extension appointment for a half-time secondment to the International Livestock Research Institute in Debre Zeit and Addis Ababa, Ethiopia. There, he conducted work to solve toxicological constraints to feeding livestock tree legumes and mycotoxin-contaminated grains. Now a full-time Cornell faculty member again, Brown has taught a graduate class in nutritional toxicology, introductory animal science courses, and a new comparative nutrition course. In the course of his research career, Brown has made a number of discoveries concerning the nutrition and toxicology of sheep, goats, cattle, mice, rabbits, swine, black rhinos, horses, and white tail deer and has conducted several studies with humans concerning toxicology and body composition control.

Currently, Brown has an active USAID-funded program in Haiti conducting investigations aimed at eliminating aflatoxin from the production chain behind the local manufacture of ready-to-use therapeutic foods for malnourished children and HIV patients.

Contributors

G. Eric Bradford
(deceased)
Department of Animal Science
University of California
Davis, California

Mindy Brashears
Department of Animal and Food
    Sciences
Texas Tech University
International Center for Food
    Industry Excellence
Lubbock, Texas

Dan L. Brown
Cornell University
Ithaca, New York

Dustin J. Burnett
Tufts University–Friedman School
    of Nutrition Science and Policy
Frances Stern Nutrition Center
Tufts–New England Medical Center
Boston, Massachusetts

Johanna T. Dwyer
Tufts University–Friedman School
    of Nutrition Science and Policy
Frances Stern Nutrition Center
Jean Mayer USDA Human
    Nutrition Research Center on Aging
Tufts–New England Medical Center
Boston, Massachusetts

Donald C. Erbach
Engineering/Energy
U.S. Department of Agriculture
Agricultural Research Service
Beltsville, Maryland

John W. Finley
Department of Food Science
Louisiana State University
Baton Rouge, Louisiana

Barbara Elaine Golden
International Child Health
Department of Child Health
University of Aberdeen
Royal Aberdeen Children's Hospital
Aberdeen, Scotland

Miguel I. Gómez
Department of Applied
    Economics and Management
Cornell University
Ithaca, New York

Monika Grillenberger
Max Rubner–Institut
Federal Research Institute of
    Nutrition and Food
Karlsruhe, Germany

Summer Hamide
Schools of Law and Public Health
University of California, Los Angeles
Los Angeles, California

Larry W. Harrington
CGIAR Challenge Program on
    Water and Food and
Department of Crop and Soil Sciences
Cornell University
Ithaca, New York

Gail G. Harrison
Department of Community
  Health Sciences
UCLA School of Public Health
UCLA Center for Health Policy
  Research
Los Angeles, California

Peter R. Hobbs
Department of Crop and Soil Sciences
Cornell University
Ithaca, New York

Cindy M. Imai
Tufts University–Friedman School
  of Nutrition Science and Policy
Frances Stern Nutrition Center
Tufts–New England Medical Center
Boston, Massachusetts

Roberto César Izaurralde
Joint Global Change Research Institute,
Pacific Northwest National
  Laboratory, and
University of Maryland
College Park, Maryland

Jeffrey Stanton Kellam
Presbytery of Geneva
Owego, New York

R. Lal
Carbon Management and
  Sequestration Center
The Ohio State University
Columbus, Ohio

Clark Spencer Larsen
Department of Anthropology
The Ohio State University
Columbus, Ohio

Paul E. McNamara
Department of Agricultural and
  Consumer Economics
University of Illinois at
  Urbana–Champaign
Urbana, Illinois

Julie M. Meeks Gardner
Caribbean Child Development Centre
The University of the West Indies,
  Mona
Kingston, Jamaica

Buford L. Nichols
U.S. Department of Agriculture
Agricultural Research Service
Children's Nutrition Research
  Center and
Department of Pediatrics
Baylor College of Medicine
Houston, Texas

Charles F. Nicholson
Department of Applied
  Economics and Management
Cornell University
Ithaca, New York

Kevin R. Pond
Department of Animal and Food
  Sciences
Texas Tech University
Lubbock, Texas

Wilson G. Pond
Department of Animal Science
Cornell University
Ithaca, New York

Roberto Quezada-Calvillo
CIEP–Facultad de Ciencias Quimicas
Universidad Autonoma de San
  Luis Potosí
San Luis Potosí, Mexico

Stacey Rosen
Economic Research Service
U.S. Department of Agriculture
Washington, D.C.

Shahla Shapouri
Economic Research Service
U.S. Department of Agriculture
Washington, D.C.

Tyler Stephens
Alberta Agriculture and Food
Agriculture Research Division
Agriculture Centre
Lethbridge, Alberta, Canada

B. A. Stewart
Dryland Agriculture Institute
West Texas A&M University
Canyon, Texas

Duane E. Ullrey
Comparative Nutrition Group
Departments of Animal Science
    and Fisheries and Wildlife
Michigan State University
East Lansing, Michigan

Susan P. Walker
Tropical Medicine Research Institute
The University of the West Indies,
    Mona
Kingston, Jamaica

Wallace W. Wilhelm
U.S. Department of Agriculture
Agricultural Research Service
Lincoln, Nebraska

Section I

Food and Agriculture
in Human History

1 Emergence and Evolution of Agriculture:
The Impact on Human Health and Lifestyle

Clark Spencer Larsen

CONTENTS

ABSTRACT

Within a remarkably short period at the close of the Ice Age, humans began to domesticate plants and animals. The cause of the shift from hunting and gathering to agriculture was complex. The outcome and implications for the health and well-being of humans have been debated for centuries. Some authorities regard it as the great leap forward, and certainly, the general public would agree. This chapter addresses the traditional model of the outcome of agriculture for human populations by reviewing what physical anthropologists have learned from ancient skeletal remains around the world. In general, this record suggests that the foraging-to-farming transition occasioned a decline in health but was accompanied by a decline in workload, at least in some settings. That is, there is strong evidence for decrease in dental and oral health, increase in infection and infectious disease, and change in skeletal and cranial morphology. The agricultural revolution that began 10 to 12 thousand years

3

ago set the stage for health and other quality-of-life issues that we are experiencing in the twenty-first century, brought about by the continuation of dietary and nutritional changes, population increase, and disease generally in a rapidly changing world.

INTRODUCTION

Imagine for a moment what our lives in the twenty-first century would be like without domesticated plants and animals,[1] especially the kind that provide the foods we eat on a daily basis. These provide all form and substance of everything we eat, ranging from the tossed salad at lunch to the baked chicken at dinner. Bottom line: today, human beings are *completely* dependent on domesticated species of plants and animals—thousands of them—for the food we eat.

It has not always been this way. In fact, domestication did not start until about 10 to 11 thousand years ago. Prior to that, *all* foods were nondomesticated. All foods were caught, trapped, hunted, fished, or gathered. The dramatic shift in the manner in which food was acquired, processed, and consumed was a huge turning point in human evolution, the shift from foraging (or hunting and gathering[2] as many anthropologists call it) to domestication, especially farming, was a transformative event in human evolution, along with bipedalism, speech, dependence on material culture, and other features that are uniquely human.

Growing up in a small town in southeastern Nebraska in the heart of the corn belt, I had been raised with the notion that agriculture was important in the lives of humans, and for the better. After all, corn was the centerpiece of the economy that fueled my state. When I went off to college to study as an anthropology major, in my archaeology class I was thrilled to read a textbook written by one of the world's leading archaeologists, Robert J. Braidwood, that agriculture was indeed central to the success of humanity. Agriculture, in his view, relieved humans from the drudgery of life before agriculture for the hunter-gatherer lifestyle was "a savage's existence, and a very tough one … following animals just to kill them to eat, or moving from one berry patch to another [and] living just like an animal" (1967, p. 113). From this perspective, to be a forager was to be deprived.

Does this characterization of the shift from foraging to farming present an accurate picture? Study of human remains from archaeological settings—a subdiscipline of anthropology called *bioarchaeology*—has revealed important insights about the past, especially with regard to testing hypotheses about the role of agriculture (and domestication in general) in quality of life and well-being (see Larsen 1997). In this chapter, I talk about early agriculture, its origins, and what has been learned about past health and lifestyle based on the study of ancient skeletons.

THE CONTEXT FOR EARLY AGRICULTURE

At the end of the Pleistocene epoch—10 to 12 thousand years ago—Earth's climate became relatively warmer and wetter compared to the cold, dry climate of the preceding 2 million years (Smith 1998). This period of global warming marked the beginning of the Holocene, the geological epoch that we live in today. It is during this climate shift that people began to control the growth cycles of some plants and animals, the process that anthropologists call *domestication*. To be sure, there are some groups in the world today that still hunt and gather, but these are a very small minority of the human population, well under 0.1% of the more than 6 billion Earth inhabitants. It is amazing to think that for nearly all of human evolutionary history, humans had eaten plants of all kinds, but that they had never planted and then harvested them. It is even more amazing to see how rapidly the transition from foraging to farming took place—an eyeblink in human evolution!

We know that the change to farming began quickly, but this is not to say that it happened overnight. Rather, it evolved over a period of centuries, but still fast in light of the 6 to 7 million years that humans and human-like ancestors were committed to foraging and hunting prior to the Holocene. In addition to when plant domestication first took place, authorities have developed a reasonably good understanding of where it occurred. In this regard, it appears not to have been a one-time event, then spreading around the world from a single center. Rather, plant domestication started as an independent event in at least 10 or 11 places around the world, including Asia, Africa, North America, and South America (Smith 1998). It was slow to develop in some of these places, taking several thousand years to gain a dominant place in diet, but a matter of decades in other localities. From these "primary" centers of domestication, it then spread from place to place. With the exception of Australia, every habitable continent saw the change begin to take place by at least by 5000 years ago.

Among the earliest centers of domestication was a region of southwestern Asia known as the Levant (Smith 1998). In the Levant, people began to harvest and process the wild ancestors of wheat, barley, and rye. These grains began to supply food for the growing populations that were beginning to form permanent and semipermanent villages. Around 11,000 years ago, these plants were fully domesticated. Within a few thousand years, agriculture—especially cultivation of wheat, barley, and rye—was well in place across the Fertile Crescent, a large region of grasslands and open woodlands of the eastern Mediterranean region. This was accompanied by the appearance of numerous small villages, some which grew into towns and cities, such as Jericho and Çatalhöyük in the Palestine territories and Turkey, respectively. Plant domestication may have been as early in China (millet and rice) as in the Levant. Other early domesticated plants are documented in Mexico (bottle gourds, 10,000 years ago; corn, 6,300 years ago); New Guinea (taro and banana, 7,000 years ago); eastern North America (squash, sunflower, and goosefoot, 6,000 years ago); South America (potatoes, sweet potatoes, and manioc, 5,250 years ago); and Africa below the Sahara Desert (sorghum and yams, 4,500 years ago).

What explains this dramatic change in how humans went about getting food? Some authorities believe that the disappearance of very large animals—such as the now-extinct mammoths—at the end of the Pleistocene created circumstances whereby people had to turn to something else for food, forcing them to domesticate plants and animals. This explanation seems rather far-fetched, however, especially given the abundance of other wild animals (and a variety of plants) available for consumption (Smith 1998). Alternatively, the foraging-to-farming transition may have been motivated by humans' seeking a food that could be more predictably acquired and stored, thereby reducing risk of not having enough to eat. Domesticated plants fit the bill in terms of this kind of predictability (Winterhalder and Goland 1993, Smith

1998). Although the quest for a more predictable food source may have been a factor in explaining agricultural origins, archaeologists are learning that the farming transition was likely not monocausal. Instead, it involved the confluence of a number of factors, including climate change, the evolution of new plants and animals adapted to the warmer and wetter Holocene landscapes, availability of water, and local knowledge of wild plants and animals whose growth cycles were well understood (Smith 1998).

Whatever the "root" cause of plant domestication, the implications of the farming transition are large and numerous: It was a new way of living, a new way of settling the landscape, and exposure to new foods with dramatically different nutritional content, all of which led to fundamental changes in health and lifestyle. For example, diets became generally less varied, more focused on fewer foods, and reduced consumption of meat and availability of protein and key micronutrients (e.g., iron; Larsen 1995, 2003).

OUTCOMES FOR HUMANITY

As the opening of this chapter indicated, most assume that the shift from foraging to farming was a positive event in human history. Indeed, by some measures it was the "great leap forward." If we measure success for humanity by the population size before and after agriculture, the event was a wonderful success. That is, at 10 thousand years ago, the world was populated by no more than 2 or 3 million people. By 2 thousand years ago, population had increased to 250 to 300 million people. The first billion mark was hit in the mid-nineteenth century, and today it is well over 6 billion and counting. Agriculture also created the economic basis for the rise of complex societies and the world's great civilizations. These certainly are measures of success, but there are other outcomes for humanity that suggest that the transition to farming was a mixed blessing. For example, as population increased, so did competition for increasingly fewer resources (e.g., arable land for raising crops). As a result, organized warfare developed. While the evidence for interpersonal violence has great antiquity, the level of violence in early humans is nothing in comparison to what plays out in the early civilizations of southwest Asia, Central America, or South America. There were also environmental consequences, including degradation of landscapes (e.g., van der Leeuw 1998, Krech 1999). Dense human settlement has contributed to soil erosion, which inhibits production of food. Overgrazing by sheep and goats also contributed to soil erosion. Exploitation of forest for wood for fuel and construction rapidly stripped vast swaths of landscapes wherever agriculture, towns, and cities arose. Finally, anthropologists have also shown how human population size has impacted diversity of animal species around the world, apparently beginning thousands of years ago but accelerating today (McKee 2003).

One of the best records for gauging the impact of the agricultural transition is from human remains—skeletons and teeth—recovered from sites around the world. Human remains provide a fund of data on health and lifestyle, and the study of this important record gives us a detailed picture of what life was like in the Holocene, especially with relation to its impact on quality of life and well-being.

HEALTH AND THE AGRICULTURAL REVOLUTION

Two key things stand out in the shift from foraging to farming that have implications for health. First, human population size increased wherever the event occurred. Moreover, these populations shifted from seasonal (and sometimes year-round) mobility to living in semipermanent or permanent communities. In other words, preagricultural populations were small and dispersed, whereas agricultural populations were larger, more sedentary, and living in closer, more crowded conditions. Second, the quality of diet changed. Prehistoric foragers generally ate a more varied diet, and farmers ate an increasingly narrow diet. As these plants became more and more central to diet, the nutritional quality of diet declined. Thus, there are both indirect (change in pattern of settlement) and direct (change in nutritional quality) consequences of the foraging-to-farming transition.

In terms of the impact of declining mobility and population concentration, a range of skeletal indicators reveals that in many areas of the world there was an elevation in the prevalence of pathology associated with infection and infectious disease (Cohen and Armelagos 1984, Steckel and Rose 2002, Walimbe and Tavares 2002, Oxenham et al. 2005, Cohen and Crane-Kramer 2007, and others; see reviews in Cohen 1989, Larsen 1995, 2003). In addition, while evidence of chronic infectious disease was certainly present in early foragers, perhaps even with great antiquity (Kappelman et al. 2008), it is virtually nonexistent. Skeletal evidence of chronic infectious disease—such as syphilis, tuberculosis, and leprosy—are primarily associated with agricultural-based populations living in close, densely crowded communities. This appears to be the case in large part because it creates the conditions conducive to the spread of infectious disease. Most pathogens—bacteria and viruses—causing infectious disease require ready transport from host to host to survive. Close proximity to humans and marginal sanitation are perfect for these kinds of circumstances. In a wide range of archaeological settings, limb bones of skeletons display characteristic lesions called *periosteal reactions*. These lesions give the surface of the bone a rough texture and sometimes lead to noticeable elevation and expansion of the bone. The lesions are caused by local infections, such as that resulting from an infected cut or abrasion. In crowded, reduced-sanitation conditions, there is a much greater chance for bacteria (e.g., *Staphylococcus aureus*) to spread to the wound, infecting the limb, including the bone. On the other hand, any injury to the outer surface of bone can lead to periosteal reactions. However, there is a clear pattern of general increase in early agriculturalists compared to their hunter-gatherer forebears. Most of these lesions documented by anthropologists are nonspecific. We really do not know the specific cause and can only conclude that their increase signals a decline in health. Some lesions, however, are specific insofar as the pattern of infection is characteristic of a specific infectious disease. For example, in eastern North America, tibia (lower leg) bones are swollen and bowed, and skulls have distinctive cavitations that are symptomatic of a group of diseases called *treponematosis*, which includes venereal syphilis, nonvenereal (endemic) syphilis, and yaws. Various late prehistoric agricultural societies pre-dating the arrival of Columbus have lesions characteristic of a nonvenereal form of treponematosis. It is only during the early contact era soon following the arrival of Columbus that the disease became venereal. While these diseases were not caused by eating domesticated plants, they did become prevalent when the conditions were established that would promote the maintenance and spread of the pathogens that caused them, namely, close, crowded, unsanitary living conditions associated with permanent or semipermanent communities that focused their food production on crops. It is these circumstances that set in motion the later evolution of pathogens and spread of contagious diseases caused by them, such as smallpox, measles, whooping cough, black death, leprosy, and other diseases, some of which are quite recent (e.g., HIV/AIDS, severe acute respiratory syndrome [SARS], Ebola).

In addition to pathogens causing infectious disease, in many settings involving sedentary living in settlements, water sources are also easily contaminated by parasites. Humans who drink water containing parasites, such as hookworm, become infected. These parasites can result in severe anemia, resulting in a range of health problems, including decreased cognitive abilities and work capacity (various authors mentioned in Beard and Stoltzfus 2001). In a range of settings, prehistoric agricultural populations show pathology associated with iron deficiency anemia called *porotic hyperostosis*. Iron deficiency anemia causes the expansion of the region of the skeleton involving red blood cell production. As a result, the flat bones of the skull (and sometimes eye orbits) become more porous. The iron deficiency is caused by parasitic infection, consumption of iron-poor foods, or consumption of foods containing phytate (e.g., corn), which inhibits iron metabolism (Larsen 1997).

There are other skeletal indicators that show a decline in health in early agriculturalists. Dental caries (tooth decay) is a progressive disease process involving the demineralization of the tooth enamel and underlying structures (see Larsen 1997, Hillson 2008, for discussion of archaeological, historical, and ethnographic settings). It is caused by acids produced as a by-product of the metabolism of dietary carbohydrates (sugars especially) by naturally occurring oral bacteria (e.g., *Streptococcus mutans*). Most profound in this regard, sugars and other carbohydrates (e.g., starches) generally are cariogenic, resulting in an increase in tooth decay and various other oral problems that give way to infections and tooth loss. A trend of increasing frequency of dental decay with agriculture has been documented worldwide (various authors as mentioned in Cohen and Armelagos 1984, Steckel and Rose 2002, Cohen and Crane-Kramer 2007). These increases have been associated with various grains, but corn appears to be exceptionally cariogenic (e.g., Larsen et al. 1991). In contrast, early rice producers in East Asia show very little or no increase in caries relative to foragers (e.g., Domett 2001, Pietrusewsky and Douglas 2002, Oxenham 2006, Oxenham et al. 2005, Temple and Larsen 2007; although see Pechenkina et al. 2002). Rice-based diets appear to be less cariogenic than those involving other staple cereals (Sreebny 1983).

Perhaps the most dramatic impact of agriculture is seen in the negative effects that the transition to farming had on growth and development. The presence of enamel defects caused by growth disruption—a pathology called *enamel hypoplasia*—shows a general pattern of increase in frequency in comparison of prehistoric foragers and farmers. Enamel hypoplasia, areas of enamel deficiency usually appearing as horizontal lines or grooves, is a nonspecific stress indicator, arising from either disease or poor nutrition or a combination of both during the years of enamel development (before birth to age 12). In addition, anthropologists have documented reduced rates of growth as estimated by comparison of lengths of limb bones of children at specific ages and adult heights (various mentions in Cohen and Armelagos 1984, Steckel and Rose 2002, Cohen and Crane-Kramer 2007). Thus, early farmers tend to be shorter as both children and adults than early foragers or in comparison with modern populations with abundant nutrition. A variety of evidence gathered by nutrition scientists indicates that meat provides the only full complement of essential amino acids necessary for growth and development, not to mention a range of micronutrients (various authors as mentioned in Demment and Allen 2003). The decline in growth rate and adult height almost certainly reflects the reduced availability of high-quality nutrition associated with the dependence on plant domesticates, which lack essential amino acids and other key nutrients.

EARLY FARMING AND LIFESTYLE

One of the big debates among anthropologists (and any scholar interested in the foraging-to-farming transition) is how agriculture, once in place, improved (or not) our quality of life, including workload. Simply put, did agriculture and the farming lifestyle require more work or less work than foraging? Braidwood (1967) obviously believed that the transition occasioned a decline in workload as food sources derived from crops and domesticated animals became more predicable. About the time that Braidwood was writing about early farmers, Richard Lee, a cultural anthropologist, had begun a long-term ethnographic study of the Ju/'hoansi (!Kung) of southern Africa. He lived with this group for years, documenting the foods they ate and the amount of work time versus leisure time. Much to his surprise, he discovered that for this particular foraging group, they had loads of leisure time, and by inference, perhaps hunting and gathering may not have been all that tough (Lee 1968).

This research created a firestorm in the anthropological community and importantly set up a series of hypotheses and questions, some of which are being addressed by physical anthropologists. How can we tell from ancient skeletal remains how hard (or easy) past lifeways were? Is it possible to reconstruct workload and activity levels in the past? Indeed it is, and one avenue for understanding past lifestyles is via the study of bone structure and its links to behavior. When a person is alive, that person's bone tissue is highly plastic, especially when the bones are growing and maturing prior to adulthood. In the nineteenth century, the German anatomist Julius Wolff observed that when bone is "stressed," it increases where it is needed to resist stress. On the other hand, where stress is reduced, bone is taken away. Thus, a high level of physical activity over the course of a person's lifetime will result in bigger, stronger bones. Similarly, at the population level, a group of people who are physically active will have skeletons that are more robust than a group of people who are less physically active. Indeed, living human groups show this association between activity level and bone size. Anthropologists have also applied the science of biomechanical engineering to measure the strength of bones. Engineers measure the strength of building materials—like the "I" beams that are used to build the infrastructure of a building or bridge—based on the premise that a cross section that has material placed furthest from its center is stronger than a cross section where the material is close to the center (Ruff 2000). Physical anthropologists have modified the engineering measurements for the study of strength of limb bones in ancient skeletons. In fact, over the course of human evolution and especially since agriculture first began, there has been a reduction in bone size and bone strength (Larsen 1997). These findings strongly suggest that the foraging-to-farming transition resulted in a decline in workload. This finding is consistent with a reduction in osteoarthritis, a disorder of the joints of the skeleton resulting from excessive wear and tear owing to lifestyle and physical activity.

THE CHANGING FACE OF HUMANITY

One of the biggest misconceptions about the evolution of recent humans, say within the last 20,000 years or so, is that once humans became anatomically modern, their evolution came to an end. There is a growing body of evidence based on the study of skulls dating to the last 20,000 years, and especially the last 10,000 years, that there has been continued morphological change in the human cranium. One of the first to observe this was the eminent British anatomist and physical anthropologist, Sir Arthur Keith (1916). Using skulls from the last several thousand years in Britain, he documented a reduction in the size of the face and jaws, which he related to eating soft foods, especially cooked cereal grains. That is, following what is now known as Wolff's law, by which bone is placed where it is needed and taken away where it is not, facial bones are reduced in size.

Since the early nineteenth century, many other anthropologists have documented a reduction in the robusticity of the faces and jaws globally (e.g., Carlson and Van Gerven 1979, Larsen 1982, Walimbe and Tavares 2002, and others). Moreover, these changes are often accompanied by a general shortening and rounding of the skull. Physical anthropologists David Carlson and Dennis Van Gerven (1979) developed their *masticatory-functional hypothesis* to explain these changes. They offered that the change in skull shape and gracilization of the face and jaws was a response to decreased demands placed on the chewing muscles (temporalis and masseter) as people shifted from eating hard foods (nondomesticated plants) to soft foods (domesticated plants). Importantly, the shift to eating domesticated plants was also accompanied by the invention of ceramics, used for cooking food into soft mushes. There is now an abundant body of experimental research on laboratory animals to support their hypothesis. In control studies comparing animals fed hard-textured foods with others fed soft-textured foods, the ones with the harder diets had bigger jaws and faces than the ones with softer diets (e.g., Ciochon et al. 1997 and many others). Clearly, bone is responsive to the mechanical environment, including bone tissue associated with chewing.

The reduction in the size of the jaws has also had an impact on our teeth. One question I ask my anthropology classes is how many had to have orthodontic treatment ("braces") when they were children. In every instance, at least half of the students answer in the affirmative. This informal polling represents a cross section of the public and reflects the fact that malocclusions and malaligned teeth and jaws affect a huge portion of our population. Yet, when I look at ancient skulls, malocclusions, crooked and crowded teeth (the so-called wisdom teeth are a great example), are rare. In other words, if orthodontists were to travel back in time, they would find themselves with very little to do, at least with respect to fixing crooked teeth.

What explains this rapid growth in the incidence of malaligned teeth in the modern world? The answer appears to lie in the relationship between tooth size and jaw size. Tooth size has a much higher heritability than jaw size. That is, the teeth have a greater genetic component than bone. Teeth have decreased in size, but not nearly to the extent of bone (jaw) size in the last 10,000 years. This disjunction in reduction of tooth and jaw size has resulted in a kind of disharmony between the teeth and the supporting bone. That is, people have more crowded teeth today than in the past, at least in part, because of the greater reduction in the size of the jaws than in the size of the teeth. By all accounts, our diet is continuing to get softer. Thus, the incidence of malocclusions will likely continue to grow. Orthodontists will enjoy job security for the foreseeable future.

TRADE-OFFS AND TRANSFORMATIONS: SETTING THE STAGE FOR THE PRESENT

The skeletal record from past human populations shows a similar story of a general decline in health in early agriculturalists relative to their foraging predecessors. But, this is not to leave the impression that the shift was a full-blown disaster—far from it. On the negative side, there is clear evidence of increased infections and poorer health generally. However, in nearly every place where humans made the shift to agriculture there was a dramatic increase in population size, largely shaped by increase in birth rate (Bocquet-Appel and Naji 2006). This remarkable demographic transition is now well documented in the archaeological record, from European and North American settings especially. Importantly, this transition is an evolutionary success story in increased numbers and in those who survive to reproductive age. New genetic evidence indicates that this rapid increase in human population has also resulted in increase in human adaptive evolution in the last 10,000 years, far more than once thought (Hawks et al. 2007).

Agriculture provided the economic context for the rise of civilization and eventually the technology that we currently enjoy. The downside, of course, is that large population and its concentration provide the essential breeding ground for the maintenance and transport of pathogens from host to host, eventually infecting many thousands, if not millions, of people. Examples in recent history abound, such as the 1918 influenza epidemic that resulted in the deaths of millions globally. Today, we are seeing an alarming increase in new infectious diseases, fueled by population increase and human-to-human transfer of rapidly evolving pathogens. In addition, some of the old diseases have reemerged, such as tuberculosis, resulting in the deaths of some 2 to 3 million people a year.

The biological challenges facing humans in the early twenty-first century have their roots in the agricultural revolution of the Holocene. There are many lessons in the record that I have described in this chapter. Key among them is that a large and dense population provides the basis for infectious disease and its spread. If we are to set a better course for the future, it will be essential to get the staggering growth of our numbers under control. Moreover, the shifts in nutrition reinforce the notion that focus on a limited number of foods places populations at risk. Attention has to be given to providing foods that are diverse, providing essential nutrients that are missing in a narrow-range diet deficient in micronutrients. Finally, the Holocene dietary changes also set the stage for technology that has resulted in recent years with a strong focus on high-fat, high-carbohydrate diets. This event, combined with technology that frees us of labor (e.g., the automobile), has engendered a worldwide obesity epidemic affecting every nation on the globe. Simply, we are seeing the rapid increase in fatter children and fatter adults around the world, both in developed and developing nations (Popkin 2004). This, too, is a risk for our future, resulting in an increase in diseases, such as coronary heart disease, osteoarthritis, osteoporosis, and disability and creating a health crisis that is sure not to go away unless we can create a new culture of improved nutrition and diet. There are signs of improvement and solutions to difficult issues, such as public education about diet, but we are a long way from a solution. A solution will need to be developed, taking us beyond the heritage of the agricultural revolution.

CONCLUSIONS AND STUDY TOPICS

The key point of this chapter was to outline and discuss the impact of the transition from hunting and gathering to agriculture, a process that began around 10 to 12 thousand years ago. Eventually, the process resulted in most human populations being dependent on domesticated plants (and animals). The key findings by anthropologists about the impact on human biology are the following:

1. Rapid and remarkable increase in population size, resulting in crowded and increasingly deteriorating living conditions
2. Less varied diet and hence poorer nutrition
3. Increase in infection and infectious disease
4. Increase in iron deficiency anemia
5. Increase in dental decay (caries)
6. Poorer growth and development in childhood and reduced adult height
7. Reduction in bone size and robusticity, reflecting general decreased activity and workload;
8. Decrease in osteoarthritis (although this is highly variable and age dependent)
9. Reduced face and tooth size, increase in dental crowding, and alteration in shape of skull, reflecting decrease in chewing demand as humans shifted to eating softer foods
10. Patterns of activity and dietary intake that set the stage for the problems facing human beings in the twenty-first century (e.g., obesity, coronary heart disease, osteoporosis, and disability general at all ages)

ACKNOWLEDGMENTS

I thank Wilson Pond for inviting me to contribute to this book. The discussion in this chapter is based on a life-long interest in the history of the human condition, especially major adaptive shifts in recent human evolution, including the foraging-to-farming transition. I also thank the National Science Foundation, the National Endowment for the Humanities, the National Geographic Society, the L. S. B. Leakey Foundation, and the St. Catherines Island Foundation for research funding.

NOTES

1. Domesticated plants and animals are new varieties or species of plants and animals that evolved from existing wild varieties or species through either active or incidental selection by humans (Smith 1998, Winterhalder and Kennett 2006).
2. Hunters and gatherers obtain all their food resources from wild foods exclusively, foods that are not directly managed by humans (see Winterhalder and Kennett 2006). Hunter-gatherers today typically eat a mix of domesticated and nondomesticated foods.

REFERENCES

Beard, J. and R. Stoltzfus. 2001. Iron-deficiency anemia: reexamining the nature and magnitude of the public health problem. *Journal of Nutrition* 131 (Supplement 2).

Bocquet-Appel, J.-P. and S. Naji. 2006. Testing the hypothesis of a worldwide Neolithic demographic transition: corroboration from American cemeteries. *Current Anthropology* 47:341–365.

Braidwood, R.J. 1967. *Prehistoric Men*, 7th ed. Glenview, IL: Scott, Foresman.

Carlson, D.S. and D.P. Van Gerven. 1979. Diffusion, biological determinism, and biocultural adaptation in the Nubian Corridor. *American Anthropologist* 81:561–580.

Ciochon, R.L., R.A. Nisbett, and R.S. Corruccini. 1997. Dietary consistency and craniofacial development related to masticatory function in minipigs. *Journal of Craniofacial Genetics and Developmental Biology* 17:96–102.

Cohen, M.N. 1989. *Health and the Rise of Civilization*. New Haven, CT: Yale University Press.

Cohen, M.N. and G.J. Armelagos (Eds.). 1984. *Paleopathology at the Origins of Agriculture*. Orlando, FL: Academic Press.

Cohen, M.N. and G. Crane-Kramer (Eds.). 2007. *Ancient Health: Skeletal Indicators of Economic and Political Intensification*. Gainesville: University Press of Florida.

Demment, M. and L. Allen. 2003. Animal source foods to improve micronutrient nutrition and human function in developing countries. *Journal of Nutrition* 133 (Supplement 2).

Domett, K.M. 2001. *Health in Late Prehistoric Thailand*. BAR International Series 946. Oxford, U.K.: Archaeopress.

Hawks, J., E.T. Wang, G.M. Cochran, H.C. Harpending, and R.K. Moyzis. 2007. Recent acceleration of human adaptive evolution. *Proceedings of the National Academy of Sciences* 104:20753–20758.

Hillson, S. W. 2008. The current state of dental decay. In *Technique and Application in Dental Anthropology*, J.D. Irish and G.C. Nelson, Eds. Cambridge, U.K.: Cambridge University Press, pp. 111–135.

Kappelman, J., M.C. Alcicek, N. Kazanci, M. Schultz, M. Ozkul, and S. Sen. 2008. First *Homo erectus* from Turkey and implications for migrations into temperate Eurasia. *American Journal of Physical Anthropology* 135:110–116.

Keith, A. 1916. Is the British facial form changing? Summary of paper presented to the British Association for the Advancement of Science. *Nature* 98:198.

Krech, S., III. 1999. *The Ecological Indian: Myth and History*. New York: Norton.

Larsen, C.S. 1982. *The Anthropology of St. Catherines Island: 3. Prehistoric Human Biological Adaptation.* Anthropological Papers of the American Museum of Natural History 57 (part 3). New York: American Museum of Natural History.

Larsen, C.S. 1995. Biological changes in human populations with agriculture. *Annual Review of Anthropology* 24:185–213.

Larsen, C.S. 1997. *Bioarchaeology: Interpreting Behavior from the Human Skeleton.* Cambridge, U.K.: Cambridge University Press.

Larsen, C.S. 2003. Animal source foods and human health during evolution. *Journal of Nutrition* 133:1S–5S.

Larsen, C.S., R. Shavit, and M.C. Griffin. 1991. Dental caries evidence for dietary change: an archaeological context. In *Advances in Dental Anthropology*, M.A. Kelley and C.S. Larsen, Eds. New York: Wiley-Liss, pp. 178–202.

Lee, R.B. 1968. What hunters do for a living; or, how to make out on scarce resources. In *Man the Hunter*, R.B. Lee and I. DeVore, Eds. Chicago: Aldine, pp. 30–48.

McKee, J.K. 2003. *Sparing Nature: The Conflict between Human Population Growth and Earth's Biodiversity.* New Brunswick, NJ: Rutgers University Press.

Oxenham, M. 2006. Biological responses to change in prehistoric Viet Nam. *Asian Perspectives* 45:212–239.

Oxenham, M.F., N.K. Thuy, and N.L. Cuong. 2005. Skeletal evidence for the emergence of infectious disease in Bronze and Iron Age northern Vietnam. *American Journal of Physical Anthropology* 126:359–376.

Pechenkina, E.A., R.A. Benfer Jr., and W. Zhijun. 2002. Diet and health change with the intensification of millet agriculture at the end of Chinese Neolithic. *American Journal of Physical Anthropology* 117:15–36.

Pietrusewsky, M. and M.T. Douglas. 2002. *Ban Chiang: A Prehistoric Village Site in Northeast Thailand I: The Human Skeletal Remains.* Philadelphia: University of Pennsylvania Press.

Popkin, B.M. 2004. The nutrition transition: an overview of world patterns of change. *Nutrition Reviews* 62:S140–S143.

Ruff, C.B. 2000. Biomechanical analyses of archaeological human skeletons. In *Biological Anthropology of the Human Skeleton*, M.A. Katzenberg and S.R. Saunders, Eds. New York: Wiley-Liss, pp. 71–102.

Smith, B.D. 1998. *The Emergence of Agriculture.* New York: Scientific American Library.

Sreenby, L.M. 1983. Cereal availability and dental caries. *Community Dentistry and Oral Epidemiology* 11:148–155.

Steckel, R.H. and J.C. Rose (Eds.). 2002. *The Backbone of History: Long-Term Trends in Health and Nutrition in the Americas.* New York: Cambridge University Press.

Temple, D.H. and C.L. Larsen. 2007. Dental caries prevalence as evidence for agriculture and subsistence variation among the prehistoric Yayoi of Japan: biocultural interpretations of an economy in transition. *American Journal of Physical Anthropology* 134:501–512.

van der Leeuw, S.E. 1998. Introduction. In *The Archaeomedes Project: Understanding the Natural and Anthropogenic Causes of Land Degradation and Desertification in the Mediterranean Basin*, S.E. van der Leeuw, Ed. Luxembourg: European Communities.

Walimbe, S.R. and A. Tavares. 2002. Human skeletal biology: scope, development and present status of research in India. In *Recent Studies in Indian Archaeology*, K. Paddayya, Ed. New Delhi: Munshieam Manoharolal, pp. 367–402.

Winterhalder, B. and C. Goland. 1993. On population, foraging efficiency, and plant domestication. *Current Anthropology* 34:710–715.

Winterhalder, B. and D.J. Kennett. 2006. Behavioral ecology and the transition from hunting and gathering to agriculture. In *Behavioral Ecology and the Transition to Agriculture*, D.J. Kennett and B. Winterhalder, Eds. Berkeley: University of California Press, pp. 1–21.

2 The Evolving Knowledge of Nutrition

Buford L. Nichols and Roberto Quezada-Calvillo

CONTENTS

ABSTRACT

In Chapter 1, the origins of agriculture were traced to a period of about 10,000 years ago when the hunter-gatherer lifestyle was replaced by farming. This "green revolution" occurred in many locations and among many peoples. The diversity of foods and food compositions before and after the first green revolution suggest that fundamental genomic mechanisms exist to digest and utilize the broad spectrum of human diets (Chapters 1, 3, 4). Some of the genes that express digestive enzymes, such as amylase, are about 3.5 billion years old, as old as the first living cells. Other genes, such as enterokinase, which regulates protein digestion, are as young as 0.5 billion years old (Hedges et al., 2004). These digestive genes are part of the genome of all living organisms and are essential for maintenance, growth, and reproduction (de Duve, 2007). Their generic nature permits the present and future diversity of agriculture and food availability. As pointed out in Chapters 1 and 22, the growth of global population is already placing strains on food availability and diversity. It is my contention that greater understanding of basic human food needs, when coupled to understanding of the spectrum of food genomes, can continue to evolve and sustain global population growth and health through a new green revolution.

WHAT IS NUTRITION?

Entropy is a universal physical quantity that defines the second law of thermodynamics: Energy dissipates, maximizing the disorder in the universe. However, living organisms defy the decay into equilibrium with the environment by feeding on negative entropy and decreasing their disorder. By living, the organism maintains itself in a stationary or low level of entropy (Schrödinger, 1956, p. 73). Nutrition is the process by which the organism is continually consuming negative entropy from the environment. In humans, the ingested negative entropy consists of foods. Here, we examine the genomic history of the processes of digesting the negative entropy contained in food macronutrients: proteins, fats, and carbohydrates. In plants, the most powerful supply of negative entropy is from sunlight. All living organisms concentrate a "stream of order" from the environment to escape the atomic chaos of entropy and display the power of maintaining self and expressing orderly events (p. 75). These interwoven events are guided by genetic mechanisms that are completely at odds with the "probabilistic mechanisms" of physics to ensure the living paradigm of orderliness. Growth and reproduction are due to an "order-from-order" principle (p. 78). The genome thus defies the disorder of the physical universe. A fundamental difference between the physical and biological universes is the harmony that bridges the genome of single with multicellular organisms. By defying the laws of physics, living cells were described as "the finest masterpiece ever achieved along the lines of the Lord's quantum mechanics" (p. 83).

WHAT IS THE HISTORY OF LIFE?

Planet earth is thought to have formed about 4.5 billion years ago (bya). The common ancestor of contemporary life forms populated the earth about 3.5 bya (Pollard et al., 2008, p. 17). Biochemical features stored in all present life forms suggest that this primitive microscopic cell had about 600 genes encoding DNA, protein synthetic machinery, and a plasma membrane and with mechanisms for digesting polymeric molecules and assimilating negative entropy from the environment. Over about 1.7 bya, distinctive living species evolved. On the basis of evolutionary records, preserved in their genomes, living organisms are divided into three primary domains: archaea, bacteria, and eukarya. Genomic diversification evolved by mutation, duplication, and divergence, and lateral transfer of DNA (p. 19). Photosynthesis originated about 3 bya by symbiosis by two different families of bacteria. About 2.5 bya, a lateral transfer event brought the two genomes for photosynthesis together in cyanobacteria (blue-green algae). Sunlight energized the photosynthetic structures to activate a proton gradient used to synthesize adenosine triphosphate (ATP) and the many carbon compounds that living organisms required for negative entropy. About 2.4 bya, cyanobacteria produced most of the oxygen in the Earth's atmosphere as the product of photosynthesis. This increase in oxygen revolutionized the chemical environment for all other species of organisms.

Genomic history indicates that the present eukaryotic lineages diverged between 2 and 1 bya (Hedges et al., 2004; Figure 2.1). Cell surface membranes characterize both eukaryotes and prokaryotes; however, internal compartmentalization is lacking in prokaryotes. The external membrane of prokaryotes separates digestion of macromolecules outside the cell from internal machinery requiring these nutrients. These primitive cells exert digestion by exporting enzymes, either by secretion or extension from the cell surface, to hydrolyze complex organic structures. The products of digestion are then processed and ingested through the membrane for metabolic processes (de Duve, 2007). Compartmentalization of internal cell structures is thought to have evolved by the regional segregation of digestive enzymes on the external surface of the plasma membrane, a feature persistent in present-day bacteria. Invagination of

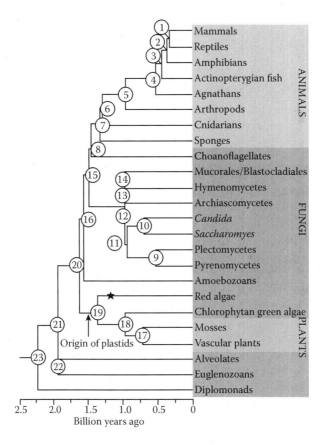

FIGURE 2.1 A timescale of eukaryote evolution. The times for each node are taken from the genomic divergence times, except for nodes 1 (310 Ma [million years ago]), 2 (360 Ma), 3 (450 Ma), and 4 (520 Ma), which are from the fossil record; nodes 8 (1450 Ma) and 16 (1587 Ma) are phylogenetically constrained and are the midpoints between adjacent nodes. Nodes 12 to 14 were similar in time and therefore are shown as a multifurcation at 1000 Ma; likewise, nodes 21 and 22 are shown as a multifurcation at 1967 Ma. The star indicates the occurrence of red algae in the fossil record at 1200 Ma, the oldest taxonomically identifiable eukaryote. Plastid (photosynthesis organelles) origin is shown with an arrow. (From Hedges, S.B., Blair, J.E., Venturi, M.L., and Shoe, J.L., *BMC Evol Biol.* 2004 Jan 28;4:2. Open Access article available at http://www.biomedcentral.com/1471-2148/4/2.)

the segregated domains of plasma membrane likely generated the membrane-bound organelles of eukarya.

One hypothesis is that cells from the prokaryotes joined in a symbiotic relationship to generate the first eukaryotes about 2.7 bya. Later, lateral transfer of the genome was surrounded by plasma membrane to become the eukaryotic nucleus. Genomic evidence has established that eukaryotes acquired mitochondria about 1.8 bya when a protobacterium became symbiotic. The bacterial genome thus contributed molecular machinery for ATP synthesis by oxidation phosphorylation. Over many centuries of evolution, most symbiotic bacterial genes moved into the host cell nucleus.

The acquisition of chloroplasts began when the cyanobacterial symbiant brought photosynthetic machinery into a primitive algal cell that had mitochondria by lateral transfer about 1.6 bya. Symbiosis evolved into interdependence when the chloroplast genes were assimilated into the nuclear genome. About 2 bya, the algal and plant branches of eukarya evolved independent strategies for multicellular existence. Further increases in organism cell number occurred about 1.5 bya (Hedges et al., 2004). Fossils confirm that animals had evolved multicellular structures by 0.6 bya. These primitive metazoans (multicell organisms) had mouth, intestine, and sensory structures. Evolution of genes for intercellular adhesion proteins pre-dated the metazoan animals. The early metazoan animals resemble contemporary animal embryos in appearance. In this period, animals diverged into the three subdivisions: mollusks, annelid worms, bracheopods, and platyhelminths (~1.3 bya); arthropods and nematodes (~1 bya); and echinoderms and chordate (~0.5 bya, humans at ~ 0.06 bya) (Hedges et al., 2004; Pollard et al., 2008, p. 17).

GENOMIC HISTORY OF DIGESTION

The full scope of nutrient metabolism is discussed in Chapters 5–8. Here, we review the evolution of the major digestive processes that demark the boundary between human negative entropy intake (food) and nutrient utilization (metabolism). The macronutrients are viewed as purveyors of essential food micronutrients. We trace the genomic history (phylogenies; Huerta-Cepas et al., 2007) of the major human digestion enzymes. As described, digestion was well established in the prokaryotic organisms at 3.5 bya and advanced to a compartmentalized system in early eukarya by 2 bya. In fossil metazoan animals, a gastrointestinal tract was evidenced by 0.6 bya.

PROTEIN DIGESTION

The central dogma of molecular biology is that DNA is transcribed into RNA, and this RNA is translated into proteins (Pollard et al., 2008, p. 251). Thus, proteins are most tightly controlled by the genome. This dogma has significance for the protein substrates and the digesting enzymes converting food into oligopeptides and free amino acids for absorption. The synthesis of all proteins is called *translation* because of the conversion of the genetic code into amino acids in the peptide chain by messenger RNA (mRNA). Small transfer RNA (tRNA) is the purveyor of specific amino acids in response to successive codons within the mRNA. The codons are made up of three nucleic acids whose sequence is transcribed from genomic DNA.

The genetic code converting the information from codons into amino acid sequence is almost universal. Thus, the four nucleotides in the genome are expanded to 64 different triplet codons, resulting in 20 specific amino acid translations. One codon specifies the start, and three specify the stop of coding. The mRNA translation takes place in ribosomes in the cytoplasm of prokaryotes and the endoplasmic reticulum (ER) of eukaryotes. Resulting soluble proteins are folded by mechanisms encoded in the sequence but are sensitive to cytoplasmic physical conditions. Transmembrane protein folding is frequently assisted by molecular chaperones that inhibit aggregation and assist sorting to cellular organelles and membrane domains such as the lumen of the digestive tract. Many cell surface proteins are glycosylated in the Golgi as a processing step; more than 200 enzymes orchestrate the addition of sugar residues to peptides. Glycoproteins are important for cell adhesion and are highly resistant to digestion.

Adult humans consume about 50 g/d of protein. About 90% of this intake is digested to absorbable peptides and amino acids in the upper intestine (Erickson and Kim, 1990; see Chapter 7). A series of different peptidases participate in this digestion process (Erickson and Kim, 1990). Gastric pepsin is an endoprotease hydrolyzing an aspartic residue in the peptide sequence (Whitcomb and Lowe, 2007). This human chromosome 11 paralogous gene with multiple isoforms is limited in expression throughout animal phylogeny, suggesting a common ancestor before differentiation of the arthropods about 1.5 bya (Benson et al., 2008). A bovine homologue, renin, is an enzyme used for making cheese by precipitation of milk caseins.

The pancreatic peptidases include trypsin, chymotrypsin, and elastin (Whitcomb and Lowe, 2007). The trypsins are a family of secreted serine endoproteases coding from nine genes on chromosome 7 and one on chromosome 9. The genes are embedded within T-cell receptor beta (TCR-B) loci of both chromosomes. The multiple-protein isoforms from chromosome 7 trypsins have redundant secreted activities; however, that from chromosome 9 is more resistant to inhibitors (Benson et al., 2008). An association with TCR-B is present throughout mammals from about 0.2 bya, but the gene itself is rooted in many bacterial species and thus was present from as long as 1.1 bya. Chymotrypsin is another secreted pancreatic serine endoprotease. It is coded from a duplicated gene located on chromosome 16 and is a neighbor of two more pancreatic elastase serine endoprotease genes on chromosome 1. The phylogeny of these chymotrypsin and elastase genes is parallel to that of trypsin, with a secreted presence in bacteria and conservation of associated genes on chromosomes on mammalian species. The pancreas also produces a family of secreted carboxypeptidase exoproteases. Carboxypeptidase A is expressed as two isoforms and cleaves terminal aromatic amino acids; carboxypeptidase B cleaves terminal aliphatic amino acids. Both are zinc-requiring metalloproteases. These reside on human chromosome 7 within a cluster of four genes, two of which are specific for the pancreas. The genomic context of these human pancreatic proteolytic enzymes is conserved within rodent chromosomes, which diverged about 0.3 bya, and the phylogenic roots extend back to bacteria that evolved about 2 bya. These secreted digestive gene products are likely rooted in the genome of the common ancestor of all cell lineages (de Duve, 2007).

The final steps of digestion of food proteins to absorbable peptides and free amino acids take place at the lumenal membrane of small intestinal enterocytes (Sterchi and Woodley, 1980a, 1980b; Sterchi, 1981; Rawlings et al., 2008). The lumenal brush border membrane anchors a series of peptidases, most of which hydrolyze terminal amino acids (Table 2.1). Enterokinase has the specialized function of activating the secreted pancreatic proteases described. It has a younger phylogeny than the remaining enzymes. The remaining peptidases are expressed in many tissues, including T cells, and are identified by CD antigen numbers. These membrane enzymes often play a role in the regulation of peptide hormone blood levels and are targets for pharmacologic inhibitors. γ-Glutamyl peptidase is a key enzyme in the glutathione cycle and plays a critical role in xenobiotic detoxification. The last five enzymes diverged before bacterial emergence (2.5 bya) and likely are descendants from membrane-bound digestive genes expressed by the first common ancestor of life (de Duve, 2007).

LIPID DIGESTION

A surrounding membrane contributes to the defiance of entropy by the living cell. This is a planar structure composed of phosphodiglycerides oriented to display hydrophilic PO_4 external extensions coating a core of hydrophobic diglyceride tails (Pollard et al., 2008, p. 113). Other lipids are also embedded into the external faces, and hydrophobic membrane proteins and transporters traverse the lipid core. The membrane is stable and relatively impermeable to ions and electrons. It is believed that the earliest life forms evolved the lipid membrane to decrease local entropy within a living cell about 3.5 bya. All sequenced genotypes have conserved enzymes that catalyze the synthesis of coenzyme A and mevalonate in the synthetic pathway to phosphodiglycerides and cholesterol (Friesen and Rodwell, 2004). By contrast, only eukaryotes synthesize neutral triglycerides, which are stored as intracellular hydrophobic droplets (Turkish and Sturley, 2007). These membrane and stored lipid classes are major contributors to food negative entropy. The disorders of atherosclerosis and obesity are thought to be influenced by quality and quantity of lipids in foods (see Chapter 6). The glycerides are not primary products of DNA transcription, as are proteins, but are products of regulated multienzyme metabolic pathways; the consequence is the synthesis of a family of di- and triglycerides with fatty acids ranging from 14 to 20 carbons in length with variable degrees of saturation.

The adult Western diet contains about 100 g/day of fat, of which more than 90% exists as triglycerides. Virtually all food triglycerides and diglycerides require lumenal small intestinal digestion before more than 95% absorption (Lowe 1997, 2002). This is accomplished by a series of lipase enzymes. Lipases are esterases that can hydrolyze acyltriglycerides into di- and monoglycerides, glycerol, and free fatty acids at a water-lipid interface. Gastric lipase is the initial digesting activity hydrolyzing position 3 (sn-3) ester linkages of the glycerides. This is a developmentally important enzyme for normal young human infants, who have a physiological delay in maturation of pancreatic lipase. The human enzyme is transcribed from chromosome 10, where it resides within a cluster of five paralogous genes. This gastric lipase chromosomal grouping is conserved in rodents. The gastric expressed gene is only found

TABLE 2.1

Selected Membrane-Bound Peptidases Active in Food Digestion in Human Small Intestine

Name	Abbreviation	Chromosome	Subunits	Functions	Divergence	EC
Enterokinase	ENTK	21q21	Yes	Activates pancreatic proteases	Bacteria ~ 2 bya	3.4.21.9
Dipeptidyl-peptidase IV	DPP IV	2q24	No	Degrades GLP-1 **CD26**	Archaea ~ 3.5 bya	3.4.14.5
Aminopeptidase N	AMPN	15q25	Yes; Zn	Degrades enkephalins **CD13**	Archaea ~ 3.5 bya	3.4.11.2
Aminopeptidase A	APA	4q25	Yes; Zn	Degrades ACE **CD249**	Archaea ~ 3.5 bya	3.4.11.7
X-Pro aminopeptidase 2	XPNPEP2	Xq25	Yes; Mn isoforms	Degrades bradykinin	Archaea ~ 3.5 bya	3.4.11.9
γ-Glutamyl transpeptidase	GGT1-6	20p11.1, 20q11.22, 22q11.21, 22q11.23	Yes; isoforms	Degrades GSH **CD224**	Archaea ~ 3.5 bya	2.3.2.2

Note: Named lymphocyte antigens are shown in **bold**. bya, billion years ago. GLP-1, glucagon-like peptide 1; Enkephalins, endogenous opioids; ACE, angiotensin1-converting enzyme; GSH, L-gamma-glutamyl-L-cysteinyl-glycine.

in eukaryotes. Pancreatic triglyceride lipase (PTL) is the second and major lipase that hydrolyzes all sn-1 and sn-3 esters of acylglycerides but not membrane lipids. Pancreatic lipase is also on human chromosome 10 within a locus of four paralogs whose relationship is conserved in rodents. One of these paralogous genes produces an 80% homologous pancreatic lipase-related protein (PLRP2), which hydrolyzes acyltriglycerides and all membrane lipids. Both pancreas-expressed genes are only found in eukaryotes. Both proteins have two domains: N-terminal and C-terminal. The N-terminal is associated with *interfacial activation*, the process of becoming active at the lipid-water interface. The function of the C-terminal is to mediate interaction with lipids. Many food components inhibit pancreatic lipase, and the pancreas secretes colipase, which conserves activity by functioning as a PTL cofactor. Colipase binds to the bile-salt covered triacylglycerol interface thus allowing the PTL enzyme to anchor itself to the water-lipid interface. Colipase gene is located on chromosome 6 and only found in eukaryotes. The locus is conserved on the mouse chromosome. In mouse knockout (KO) studies, PTL deficiency was asymptomatic, but colipase deficiency was associated with steatorrhea.

Bile salt-dependent lipase is also secreted by the pancreas. This enzyme has broad substrate specificity for all fat and membrane lipids. It is also secreted in human milk. This enzyme codes from human chromosome 9 and is conserved in archael and bacterial genomes. Phospholipase A2 (PLA2) is a secretory pancreatic enzyme that cleaves the sn-2 position of the glycerol backbone of membrane phospholipids. It is clustered with two paralogs on chromosome 1 and is also expressed on placenta, synovial membranes, and platelets. PLA2 is only expressed in eukaryotes. Knockout of the last two specific genes in mice was not associated with steatorrhea, suggesting that the overlapping substrate specificities of the various lipases are redundant, and that individual deficiencies are compensated by the remaining lipases.

CARBOHYDRATE DIGESTION

The carbohydrates are a major source of negative energy in the human diet; some rural agricultural workers consume more than 500 g/day (Robayo-Torres et al., 2006, and see Chapter 5). The amount digested in the small bowel is dependent on the species of carbohydrate fed; the range is from more than 95% down to less than 30% with digestion-resistant starches (described in the section α-Glucosidase Digestion of Starch). Food carbohydrates exist as glycosides bound to membrane proteins and lipids and as sugar units of disaccharides and glucose polymers. In the first category, the carbohydrates provide stability to the lipid-water plasma membrane (Pollard et al., 2008, p. 113). The second case provides stored energy-rich foods that ensure adequate glucose for prandial metabolism.

Milk Sugar

Lactose is the principal carbohydrate in milk (Robayo-Torres et al., 2006). It is a disaccharide composed of glucose and galactose linked as 1-β-D-galactopyranosyl-4-α-D-glucopyranose. *N*-Acetyllactosamine synthase is highly conserved, with seven paralogous genes on various chromosomes and is a component of lactose synthase along with α-lactalbumin (chromosome 12). The *N*-acetyllactosamine synthase is a highly conserved gene in all genomes. In contrast, the α-lactalbumin gene is limited to mammalian genomes but has an ancestral root as a lysosomal enzyme gene. N-Acetyllactosamine synthase is expressed in seven isoforms and plays a crucial role in protein N-glycosylations. As lactose synthase complex, these two genes are central to human and bovine lactation, for which lactose production drives the volume of milk produced.

Small intestinal mucosal lactase is the hydrolase that digests lactose to the monosaccharide units. The lactase protein is an internally duplicated enzyme that belongs to the glycosyl hydrolases family GH 1 (Benson et al., 2008). The enzyme is bound at the C-terminal to enterocyte lumenal plasma membrane and has both lactase and phlorizin hydrolase activity. The second activity also hydrolyzes β-glucosides of lipids and micronutrients such as pyridoxine-5′-β-D-glucoside and other glycosylated phytochemicals. The enzyme has a glutamic acid proton donor and glutamic acid nucleophile and a $(\beta/\alpha)_8$ barrel structure. In the human, the gene for lactase activity is downregulated at about 4 years of age, resulting in symptomatic lactose intolerance. A mutation in regulation of this gene permits lactase persistence in adults, and onset of the mutation is correlated with the development of dairy cattle about 10,000 years ago (see Chapter 1). Symptoms of lactose intolerance are treated by elimination of lactose in the diet or by oral lactase enzyme supplementation (Robayo-Torres et al., 2006).

Table Sugar

Sucrose is a disaccharide composed of glucose and fructose: α-D-glucopyranosyl β-D-fructofuranoside (Robayo-Torres et al., 2006). The sweetness, for which it is favored in candy and pastries, arises from the fructose unit. It is hydrolyzed to monosaccharides by small intestinal membrane-bound sucrase-isomaltase (SI) activity. The two subunits are both glucohydrolase family 31 α-glucosidase activities that play a prominent role in the digestion of starch. Given that sucrose only became available as a cultivated crop about 4000 years ago, the role in starch digestion is likely its primitive function. SI is bound to the membrane by the isomaltase containing N-terminal. The human SI gene is located on chromosome 2 in a context conserved among rodents. The family GH 31 genome extends throughout the archaea, bacteria, and plant kingdoms (Benson et al., 2008). In rodents, SI intestinal activities are low until the time of weaning to a starch-based diet. A human disorder, congenital sucrase-isomaltase deficiency (CSID) has been discovered with clinical symptoms similar to lactose intolerance. These patients are relieved if sucrose is removed from the diet or an oral supplement of sucrase enzyme is fed with the sugar. Some CSID patients also have symptoms when fed starch (Robayo-Torres et al., 2006).

Starch Digestion

The plant kingdom converts radiant to chemical negative energy by fixation of atmospheric CO_2 through photosynthesis by leaf chloroplasts (Quezada-Calvillo et al., 2007a, 2007b). The immediate product of carbon fixation by chloroplasts is starch, which is synthesized in light and degraded in dark cycles. The disaccharide sucrose is produced in leaf cytosol from starch-derived adenosine diphosphate (ADP)-glucose. Sucrose is then transported from leaves to amyloplasts in reproductive tissues, where it is converted to storage molecules composed of thousands of polymeric glucose units. Starches with mostly linear α-1,4 glucose linkages are known as amyloses, and those with a mixture of α-1,4 and α-1,6 branched linkages are amylopectins. In the plant, these are stored as chemical energy within semicrystalline granules whose glucose units become available during reproduction. Plant cell walls are composed of β-linked glucoside (cellulose) and nonglucose polymers that are poorly digested by the human small intestine; these carbohydrates are classed as dietary fibers. These polymers are not primary products of DNA transcription but are products of regulated multienzyme metabolic pathways; the product is synthesis of a family of starch and fiber polymers varying in size.

Starches are a gift of the vegetable kingdom to animal diets and a major food source of negative entropy. However, animal digestion is made complicated by the hundreds of botanical varieties of food starches. In contrast to the digestion of sucrose by a single mucosal enzyme, starch digestion requires a committee of six enzymes. The multiplicity of animal starch-digesting enzymes mirrors the multiplicity of the starch synthetic enzymes of plants (Quezada-Calvillo et al., 2007a, 2007b). Amylase was the first enzyme ever identified by biochemists. The activity is increased by the process of malting by sprouting grain. The same enzyme activity is found in animal salivary and pancreatic secretions. The product maltose was first discovered in malted grains, and mucosal maltase activity was found by brewers. Four different membrane-bound maltase enzyme activities of the small intestine can be identified. Two are associated with SI activities and two lack any other identifying activities (maltase-glucoamylase, MGAM). Because the four mucosal maltase enzymes hydrolyze the nonreducing end of all α-1,4 glucose oligomers to free glucose, here they are referred to as α-glucosidase activities (Table 2.2).

Amylase Solubilization of Starch

The Amylase gene (AMY) is highly represented in 615 species of bacteria and vertebrates. It is α-endoglucosidase and a member of family GH 13 (Stam et al., 2006; Benson et al., 2008). The gene is found in some archaea, suggesting differentiation before 2.5 bya. The human gene is on chromosome 2, where six paralogs are found (Quezada-Calvillo et al., 2007a). The grouping with neighboring genes is conserved in rodents. The secreted enzyme has a glutamic acid proton donor and aspartic acid nucleophile and a $(\beta/\alpha)_8$ barrel structure. It requires bound calcium and chloride ions for activity. The catalytic action is on internal α-1,4 linkages of starch granules, and the hydrolyzed products are a mixture of soluble maltodextrins. When there is no remaining hydrolysis possible, the produced oligomers are referred to as α-limit dextrins. The amylase activities produce very little free glucose during hydrolysis of starch granules (Quezada-Calvillo et al.). The redundant nature of the root gene expression of this secreted protein suggests that it coded digestion by the most primitive cells (de Duve, 2007).

α-Glucosidase Digestion of Starch

The most active mucosal α-glucosidase is the C-terminal of MGAM, which is located on human chromosome 7. This gene is in family GH 31 along with the complimentary gene SI (as discussed in the starch digestion section). The genes for GH 31 are expressed in 491 species, including archaea and bacteria (Sim et al.,

TABLE 2.2
Glucosidases Active in Food Carbohydrate Digestion in Human Small Intestine

Name	Abbreviation	Chromosome	Subunits	Functions	Divergence	EC
Lactase	LCT	2q21	Yes	Digests lactose	Archaea ~ 3.5 bya	3.2.1.108 3.2.1.23
-phlorazin hydrolase	LCT	2q21	Yes	Digests glycosides	Archaea ~ 3.5 bya	3.2.1.62
Amylase, pancreatic	AMY2A	1p21	Yes; Ca, Cl	Solubilizes 1,4 starch	Archaea ~ 3.5 bya	3.2.1.1
Amylase, salivary	AMY1A-1C	1p21	Yes; Ca, Cl	Solubilizes 1,4 starch	Archaea ~ 3.5 bya	3.2.1.1
Maltase	MGAM	7q34	Yes; isoforms	Degrades 1,4 oligomers	Archaea ~ 3.5 bya	3.2.1.20
-glucoamylase	MGAM	7q34	Yes; isoforms	Degrades 1,4 polymers	Archaea ~ 3.5 bya	3.2.1.3
Sucrase	SI	3q25	Yes; isoforms	Degrades 1,4 oligomers	Archaea ~ 3.5 bya	3.2.1.48 3.2.1.20
-isomaltase	SI	3q25	Yes; isoforms	Degrades 1,4, 1,6 oligomers	Archaea ~ 3.5 bya	3.2.1.10 3.2.1.33 3.2.1.20

bya, billion years ago.

2008; Figure 2.2). It is not possible to differentiate GH 31 substrate specificities from genome sequences (Benson et al., 2008). Both enzymes are internally dupli- cated, have two catalytic sites, and are bound to the enterocyte membrane at the N-terminal (Quezada-Calvillo et al., 2007a, 2007b). MGAM and SI both have two aspartic acid proton donors and aspartic acid nucleophiles in a $(\beta/\alpha)_8$ barrel struc- ture. Both enzymes are exoglucosidases that can hydrolyze granular starch, but the rate of glucose production is amplified fivefold if starch is also treated by AMY. The rate of α-glucogenesis from maltodextrins and α-limit dextrins by MGAM is tenfold greater than SI, but MGAM is inhibited by lumenal substrates and SI is not. This suggests that although sharing the same substrates, MGAM accelerates glucogen- esis on low-starch diets, and SI constrains glucogenesis after higher starch intakes (Quezada-Calvillo et al., 2007b).

The concept of glycemic index (GI) of foods was introduced in the 1980s (Englyst et al., 2007). The GI is calculated by comparing the rise of blood glucose of fasting

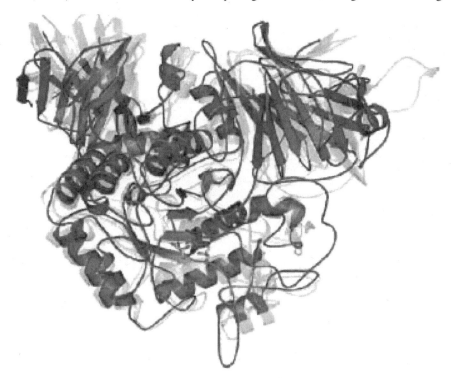

FIGURE 2.2 *A color version of this figure follows page 198.* Phylogenetic conservation of GH 31 enzyme sequences and structures. The superposition of the human N-terminal of human maltase-glucoamylase (MGAM) (red), archael YicI (cyan), and bacterial MalA (yel- low) structures is shown within a conserved $(\beta/\alpha)_8$ barrel structure. Glycerol molecules that indicate locations of glucose-binding areas in the active-site MGAM are represented as stick figures. Note the conservation of the tertiary protein structures, which diverged over more than 2.5 billion years. (From Sim, L., Quezada-Calvillo, R., Sterchi, E.E., Nichols, B.L., and Rose, D.R., *J Mol Biol.* 2008 Jan 18;375(3):782–792. With permission.)

individuals after a test meal with the rise after a glucose feeding. *Resistant starches* (RSs) have been defined as "the sum of starch and products of starch degradation not absorbed in the small intestine of healthy individuals." An in vitro analysis of rates of digestion of foods to glucose has been developed, and it is reported that rapidly available glucose (RAG) correlates with the GI (Englyst et al.). The nature of slowly available glucose (SAG) in the assay has been investigated with various starches; granular, supramolecular, and molecular structures of botanical starch species all determine the ratio of RAG to SAG (Zhang et al. 2006a, 2006b, Ao et al. 2007a, 2007b). Increasing the number of branches and reducing the length of chains increases SAG (Ao, 2007a, 2007b). In vitro mucosal α-glucogenesis of mice is slowed by experimental molecular adjustments of starch structure that are only partially restored by amylase activity (Quezada-Calvillo et al., 2007a, 2007b). Further investigations are needed to clarify the starch chemistry and enzyme molecular mechanisms behind resistant food starches because of suspected roles of starch digestion in glucose disorders such as type 2 diabetes and obesity.

SUMMARY

- The study of genomics leads to the conclusion that the human shares a common ancestry with other life-forms.
- The genomic roots of some human genes for secreted and membrane-bound digestive proteins were present in archael organisms about 3.5 bya. Others appeared when bacteria diverged from archaea about 2.5 bya. Others only appeared when eukarya appeared about 1 bya.
- By the time of divergence of the human from rodent genomes at 0.06 bya, the loci of digestive genes in chromosomal neighborhoods became fixed.
- The enzymes expressed by genes most essential for digestion and assimilation of negative energy often have redundancies at genome and activity levels for the important digestive enzymes, such as γ-glutamyl transpeptidase, which ensure normal phenotypes in the face of individual coding mutations. Where no genomic redundancy exists, as with dipeptidyl peptidase IV, mutated genes can result in clinical disorders.
- Some may express reservations about the religious implications of the genomic roots of our nutritional heritage. In the words of Francis Collins, director of the National Institutes of Health (NIH) Genome Project, "The God of the bible is also the God of the genome" (Collins, 2006, p. 211; also see Chapters 3 and 4).
- The online GenBank of the NIH Genome Project presently contains over 65 billion nucleotide bases from more than 61 million individual sequences. Over 240,000 named species are represented, and only about 16% of the sequences are of human origin. The number of complete sequences of genomes of food plants and animals is rapidly expanding (Benson et al., 2008).
- Agriculture production was critical in the past 10 to 12 thousand years as food crops were developed to provide negative entropy through phenotype selection. This process supported the growth of civilizations (see Chapters 1, 19–22).

- Given a better precision in understanding of human negative energy needs and with the expanding genomic understanding of crop phenotypes, we can envision future advances in genetically characterized foods that satisfy expanding global human needs for quality and quantity and sustain agricultural ecology (see Chapters 1, 14–22).
- Genetically modified organisms (GMOs) are already accepted in much of the world and hold the promise to resolve much of the existing discrepancies between food availability and human negative energy needs (see Chapters 20–22).

REFERENCES

Ao, Z., Quezada-Calvillo, R., Sim, L., Nichols, B.L., Rose, D.R., Sterchi, E.E., and Hamaker, B.R. Evidence of native starch degradation with human small intestinal maltase-glucoamylase (recombinant). *FEBS Lett.* 2007a May 29; 581(13):2381–2388.

Ao, Z., Simsek, S., Zhang, G., Venkatachalam, M., Reuhs, B.L., and Hamaker, B.R. Starch with a slow digestion property produced by altering its chain length branch density, and crystalline structure. *J. Agric Food Chem.* 2007b May 30; 55(11):4540–4647.

Benson, D.A., Karsch-Mizrachi, I., Lipman, D.J., Ostell, J., and Wheeler, D.L. GenBank. *Nucleic Acids Res.* 2008 Jan; 36 (Database issue):D25–D30.

Collins, F.S. *The Language of God: A Scientist Presents Evidence for Belief.* New York: Free Press Simon & Schuster, 2006.

de Duve, C. The origin of eukaryotes: a reappraisal. *Nat Rev Genet.* 2007 May; 8(5):395–403.

Englyst, K.N., Liu, S., and Englyst, H.N. Nutritional characterization and measurement of dietary carbohydrates. *Eur J Clin Nutr.* 2007 Dec; 61 (Suppl. 1):S19–S39.

Erickson, R.H. and Kim, Y.S. Digestion and absorption of dietary protein. *Annu Rev Med.* 1990; 41:133–139.

Friesen, J.A. and Rodwell, V.W. The 3-hydroxy-3-methylglutaryl coenzyme-A (HMG-CoA) reductases. *Genome Biol.* 2004; 5(11):248.

Hedges, S.B., Blair, J.E., Venturi, M.L., and Shoe, J.L. A molecular timescale of eukaryote evolution and the rise of complex multicellular life. *BMC Evol Biol.* 2004 Jan 28; 4:2.

Huerta-Cepas, J., Dopazo, H., Dopazo, J., and Gabaldón, T. The human phylome. *Genome Biol.* 2007; 8(6):R109.

Lowe, M.E. Structure and function of pancreatic lipase and colipase. *Annu Rev Nutr.* 1997; 17:141–158.

Lowe, M.E. The triglyceride lipases of the pancreas. *J Lipid Res.* 2002 Dec; 43(12):2007–2016.

Pollard, T.D., Earnshaw, W.E., and Lippincott-Schwartz, J. Evolution of life on earth. In *Cell Biology*, 2nd ed. Philadelphia: Saunders Elsevier, 2008, pp. 17–28.

Quezada-Calvillo, R., Robayo-Torres, C.C., Ao, Z., Hamaker, B.R., Quaroni, A., Brayer, G.D., Sterchi, E.E., Baker, S.S., and Nichols, B.L. Luminal substrate "brake" on mucosal maltase-glucoamylase activity regulates total rate of starch digestion to glucose. *J Pediatr Gastroenterol Nutr.* 2007a Jul; 45(1):32–43.

Quezada-Calvillo, R., Robayo-Torres, C.C., Opekun, A.R., Sen, P., Ao, Z., Hamaker, B.R., Quaroni, A., Brayer, G.D., Wattler, S., Nehls, M.C., Sterchi, E.E., and Nichols, B.L. Contribution of mucosal maltase-glucoamylase activities to mouse small intestinal starch alpha-glucogenesis. *J Nutr.* 2007b Jul; 137(7):1725–1733.

Rawlings, N.D., Morton, F.R., Kok, C.Y., Kong, J., and Barrett, A.J. MEROPS: the peptidase database. *Nucleic Acids Res.* 2008 Jan; 36 (Database issue):D320–D325.

Robayo-Torres, C.C., Quezada-Calvillo, R., and Nichols, B.L. Disaccharide digestion: clinical and molecular aspects. *Clin Gastroenterol Hepatol.* 2006 Mar; 4(3):276–287.

Schrödinger, E. *What is life? and Other Scientific Assays.* New York: Doubleday Anchor, 1956, pp. 1–88.

Sim, L., Quezada-Calvillo, R., Sterchi, E.E., Nichols, B.L., and Rose, D.R. Human intestinal maltase-glucoamylase: crystal structure of the N-terminal catalytic subunit and basis of inhibition and substrate specificity. *J Mol Biol.* 2008 Jan 18; 375(3):782–792.

Stam, M.R., Danchin, E.G., Rancurel, C., Coutinho, P.M., and Henrissat, B. Dividing the large glycoside hydrolase family 13 into subfamilies: towards improved functional annotations of alpha-amylase-related proteins. *Protein Eng Des Sel.* 2006 Dec; 19(12):555–562.

Sterchi, E.E. The distribution of brush border peptidases along the small intestine of the adult human. *Pediatr Res.* 1981 May; 15(5):884–885.

Sterchi, E.E. and Woodley, J.F. Peptide hydrolases of the human small intestinal mucosa: distribution of activities between brush border membranes and cytosol. *Clin Chim Acta.* 1980a Mar 14; 102(1):49–56.

Sterchi, E.E. and Woodley, J.F. Peptide hydrolases of the human small intestinal mucosa: identification of six distinct enzymes in the brush border membrane. *Clin Chim Acta.* 1980b Mar 14; 102(1):57–65.

Turkish, A. and Sturley, S.L. Regulation of triglyceride metabolism. I. Eukaryotic neutral lipid synthesis: "Many ways to skin ACAT or a DGAT." *Am J Physiol Gastrointest Liver Physiol.* 2007 Apr; 292(4):G953–G957.

Whitcomb, D.C. and Lowe, M.E. Human pancreatic digestive enzymes. *Dig Dis Sci.* 2007 Jan; 52(1):1–17.

Zhang, G., Ao, Z., and Hamaker, B.R. Slow digestion property of native cereal starches. *Biomacromolecules.* 2006a Nov; 7(11):3252–3258.

Zhang, G., Venkatachalam, M., and Hamaker, B.R. Structural basis for the slow digestion property of native cereal starches. *Biomacromolecules.* 2006b Nov; 7(11):3259–3266.

Section II

Foods by Choice

3 Role of Religion, Spirituality, and Faith in Food Choices

Jeffrey Stanton Kellam

CONTENTS

ABSTRACT

The chapter examines the profound personal relationship that humans have with food. Besides providing physical sustenance, there is an abundance of evidence that one looks to food to satisfy other needs. There are specific religious laws, followed by millions of people for centuries, that concern food restrictions, consumption, production and cultivation, preparation, animal care and slaughter, fasting, feasts, and prayer. Many human emotions and behaviors are intrinsically linked to food. Eating symbolic food is a ceremoniously spiritual experience, and the use of food items as metaphors for life is commonplace in our vernacular. Faith synthesizes matters of religion, individuality, and spirituality. The importance of food on our psyche is timeless and contemporary, holistic and ever modulating, and is affected by governance and secularity. The chapter intends only to highlight key areas of this complicated subject matter.

INTRODUCTION

"Eat of the things provided, and render thanks to Allah."

—Qur'an 2.172

That admonition succinctly summarizes the timeless religious attitude toward the food that sustains both body and spirit. Even those who do not affirm belief in a deity may well be thankful in heart for food that nourishes the body, brings pleasure, and engenders community. Those who do affirm, worship, and name a divinity ordinarily consider the food that sustains and enriches life as the gift or blessing of their god or gods. When, thanks to sun and rain, crops grew and bellies were filled, thanksgiving filled the heart. It could not be contained there but had to be expressed in words, song, dance, and thank offerings (Fick, 2008). When food became scarce and hunger bordered on starvation, the words, the song, the dance became petitions to the divine for help or served as acts of appeasement or repentance.

Food is more than mere physical nutrition, no matter a person's religious beliefs. It is the stuff of comfort, community, caring, and creativity. It is a matter of conscience, morality and justice, and politics. It plays a substantial role in the rituals of daily life at or away from the table. In the realm of religion or general spirituality, "Food can be sacramental: simply hold in your hand a piece of fruit," Michael Schut wrote. "Is it not a window through which you can sense your connection to soil, farmers, sunlight, rain?" (Schut, 2002). Going further, a Jewish scholar noted a helpful progression that moves from food consumption as a biological act, a means of satisfying physical hunger, to sharing food with others as a social act (whether a simple lunch or celebratory banquet), and finally a meal that represents (or is) a solemn or sacred act (Klein, 1979). That would include the Seder for Jews and Holy Communion for Christians.

A helpful study noted some 17 ways food can be understood as more than a collection of nutrients. "Food ... may be used as a gesture or a language to communicate intentions, feelings, and attitudes" (Grimm, 1996). Of that list (see Figure 3.1), at least 9 of the 17 actions noted can relate to how food is used in spiritual or religious practices or settings. For example, a given food practice may certainly initiate and maintain personal relationships (item 1); demonstrate the nature and extent of relationships (item 2); provide a focus for communal activities (item 3); express love and caring (item 4); both proclaim the separateness of a group (item 6) and demonstrate belonging to a group (item 7); symbolize an emotional experience (item 14); display piety (item 15); and express moral sentiments (item 17). Others may apply, but those listed here are not at all a stretch for people who recognize and practice either tenetless spirituality or doctrinal religion.

FOOD AND SPIRITUALITY

A nonfiction book that spent many months at the top of various best-seller lists was entitled *Eat, Pray, Love* (Gilbert, 2006). While visiting Italy, author Elizabeth Gilbert helped prepare a birthday meal for her friend Luca Spaghetti. His birthday coincided that year with America's Thanksgiving Day, which the author described as "a day of grace and thanks and community and—yes—*pleasure*." An American friend Deborah and several others gathered at the table, and after several bottles of Sardinian wine, "Deborah introduces to the table the suggestion that we follow a nice American custom ... by joining hands and—each in turn—saying what we are most grateful for. In three languages, then, this montage of gratitude comes forth, one testimony at a time."

Beyond satisfying physical hunger, a given food practice may:

1. initiate and maintain personal and business relationships
2. demonstrate the nature and extent of relationships
3. provide a focus for communal activities
4. express love and caring
5. express individuality
6. proclaim the separateness of a group
7. demonstrate belonging to a group
8. cope with psychological stress
9. reward or punish
10. signify social status or wealth
11. bolster self-esteem and gain recognition
12. wield political or economic power
13. prevent, diagnose, and treat disease
14. symbolize emotional experience
15. display piety
16. represent security
17. express moral sentiments

FIGURE 3.1   This list is based on research reported in *Community, Nutrition, and Individual Food Behavior*, M.A. Bass, L.M. Wakefield, and K.M. Kolasa, 1979.

In that brief excerpt is found what one might describe as a *spiritual* meal, full of ritual (requisite wine, hand-holding, and personal testimony) and communion, yet by no means a formal religious feast as defined by doctrine or commanded by law. While the context of that particular meal may have been "secular," the elements of food and fellowship provided an experience that fed the spirit as well as the body.

Before exploring the relationship of food and formal religious practice, law, and ritual, it is appropriate to first explore the broad and expansive territory of *spirituality*, a term that is embraced by both the religious and nonreligious. In Mortimer Ostow's *Spirit, Mind, and Brain: A Psychoanalytic Examination of Spirituality and Religion* (Ostow, 2006), the author reminds us that the very word *spirituality* has no clear-cut meaning. The term infers transcendence, a sense of awe, including what Freud referred to as "oceanic feeling," and finds expression in one's appreciation of beauty, music, art, poetry, humor, even math and logic. Free of credo, spirituality "is neither strictly related to religion or even to belief in God." It has to do with meanings and values, ideas and ideals, the belief and ethics that a person holds, and as one psychologist suggests, "deliberately lived concern of the transcendent dimension of life" (Helminiak, 2008).

Culturally speaking, it is common to refer to "team spirit," "school spirit," or the "human spirit." It is the last that is nourished by what is often referred to as *comfort food*, that is, food that brings comfort, pleasure, security, reward, or a sense of connection to a special time, place, or person. From pizza to a favorite cheese, from ice cream to anything chocolate, certain foods feed the spirit.

Further, even the secular culture engages in its *rituals* involving food. Morning coffee or "an apple a day" may be considered food rituals. One woman told a public television producer that at the end of her stressful day at the office, it was her ritual at home to put her feet up, watch reruns of the gentle children's host "Mr. Rogers," and have a glass of wine and some cheese. "Take me out to the ball game ... buy me some peanuts and Cracker Jack" or a hot dog. When a death occurs in the neighborhood, food arrives, not just to ease the labor of planning and cooking meals, but to express friendship, support, and consolation. Family celebrations include the food rituals of birthday and wedding cakes or "Mom's apple pie," savoring favorite regional dishes, family reunion barbecues, and perhaps expressions of gratitude for the feast.

While pleasurable and comforting experiences such as those cited may be examples of spiritually uplifting food consumption, they are not necessarily understood as "religious" experiences. In considering the meaning of food in global religious history and traditions, ancient beliefs and deeply held convictions feed contemporary understanding of the gift and responsible stewardship of food.

FOOD AND RELIGIOUS BELIEF

One may distinguish general spirituality from religious belief in this way: Those who express appreciation and gratitude for food in a religious context understand food as a gift of not only the Earth and its natural resources, but as the gift of a divine Giver, or Creator, or God. A traditional Jewish prayer makes this generalization more explicit. "Blessed art Thou, O Lord our God, King of the universe, who createst many living beings with their wants, for all the means Thou hast created to sustain the life of each of them. Blessed is he who is the life of all worlds" (Hedges, 1972). One account of creation in the Torah reads, "God said, 'See, I give you every seed-bearing plant that is upon all the earth, and every tree that has seed-bearing fruit; they shall be yours for food. ... And it was so" (Genesis 1:29, 30b, Jewish Publication Society [JPS], 1962) Among even older writings, a Zoroastrian statement affirms, "We worship God (Ahura Mazda) who created the cattle, the water, and the wholesome plants" (Zend-Avesta, as quoted in Hedges, 1972).

The Hebrew Scriptures resound with concrete, down-to-earth themes that relate to food and eating (Jung, 2004). It is clear that God, whose image is that of Creator, but also Gardener, provides abundance for every creature. According to the Psalmist, God causes "the grass to grow for the cattle, and plants for people to use, to bring forth food from the earth, and wine to gladden the human heart" (Psalm 104:14, 15a; Metzger and Murphy, 1991; Metzger and Coogan, 1993). The creature's immediate welfare and future destiny depend on the stewardship of the gift of the garden, to cultivate and protect it. In response to the Creator's beneficence, the grateful creature is moved to act responsibly and obediently, to honor the gift of food by enjoying and delighting in it and by respecting the boundaries or restrictions found in sacred texts or understood from long tradition. The themes of gratitude, responsibility, and obedience are prominent in most of the major religions of the world. Offered in this chapter are only brief summaries of topics worth exploring in more depth through further reading and personal research. (See the reference resources for more detail.)

DIETARY LAWS, CEREMONIES, AND RITUALS
IN REPRESENTATIVE RELIGIONS

One of the Three Ways of Salvation in Hinduism is Karma Marga, or the Way of Works (or Actions or Duties). Strict dietary laws are lodged in a law book known as the Code of Manu. According to Manu, "Food that is always worshipped gives strength and manly vigor; but eaten irreverently it destroys both." Found in the Taittiriya Upanishad is this proclamation regarding food: "From food are produced all creatures which dwell on earth. Then they live by food, and in the end, they return to food. For food is the oldest of beings and therefore it is called panacea."

One sacramental rite at the beginning of a follower's life is a first feeding with boiled rice. Later in life, the head of the house must honor the guardian deities with portions of food fresh from the hands of the lady of the house. Until this ceremony is completed, no one eats. At life's end, ancestral spirits are honored through shraddha rites, which include food offerings of pinda (rice balls pressed into firm cakes) that help contribute to a new body for the dead (Noss, 1963). It is also required to practice hospitality by offering food to guests before members of the household eat, with the guests representing the visitation of "the Lord."

Three dimensions of Hindu dietary laws can be expressed by the words *sacrifices* (various food offerings), *austerity* (devotion, self-denial, and fasting), and *charity* (interestingly, this covers both giving food to the needy—"good karma"—and various situations where begging for food was a requirement for students). Although a primarily vegetarian diet is preferred, many Hindus today do eat meat. Other restrictions include spicy foods, mushrooms, onions, and garlic.

In some religious traditions, dietary laws are divided between foods that are allowed or lawful and those not allowed or unlawful. In Islam, lawful behavior falls under the rubric *halal* (free or allowable), while prohibited behaviors are known as *haram* (unlawful). Generally, those things that are lawful and contribute to clean living, are pure, delightful, or pleasant. However, too much of a good thing would violate another precept of Islam, moderation. The Qur'an advises: "O you who believe, forbid not the good things which Allah has made lawful for you, and exceed not the limits" (5:87). In terms of food, there are four specific prohibitions:

1. That which dies of itself (as in carrion)
2. Blood (as in that which is poured forth)
3. Flesh of swine
4. That over which any other name than that of Allah has been invoked at the time of its slaughter.

(There are specific laws about how the slaughter of animals is to be carried out, the major consideration being the complete and free flowing of blood so that poisons do not form part of the food; Ali, 1971).

Food laws are detailed extensively in the Qur'an and the traditions that flow from it. For example, the prophet Mohammed prohibited the eating of canine-toothed animals and clawed birds of prey. The tame ass is prohibited, but not the wild ass, and the mule, but not the horse. Interestingly, many foods considered halal are considered so simply because of personal likes and dislikes. One commentator writes that "a thing which may be good (*tayyib*) as food for one man or one people may not be so for another. Certain things may be good and even useful as food but their use might be offensive for others" (Ali, 1971). The example given was that of the Prophet's prohibition of eating raw onions and garlic just before going to the mosque; the odor would be offensive to others. Cooking such items alleviated the prohibition.

Other Islamic dietary considerations include the practice of hospitality and good manners, moderation even in the good and lawful things, and avoiding intoxicants of any kind.

Islam joins most other religions in the disciplines of fasting. It is of such importance that it is listed among the Five Pillars of Islam (along with belief in Allah, prayer, almsgiving, and the Pilgrimage to Mecca).

Many Islamic practices carry over from Judaism, which has dietary laws that are rooted in Torah and Talmud, tradition and code. The 11th chapter of Leviticus begins, "And the Lord spoke to Moses and Aaron, saying to them, Speak to the Israelite people thus: These are the creatures that you may eat from among all the land animals." The rest of the chapter alternates between, "These you may eat" and "The following shall make you unclean," and ends in summary: "These are the instructions concerning animals, birds, all living creatures that move in water, and all creatures that swarm on earth" (JPS, 1962). Further instruction is given in the final book of the Torah, in Deuteronomy 14, verses 4 to 21.

There remains speculation about the primary *reasons* for dietary laws. The twelfth century Jewish philosopher Maimonides suggested that the food commands were given primarily for health reasons, but most scholars would attribute the dietary laws to the Bible's stress on the need for "holiness." If God is holy, the people of God must strive for holiness, which includes building a right relationship with God. As Rabbi Klein wrote, "The Torah regards the dietary laws as a discipline in holiness, a spiritual discipline imposed on a biological activity. The tension between wanton physical appetites and the endeavors of the spirit was traditionally explained as the struggle" between good and evil inclinations—"the two forces that contend with each other for mastery of the soul" (1979). In addition, food laws encourage a reverence for life: of the beast, the earth, and the neighbor.

The word *kosher* is almost universally perceived as meaning acceptable, fit, or proper. Its popular understanding is rooted in the laws of *kashrut* found in the Leviticus passage noted. In terms of the food requirements, kosher foods are those ritually clean and fit to eat. The Hebrew word for nonkosher is *trayf*, derived from the Hebrew word for "torn," from the Levitical prohibition from eating meat that has been torn by predators. Trayf, however, has come to mean any food that is deemed not fit to eat.

Through the ages, various branches of Judaism have differed on how to interpret and incorporate the ancient dietary laws. Orthodox Jews are the strictest observers of kashrut; Conservative Jews also take the laws seriously, while Reform Jews are less likely to keep such observances, perhaps considering them outdated in modern culture. Keeping kosher in the pious Jewish household means devotion to God, showing deep reverence for life, and preserving the unity of and connection to Israel's roots as people of God.

While time, tradition, and commentary have added many layers to the Torah's original dietary laws, discussion here is necessarily limited to the following allowed and prohibited foods:

- Allowed are animals that both chew their cud and have cloven hooves (cattle, sheep, goats), along with fish that have both fins and scales. Forbidden are pork and shellfish.
- Even the permitted animals are not fit to eat unless ritually slaughtered according to kosher mandate: A specially trained, licensed person (*shochet*) with a highly sharpened knife kills the animal humanely in such a way that blood drains quickly and completely.
- Meat and dairy products are not to be mixed. A kosher kitchen has two sets of dishes, one for each, and even an accidental mixing renders vessels unclean (Williams, 2002).
- Among fowl, Leviticus lists a number of birds "you shall regard as detestable." The list includes scavenger birds of prey that would eat carrion and ingest the blood of the carcass. Traditionally, however, chickens, turkeys, geese, and ducks are allowed.
- All vegetables, grains, and almost all fruits are allowed. An exception is the grape, which is tied historically to food and drink offered to idols. Wine has come to be permitted if not physically touched by Gentiles in its making.
- No insects are allowed to be consumed. This may seem silly to the casual reader, but if one is to keep strict kosher, one must read the contents lists for products such as yogurt to avoid eating color dyes made from ground insects.

These few lines only introduce the extensive Jewish dietary laws. One further clarification is helpful, however. The terms *clean* and *unclean* have no context but for things that are within or outside the law. Again, the primary focus is on obedience, discipline, and reverence in regard to one's relationship to God and creation. "You shall be holy to me, for I the Lord am holy, and I have set you apart from other peoples to be mine" (Lev. 20:26, JPS).

Any discussion of dietary law and ritual must include an overview of two ceremonial meals among the feasts and table gatherings celebrated by Jews (the Passover Seder) and Christians (Eucharist or Holy Communion). Since among the first interactions of mother and infant is her feeding the infant, it is not surprising to find ritual meals among the common forms of religious worship (Ostow, 2006). Those meals are about being nourished in relationship to the One who gives life, at the same time establishing a profound sense of community, even family. The Hebrew Scriptures (the Torah, History, Prophets and Psalms) and the New Testament both contain many references of eating at table together: pilgrim festivals, suppers, feasts, banquets, and family meals (Juengst, 1992). One story in the book of Genesis has Jacob making a stew for his famished brother Esau. The price of the lentil stew was Esau's birthright. But most of the meals of the Bible are far more positive in tone: covenants renewed, festivals noted, weddings celebrated, homecomings hailed, and God's mighty acts commemorated. At the center is not the food, but God, often seen symbolically as the host. "Thou hast prepared a table before me" (Psalm 23).

At the time of Passover, the Seder table may be set lavishly or in other households more simply, but that evening meal always serves as the occasion to fulfill the exhortation to tell to one's children the story of the Exodus. The name of the ritual meal, *Seder*, means "order," the way through which human beings may know God and participate in God's work on Earth (Cernea, 1995). The food elements of the meal, indeed the dishes and vessels on the table, a traditional candlelit search for any stray bit of yeast-leavened bread, a pillow—all provide commentary to answer the question, "Why is this night different from all other nights?"

In story and prayers, symbolic foods, and traditional gestures, oppression surrenders to liberty. Matzoh is the bread of affliction, bitter herbs are a reminder of slavery, and Haroset (a paste made of apples, raisins, almonds, and spices) stands for the mortar used by the Hebrew slaves who did the bidding and building of their masters. The placement of those elements on plate and table is deliberate and prayerful. Wine and song add to the story that connects all Jews with their history and their destiny.

Out of the Passover meal that Jesus of Nazareth shared with his disciples on the night before his crucifixion has come a sacramental meal known as Holy Communion, the Eucharist (meaning thanksgiving), or the Lord's Supper. The theological understandings of the meal vary according to tradition, but the common elements include the sharing of bread (the body of Christ) and wine (the "new covenant" in the blood of Christ), along with the reading of the biblical story of Jesus handing down the meal and saying, "Do this in remembrance of me." The bread may be unleavened, a common loaf of any variety, or small squares of white bread neatly cut for handy distribution. The wine may well be grape juice in some traditions. In Roman Catholic theology, the elements are believed to be changed into the actual body and blood of Jesus (transubstantiation), while Protestants make no such claim for the symbolic foods but still affirm that the risen Jesus is present at the meal with his present-day followers.

While Seder is reserved for that one Passover eve annually, Christians may receive Communion daily, weekly, or as in some more traditional Reformed churches, only four times a year. Both meals have in common the retelling of table stories that have long shaped, and continue to nurture, spiritual journeys.

It is important to note here that while Christians gather for many meals together and have since the earliest days of the church's existence (as in Acts 2:46b, "they broke bread at home and ate their food with glad and generous hearts"), Jesus for the most part broke from the dietary restrictions of Judaism as a sign that his realm was entirely new. "Is not life more than food?" he once asked (Matthew 6:25 NRSV). Christians find no prescriptions for foods allowed or prohibited, although there are numerous passages in the gospels and epistles of the New Testament that tell of controversies over food and meals. From whom Jesus ate with, to whether it was permitted to eat food once offered to idols—these stories never led to specific dietary patterns or choices. It was enough to know that God would continue to provide manna in the wilderness and daily bread, and that the right response to that grace was, as always in the Judeo-Christian tradition, to share one's bread as a sign of hospitality and common blessing.

ETHICAL CONSIDERATIONS REGARDING FOOD CHOICES

It was Jesus who said that people "do not live by bread alone," but there is no denying that people do live by bread. And, bread is an issue of justice in the religious community if one is to understand the moral and ethical dimensions of food choice. It is the believer's responsibility to share food with the hungry. The ethic is summed up in the admonition of a young hiker a few years ago who walked from North Carolina to Washington, D.C., to call attention to world hunger. "If you have received a blessing (food), it is your responsibility to *be* a blessing (and share what you have with others)." He was working with CROP (Communities Responding to Overcome Poverty), the hunger prevention arm of an ecumenical agency known as Church World Service.

However, true justice requires more: the sharing of knowledge, scientific discovery, and technology that will increase food abundance and lessen the dependence of the "have nots" on the power of the "haves."

People of faith respond in several ways. One is to work in concert with others who share the blessing of daily bread to provide direct food assistance locally and globally, but also to provide tools, wells, cisterns, seeds, labor, and know-how to enable neighbors everywhere to successfully increase their own food choices. Theologian Walter Brueggemann (1996) once preached on this Hebrew text in Proverbs: "Better is a dinner with herbs where love is, than a fatted ox and hatred with it." He concluded:

> The choices of ox and herbs, of greens and beef, of love or strife, are not little family choices made in private when you go into the kitchen. They are big, far-ranging public choices concerning foreign policy and budget and land reform and dreams. We do not pick our food just before dinner. We pick our food by how we value life, and how we build policy and how we shape law, and how we arrange money, and how we permit poverty and hunger in a land of abundance.

Through the social pronouncements of church agencies and governing bodies, their constituents are learning the value of family farms and sustainable agriculture, of local farmers' produce versus what Wendell Berry (1995) called "bechemicaled factory fields," and of paying a bit more for fair trade coffee over so-called gourmet brands. David Beckmann, head of the interreligious agency Bread for the World, stated that 850 million people in the world are undernourished, and the prices of basic staples like rice, corn, and wheat are rising to a point that another 100 million people are being driven into hunger. His organization acts as both an advocate for the hungry ones outside the garden gate and as a lobbyist to politicians who help shape food policy.

Many people of faith join their voices in the prayer of Jesus, saying, "Give us this day our daily bread." They trust that bread will come and with it a sense of gratitude and responsibility, commitment to justice, and renewed compassion for local and global neighbors.

CONCLUSIONS

The life of the spirit is lived in the context of the table, and those who gather round it make many choices throughout the day regarding how that table will be set. The

A PERSONAL REFLECTION

I sat at table in the Guest House of Holy Cross Abbey, a Trappist Monastery in Berryville, Virginia, with the guest master Father Stephen at the head of the table, and six other retreatants gathered round. After Father Stephen had offered a prayer of thanksgiving to God for the food we were about to enjoy from the bounty of God's creation, he stabbed his fork into his vegetarian salad and encouraged the rest of us to pass the large platter of roast beef around the table. The meat had come from beef cattle the monks raised on their 1200-acre farm along the Shenandoah River, just yards from the Guest House.

Stephen abstained from meat, following the sixth-century rule of St. Benedict: "Except the sick who are very weak, let all abstain entirely from eating the flesh of four-footed animals." Yet, the monks' livelihood came from raising beef for others to eat. Stephen's traditions included spiritual fasting as well as observing Roman Catholic "feast days." As the lone Presbyterian at the table that day, my own eating habits were less guided by religious ritual, health issues, and even conscience than by my physical appetite.

Many things have changed in the 30 years since that meal. The monks found raising cattle to be less than profitable and even stopped baking Monastery Bread, although they use the ovens for brandy-fueled fruit cake production to support their religious vocation. And I am far more aware that the food I choose to eat not only directly impacts my health but also has profound ramifications for the health of the planet itself as well as that of my global neighbors. I also have come to value the rhythms of feasting and fasting that help pace my spiritual journey.

One thing has not changed as I look back to that monastery meal: Although from different ecclesiastical traditions that cannot yet share in the bread and wine sacramental supper we both refer to as "Holy Communion," there was a sense of spiritual union and fellowship around that dinner table, friendship created and nurtured in the context of tables set, plates passed, and dishes washed by hand. Had a Buddhist joined us around the table, he or she would have shared in the salad but not the beef. A Jew would have enjoyed the beef but passed on the pork had it been offered. And a Taoist might have cautioned us with words from the Tao Te Ching, "To take all one wants is never so good as to stop when one should."

booth at the fast food restaurant may be the setting for an entertaining time, but it is doubtful that the meal or the occasion will be nourishing or memorable. Religious and other spiritual folk consider their kitchens more than mere "filling stations" (to use Wendell Berry's term) and see their tables as places where the earth's bounty may be celebrated, where hospitality is as broad as creation is rich, and where divine justice universally demands that no one receives more than their share at the expense of another.

Frederic and Mary Ann Brussat (1996) have said it well:

> As we eat, we are linked with those who have prepared the food; we are grateful to all the plants and animals who have lived and died for this moment; we remember with love and compassion others throughout the world who hunger and thirst as we do; and we are graced with the presence of the Lord of the Universe at the table with us.

STUDY TOPICS

1. As in a journal entry, write about a particularly memorable meal you have enjoyed in the company of others. What components especially contributed to your enjoyment of the occasion? Was there in that meal some aspect of the "holy" or "sacred," however you define those terms?

2. Are ancient religious dietary laws helpful in today's postmodern culture? Name continuing benefits of such restrictions. Identify reasons people might not take those laws seriously today.

3. Do an online search for faith-based organizations that are involved in fighting world hunger. You might begin with Heifer International, Bread for the World, CROP, World Vision, or Web of Creation.

4. If you are part of a faith group, church, synagogue, mosque, or other religious organization, research the social justice statements of your faith community. Try to locate your group's statements or actions that deal with (a) sustainable agriculture, (b) genetically modified food, (c) fair trade and food, or (d) global food crisis.

5. For an activity, each fall many communities participate in an interfaith CROP Walk to raise consciousness and funds to "stop hunger." Look for notice of such a walk in your community and join in. While you walk, talk with fellow walkers about their motivation in walking, why they think the few steps they take that day will make a difference, and what local organization is benefiting from the walk.

REFERENCES

Ali, M.M. *The Religion of Islam*. Lahore, Pakistan: Ahmadiyya Anjuman Isha'at Islam, 1971.
Bass, M.W., L.M. Wakefield, and K.M. Kolasa. *Community, Nutrition, and Individual Food Behavior*. Minneapolis, MN: Burgess, 1979.
Berry, W. *Another Turn of the Crank*. New York: Counterpoint, 1995.
Brueggemann, W., edited by C.L. Campbell. *The Threat of Life: Sermons on Pain, Power, and Weakness*. Minneapolis, MN: Fortress Press, 1996.

Brussat, F. and M.A. Brussat. *Spiritual Literacy: Reading the Sacred in Everyday Life*. New York: Scribner, 1996.

Cernea, R.F. *The Passover Seder: An Anthropological Perspective on Jewish Culture*. Lanham, MD: University Press of America, 1995.

Fick, G.W. *Food, Farming, and Faith*. Albany: State University of New York Press, 2008.

Gilbert, E. *Eat, Pray, Love*. New York: Penguin Books, 2006.

Grimm, V.E. *From Feasting to Fasting, the Evolution of a Sin*. London: Routledge, 1996.

Hedges, S.G. *Prayers and Thoughts from World Religions*. Richmond, VA: John Knox Press, 1972.

Helminiak, D.A. *Spirituality for Our Global Community: Beyond Traditional Religion to a World at Peace*. Lanham, MD: Rowman and Littlefield, 2008.

Jewish Publication Society of America. *The Torah: The Five Books of Moses*. Philadelphia: Jewish Publication Society of America, 1962.

Juengst, S.C. *Breaking Bread: The Spiritual Significance of Food*. Louisville, KY: Westminster/John Knox Press, 1992.

Jung, L.S. *Food for Life: The Spirituality and Ethics of Eating*. Minneapolis: Fortress Press, 2004.

Klein, I. *A Guide to Jewish Religious Practice*. New York: Jewish Theological Seminary of America, 1979.

Metzger, B.M. and R.E. Murphy (Eds.). *The New Oxford Annotated Bible*. New York: Oxford University Press, 1991.

Metzger, B.M. and M.D. Coogan (Eds.). *The Oxford Companion to the Bible*. New York: Oxford University Press, 1993.

Noss, J.B. *Man's Religions*. New York: Macmillan, 1963.

Ostow, M. *Spirit, Mind, and Brain: A Psychoanalytic Examination of Spirituality and Religion*. New York: Columbia University Press, 2006, p. 6.

Schut, M., Ed. *Food and Faith*. Denver, CO: Living the Good News, 2002.

Williams, P.W. *America's Religions: From Their Origins to the Twenty-First Century*. Urbana: University of Illinois Press, 2002.

4 The Influence of Culture and Customs on Food Choices

*Cindy M. Imai, Dustin J. Burnett,
and Johanna T. Dwyer*

CONTENTS

ABSTRACT

This chapter emphasizes the role of food in social structure, and highlights key cultural factors that influence food choices. They include geography and environment, biology and physiology, the senses, technology, politics, economics, and social and psychological factors. A brief review of some of the staple foods used in cuisines throughout the world is provided. The cultural forces that make people choose or refuse certain foods are addressed. Religion is also a factor, and it is further discussed in Chapter 3.

INTRODUCTION

"We are who we are today because of you who came before us."

—Zulu proverb

The combined influences of the environment and other factors on cultural food choices are summarized in Figure 4.1. The quality and quantity of food available to humans and humans' food choices have changed over the course of human history. They have been influenced by several cultural factors, as mentioned throughout the chapter. In the past, humans wandered for food to obtain the nutrients needed for survival. Culture represented the means for identifying foods that were appropriate to eat. As the human population grew, environmental constraints affected the means of food acquisition, ultimately affecting food choices. Humans further adapted to their surrounding environments and began to cultivate plants and domesticate animals instead of relying on hunting and gathering. This Neolithic transition provided a stable and reliable food supply that eventually allowed for the development of civilizations. Over time, as human cultures began to have more contact with each other, additional factors—economical, political, and psychological—developed that enabled humans to further discern between foods that were "for me and my family" and foods that were "for you or strangers." Human survival depends on the acquisition of food from an ever-changing, and often hostile, natural environment. For most of recorded history, food was scarce, thus affecting human activity and food choices.

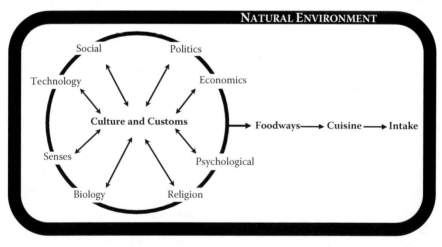

FIGURE 4.1 Factors affecting the cultural and customary influences on human food selection.

Early humans spent most of their lives hunting, gathering, and preparing food in an environment devoid of a reliable food supply. Human societies today are less isolated and more globally focused than ever before, and traveling has become easier. Highly industrialized countries are becoming more culturally diverse as immigrants arrive from rural and developing countries with new cultures, customs, and food choices (Food and Agriculture Organization of the United Nations 2004). Simultaneously, in developing countries, multinational fast-food and supermarket chains are appearing, potentially leading to more cultural homogeneity globally (Food and Agriculture Organization of the United Nations 2004). The emergence and evolution of agriculture and its impact on health and lifestyles are discussed in Chapter 1.

How do human beings choose the foods they eat? The consensus of expert opinion is that many historical and social factors as well as availability influence human food choices and consumption. Culture and customs also determine which foods are appropriate to eat, which methods are appropriate for processing foods, how foods are to be distributed throughout society, how foods are eaten, and how foods are used to promote optimal health. Table 4.1 defines key terms used throughout the chapter. Figure 4.2a illustrates the social construction of food choices; individuals often define foods by their social characteristics—what "I" eat and what "they" eat.

THE ROLE OF FOOD AND SOCIAL STRUCTURE IN FOOD CHOICES

Today, as was true in early human societies, there is still the need to acquire the essential nutrients to survive. Once an adequate and secure food supply was better achieved through agriculture and domestication of animals, greater food security allowed for more diverse food choices and food sharing both internally, within the family unit, and externally, with friends and neighbors (Fieldhouse 1996). A more secure food supply and the sharing of food are thought to have strengthened the social ties between individuals, developed a sense of trust within the community, enabled individuals to depend on one another, and allowed the community to divide and delegate tasks that were required for survival (Harris and Ross 1987, Fieldhouse 1996). Rather than spending the majority of time simply searching for food, human beings were able to build complex societies, eventually leading to the development of civilization (Ferraro 2006).

DEVELOPING THE SOCIAL IDENTITY OF FOODS

Members of different societies eventually learned to develop and prepare foods in ways that provided them with a distinctive self-identity, which either drew them closer to or separated them from their neighbors. Those who wished to associate with other societies often adopted their neighbors' food habits (Kittler and Sucher 2004). Those who wished to separate themselves from their neighboring societies often identified a certain food as a staple of that society and avoided that food to show that they were different (Harris and Ross 1987, Kittler and Sucher 2004).

As humans from various cultures continue to mix into a more global society, they are faced with unfamiliar foods, new ways to prepare familiar foods, and new ways of eating familiar foods. Even within culturally similar settings, individuals differ

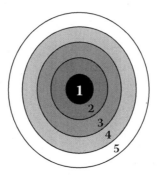

1. = What "I" eat
2. = "Family" eats
3. = "Neighbors" eat
4. = "Society" eats
5. = "They" eat

(a)

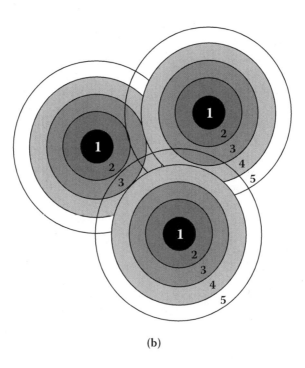

(b)

FIGURE 4.2 (a) Social construction of food choices. What *they* eat is kept far away from what *I* eat and is not acceptable for *me*. (b) Effects of acculturation on social construction of food choices. Relocation of individuals and families enables cultures to mix and share traditional foods and methods of preparation. Note that not all individuals participate in this exchange.

TABLE 4.1

Key Terms and Definitions

Terms	Definitions
Acculturation	Occurs when two or more separate cultures coexist; although one culture tends to dominate over the others, there is usually a bidirectional exchange of acceptable norms and habits; generally, younger members are more apt to adopt the norms and habits of the other culture than are the older members. Acculturation is more pronounced when different cultures share similar norms and habits (Fieldhouse 1996).
Cuisine	A tool used to distinguish one culture's food combinations from another; is a culturally transmitted set of practices used to identify acceptable (1) staple foods; (2) processing methods; (3) food combinations; and (4) characteristic flavorings (Rozin 1973; Rozin and Rozin 1981, pp. 243–252).
Culture	An unnoticeable network of ideas, beliefs, laws, and customs of everyday life that are learned and shared by a group of individuals (Fieldhouse 1996).
Custom	A routine action that is practiced out of habit and is accepted by either an individual or a group of individuals (Oxford University Press 2007).
Enculturation	The passing of culture from one generation to the next generation (Haviland et al. 2004, p. 496).
Food	An item that enters the body's systems and tissues to build, repair, and maintain life (Oxford University Press 2007).
Foodways	The shared customary habits of groups of people that relate to food and eating (Oxford University Press 2007).
Functional foods	Foods or dietary components that may provide a health benefit beyond basic nutrition (International Food Information Council 2008).
Umami	Means "deliciousness" in Japanese; is one of the five tastes; a full, meatlike taste that is used to describe the flavor of monosodium glutamate, a popular flavor enhancer (DeMan 1999).

from each other in their likes and dislikes (Douglas 1978). As humans become more adventurous and taste and choose unfamiliar foods, they may become more likely to adopt foods that meet their nutrient needs while establishing social ties with neighboring societies (Fieldhouse 1996).

GEOGRAPHY AND ENVIRONMENT AS DETERMINANTS OF FOOD CHOICES

Humans have developed a symbiotic relationship between their selection of foods and the natural environment since they evolved from their primate ancestors. Over the past 10,000 years, they have learned to utilize their environmental resources to produce a more stable and reliable food supply. However, even today, humans continue to be faced with environmental challenges as they try to meet their immediate dietary needs.

Throughout most of history, human food choices were limited to the foods that were available in the immediate environment. Climate and weather patterns varied tremendously from region to region, and thus humans migrated to hospitable environmental and geographical locales to find food (Hladik 1988; Williams 1975). The ability for humans to migrate from climate to climate, while choosing a wider variety of new plant and animal foods, helped them to overcome natural obstacles and adapt to a continually changing environment (Williams 1975).

The Neolithic revolution (see Chapter 1) occurred largely as a result of overpopulation and excessive strain on natural resources (Hillman et al. 1989). It enabled humans to settle down and grow select foods around their settlements instead of being nomads. They were able to obtain a more reliable supply of foods, although the supply was also less diverse. This decrease in food variety, coupled with the growing population, was partly responsible for an increase in biological stress, including malnutrition and infectious diseases, which were extremely common during this transition (Angel 1984, Smith et al. 1984, Cohen 1989).

STAPLE FOODS OF THE WORLD

Table 4.2 presents some of the common staple foods that are popular dietary choices in various parts of the world. Grains were domesticated and incorporated into human diets very early. Wheat, rye, barley, spelt, corn, and rice produced a readily available source of carbohydrates (Ulijaszek et al. 1991). Potatoes, taro, and cassava became commonly cultivated tubers (Lieberman 2003). For animals to be suitable for human consumption, they could not compete significantly with humans for food; they needed to provide a relatively quick and nutritious source of nutrients, and they needed to suit the environment in which they were domesticated (Harris and Ross 1987).

MODERN CHALLENGES INFLUENCING FOOD CHOICES

Today, humans are still faced with environmental constraints that influence food choices. For example, during the early part of the twentieth century in the high plains of the United States—and the steppes of Kazhakstan in the former Soviet Union in the 1950s and 1960s—the prairie was plowed and planted with grains, and little attention was paid to conservation or sustainability. This ultimately contributed to the Dust Bowl of the 1930s and led to an exodus of humans to areas of the United States with more readily available food supplies (Egan 2006). Although more attention is paid to conservation today than in the past, some argue that certain agricultural methods—deforestation and production of staple grains and livestock—and their heavy reliance on the combustion of fossil fuels and the use of other natural resources are contributing to environmental and climatic changes that may affect food choices in the future (U.S. Department of Agriculture 2004).

BIOLOGY AND PHYSIOLOGY AS DETERMINANTS OF FOOD CHOICES

Food choices must fulfill nutrient needs. Without the essential nutrients, neither human beings nor societies can survive. Human biology, homeostatic mechanisms, and the interdependence of plants, animals, and humans are covered in Chapter 1. Nutrients and their requirements and metabolism are covered in Chapters 5 through 8.

TABLE 4.2

Regions of the World and the Cultural Transformation of Foods from Nature to Cuisine

World Regions	Nature					Culture			
	Protein Sources		Carbohydrate Sources					Seasonings / Spices /	Cuisines/
	Animal	Plant	Fruit	Vegetable	Starch	Processes	Flavorings / Accompaniments		Dishes/Foods
Africa	Antelope	Almond	Banana	Artichoke	Cassava	Bake	Basil	Lemon	Bosaka
	Buffalo	Bean	Date Mango	Avocado	Fonio	Boil	Celery	Lime	Caïdou
	Camel	Chickpea	Pawpaw	Cabbage	Maize	Braise	Cinnamon	Marjoram	Chakchouka
	Chicken	Lentil	Pineapple	Eggplant	Millet, Rice	Fry	Coriander	Mint	Foutou
	Cow	Peanut	Plantain	Okra	Sorghum	Grill	Cumin	Nutmeg	Kourkouri
	Crab	Walnut	Pomegranate	Cabbage	Sweet potato	Marinate	Fennel	Onion	Mafé
	Crayfish		Prune	Spinach	Taro	Paste	Garlic	Pepper	Pepe supi
	Egg		Pumpkin	Tomato	Wheat	Stew	Ginger	Pimiento	Tajine
	Fish		Quince		Yam		Honey	Syrup	Went
	Gazelle						Jasmine	Tamarind	Yassa
	Goat								
	Lamb								
	Larvae								
	Locust								
	Monkey								
	Oyster								
	Pig								
	Rabbit								
	Snake								
	Zebra								
Asia	Abalone	Almond	Coconut	Bamboo	Arrowroot	Boil	Bonito	Nuoc mâm	Chapatti
	Chicken	Bean	Durian	Daikon	Buckwheat	Braise	Cardamom	Sesame	Halva Hincho
	Cow	Chickpea	Ginkgo	Eggplant	Lentil	Dry	Chive	Sichoun	Lap Murghi
	Crab	Lentil	Jujube	Gourd	Potato	Fry	Chutney	Star anise	buriani Nigiri
	Duck	Pulse	Kumquat	Lotus	Rice, Wheat	Grill	Cinnamon	Tamarind	zushi Noodle
	Eel	Sesame	Longan	Mushroom		Marinate	Clove	Turmeric	Peking duck
	Egg	Soya	Lychee	Seaweed		Pickle	Coriander	Vinegar	Tandoori Tom
	Fish		Mango	Spinach		Salt	Ghee	Wasabi	yam
	Lamb		Pineapple	Tomato		Steam	Ginger	Saffron Soy	
	Octopus		Rambutan	Turnip		Stir-fry	Mustard	sauce	
	Oyster								
	Pig								
	Shark								
	Shrimp								
	Squid								
	Tiger								
	Trout								
	Tuna								
	Turtle								
	Urchin								

(continued)

TABLE 4.2

Regions of the World and the Cultural Transformation of Foods from Nature to Cuisine (continued)

World Regions	Nature					Culture			
	Protein Sources		Carbohydrate Sources				Seasonings / Spices / Flavorings / Accompaniment		
	Animal	Plant	Fruit	Vegetable	Starch	Processes			Cuisines / Dishes/Foods
Australia and New Zealand	Bat	Almond	Banana	Bell pepper	Wheat	Broil	Basil	Juniper	Dinkum
	Chicken	Walnut	Lemon	Celery	Oat, Potato	Barbeque	Clove	Mint	Jack
	Cow		Pineapple	Onion	Squash	Can	Cumin	Rosemary	Kebab
	Crocodile		Passion fruit	Mushroom	Sweet potato	Microwave	Garlic	Garlic	Lamb stew
	Fish		Kiwi	Tomato			Ginger		Meat pie
	Grub								Possum soup
Europe	Bear	Almond	Apple	Artichoke	Barley	Bake	Aniseed	Nutmeg	Açorda
	Capon	Bean	Blackberry	Asparagus	Millet	Boil	Basil	Olive oil	Bread
	Chicken	Chickpea	Cherry	Beet	Oat	Braise	Caper	Oregano	Cheese
	Cow	Pistachio	Cranberry	Eggplant	Potato	Pickle	Cardamom	Paprika	Kefta
	Donkey	Walnut	Fig	Green bean	Pulse	Purée	Chervil	Parmesan	Koulibiaca
	Duck		Grape	Mushroom	Rice	Roast	Cinnamon	Parsley	Pâté
	Egg		Melon	Olive	Rye	Salt	Clove	Pepper	Puchero
	Fish		Pear	Spinach	Wheat	Sauté	Dill	Rosemary	Quark
	Goose		Pomegranate	Turnip		Smoke	Fennel	Saffron	Risotto
	Heron		Strawberry	Zucchini			Garlic	Sage	Salami
							Ginger	Sauerkraut	Slottsstek
							Mustard	Thyme	Taramasalata

Region										
North America	Alligator, Bear, Bison, Chicken, Cod, Cow, Crab, Crayfish, Egg, Fish	Lobster, Moose, Oysters, Pig, Rabbit, Reindeer, Salmon, Shrimp, Turkey, Turtle	Almonds, Beans, Lentil, Peanut, Pecans, Red bean, Soya, Walnuts	Apple, Banana, Blueberry, Cranberry, Grapefruit, Orange, Pawpaw, Pear, Pineapple	Agave, Avocado, Carrot, Eggplant, Mushroom, Onion, Radish, Tomato, Turnip, Yucca	Corn/maize, Oat, Potato, Pumpkin, Rice, Rye, Sweet potato, Wheat, Yam, Squash	Bake, Barbeque, Boil, Braise, Fry, Grill, Marinate, Microwave, Roast, Simmer	Basil, Bay leaf, Chili, Cinnamon, Cocoa, Fennel, Garlic, Mustard, Nutmeg, Oregano	Parsley, Pepper, Red pepper, Rosemary, Sage, Salt, Thyme	Chaudronnée, Chowder, Doughnut, Empanada, Enchilada, Gumbo, Johnnycake, Tortillas, Tostadifa, Waffle
South America	Alligator, Ants, Armadillo, Chicken, Cow, Crab, Egg, Goat, Iguana	Lamb, Mussel, Ox, Oyster, Pig, Rabbit, Sea cow, Shrimp	Almond, Bean, Peanut	Banana, Coconut, Guava, Mango, Orange, Papaya, Pawpaw, Pineapple, Prune	Bell pepper, Olive, Onion, Palm hearts, Tomato	Cassava, Corn/maize, Potato, Pumpkin, Rice, Sweet potato, Yam	Boil, Dry, Freeze, Fry, Grill, Roast, Stew	Caper, Cocoa, Cumin, Garlic, Lemon, Lime, Mustard	Oregano, Paprika, Parsley, Pepper, Pimiento Salt	Anticucho, Bori bori, Ceviche, Chupe, Conejo estirado, Custard, Feijoada, Matambre, Sancocho

Note: Nature provides food in the form of animals or plants. Human beings obtain the nutrients they need for survival by first finding the food in nature, then transforming it into a dish that is appropriate for consumption. Culture dictates which animals or plants are edible, which methods are to be used for preparing the food, which flavorings and seasonings are to be used, and what the dish will be called. All of these steps determine which foods are culturally appropriate. Notice that similar foods are used throughout the world. Many of the common foods originated in one region and were carried to other regions of the world through exploration and trade.

Sources: Kittler, P.G. and K.P. Sucher, *Food and Culture*, 4th ed., Belmont, CA: Thomson Wadsworth, 2004; Lang, J.H., Ed., *Larousse Gastronomique: The New American Edition of the World's Greatest Culinary Encyclopedia*, New York: Crown, 1995; Massachusetts Institute of Technology, Australian recipes, 2005, available at http://web.mit.edu/~mitsos/ www/WCA_cook2/Australian%20cooking%20class.pdf.

There is no single food that contains adequate amounts of all of the essential components necessary for life. Nutrients must therefore be obtained by eating a variety of foods. For example, rice is low in tryptophan; however, choosing rice with beans, which contains more tryptophan, produces a more "complete" protein. Culture and the daily customs of people that are passed on through generations serve to ensure that appropriate food combinations are used for obtaining the essential nutrients. Food-processing techniques that are transmitted by culture are necessary for many foods because nutrients are often not available in a readily accessible form, or food is unpalatable if it is not processed. Cassava (manioc), for example, cannot be eaten directly in its natural form. It must first be peeled, pressed, and washed to remove the chemical hydrogen cyanide before it can be eaten.

THE SENSES AS DETERMINANTS OF FOOD CHOICES

The sensory effects of food are powerful determinants of which foods will be eaten or avoided. Food preferences are established early in life and are usually carried into adulthood. Sensory and gustatory factors of foods may be used as distinguishing characteristics in a multicultural environment, especially if one culture is trying to blend into another. When two cultures mix, those in one culture may prefer not to choose foods that identify them as foreign. Figure 4.2b illustrates a multicultural environment and its effects on food choices. As societies become more global and cultures become intertwined, the youngest members of one culture are usually more likely to accept the flavors of the other culture, especially when the flavors are socially desirable and accepted by the household.

Table 4.3 summarizes some of the sensory and gustatory influences on food choices and their biological significance. The five tastes—sweet, bitter, salt, sour, and umami—are inborn. Nevertheless, some tastes are more highly preferred than others. The tastes themselves may have evolved originally because they conferred a survival advantage by allowing humans to distinguish between nutritious foods and poisonous foods. The most distinguishable characteristics of foods are sensory: how a food looks, tastes, feels, smells, and sounds. Groups relate to their cultural heritage through the types of foods and condiments they choose to eat. They also express themselves through unique cooking and serving methods (see Table 4.2). Humans use these sensory cues to choose foods that are culturally acceptable (Pelto et al. 1989).

DEVELOPING TASTE PREFERENCES

Eating patterns are adopted early in life by individuals watching others eat. Children develop food preferences by exploring the foods that are provided to them (Kittler and Sucher 2004). A clear preference for foods containing either fat or sugar, both of which increase the palatability of food, is present in both children and adults in all cultures (Drewnowski and Greenwood 1983; Mela and Catt 1996). These sensory and gustatory preferences are frequently carried into adulthood and passed on to future generations (Fieldhouse 1996).

TABLE 4.3

Sensory and Gustatory Influences on Food Choices

Sense	Chemical Compound	Sources	Biological Significance
Taste			
Sweet	Glucose	Fruits Milk Vegetables	Energy
Bitter	Glycosides	Vegetables	Toxins
Salt	Sodium	Salt	Fluid balance
Sour	Hydrogen	Acidic foods	Stimulate digestion
Umami	Monosodium glutamate (MSG)	Glutamate- containing foods	—
Smell	Ethylene dichlor (ethereal) 1,8-Cineole (camphoraceous) Pentadecanlacton (musky) Phenylethylmethylethylcarbinol (floral) Methone (minty) Formic acid (pungent) Dimethyl disulfide (putrid)	All foods	Food acceptability Stimulates digestion
Texture	Lipid Glucose Protein	All foods	Food acceptability Energy

ROLE OF OTHER SENSES IN FOOD CHOICES

The sense of smell is also important in food choice. Smell arouses the appetite, triggers the cephalic phase of digestion, and stimulates the digestive juices. Certain food smells can also cause disgust or nausea. The odor of compounds in foods helps to identify acceptable foods in foreign cultures. However, there are also cultural differences—strong odors that are acceptable in one culture may be revolting in another.

The texture of foods is also aesthetically important. The mouth feel of food is an important determinant of its acceptance (Kittler and Sucher 2004). Some cultures prefer slimy foods, whereas others prefer dry. Some prefer crunchy, others soft. Processing is often used to alter texture (Pelto et al. 1989). Take, for example, the processing of chocolate, which varies between cultures. The conching process agitates and distributes the cocoa butter within the chocolate by pressing and mixing the two together. Chocolate that is conched for a longer period of time tends to be creamier and waxier than the less-processed chocolates, which are more granular. Cultural preferences for acceptable textures determine the appropriate length of processing.

SOCIAL INFLUENCES ON FOOD CHOICES

The social influences determining food choices depend on multiple factors, including times of high and low availability of foods; the desire to maintain peace with neighbors; and influences of social symbolism. Sharing foods between communities also serves as a means of building and maintaining ties with neighbors. Finally, food choices are influenced by the social symbolism that is associated with particular foods.

Food is a tool for facilitating social relations. Social bonds may be formed and broken depending on the variety of food (Mela 1996, de Castro 1999). The number of individuals present at a meal and the quantity and quality of food eaten are directly related (de Castro 1990, 1994). Historically, humans needed to find ways to sustain themselves during times of the year when food choices were scarce and to find ways to share food to maintain social ties with neighbors. Communal meals have the advantages of providing an abundance of food choices as well as fellowship that would not be acquired if individuals were to eat on their own (Ulijaszek 2002).

Social and cultural norms of eating diversified further in the context of ecological change over time with human migrations, the emergence of agriculture, and the development of socially and economically stratified societies (Strickland and Shetty 1998, p. 346). In most societies, festivals and holidays tend to occur after the harvesting period and during the times of the year when food is abundant (Kittler and Sucher 2004). Feasts are also associated with festivals and holidays as a way of sharing food with each other and providing food for individuals who may not have an adequate supply (Pelto et al. 1989).

Each culture modifies new foods in ways that make the foods acceptable to members of its culture (see Table 4.2). The sense of belonging to a larger community is established by partaking in these shared meals (Scapp and Seitz 1998, p. 299). The components of modern meals comprise foods from various locales and cultures that can be eaten anywhere in the world, from Southern California to southern France, Japan, or Brazil. Still, certain foods remain unacceptable to a given culture or subculture and will not be added to the meal.

As long as there is an abundance of food choices, humans are hospitable and peaceful and share foods with each other (Harris and Ross 1987). However, during times of famine and wars, when individuals and families are faced with potential starvation, hoarding is common (Pelto et al. 1989). When food supplies are threatened and individuals and families are faced with potential starvation, they often eat alone to keep from disclosing the amount of food that they actually have for themselves (Pelto et al.). With severe population pressures in the environment, food becomes less abundant, and people tend to become suspicious of one another (Harris and Ross 1987). Peace between neighboring societies is often maintained by the exchange and sharing of food.

FOOD AS SYMBOLS AND INFLUENCE ON FOOD CHOICES

Foods also serve as symbols or tools of communication between individuals or groups and may be chosen for these reasons. Certain foods—processed, refined, and packaged—are associated with a higher level of prestige or purity, whereas other foods—unprocessed and raw—are less prestigious (Pelto et al. 1989, Kittler and Sucher 2004). As foods become more refined, some believe that the "dirty" parts are removed, and the "pure" parts are preserved. In actuality, less-refined foods often provide more vitamins, minerals, and fiber and may actually be more nutritious than their prestigious counterparts.

Packaging may further enhance or confuse the prestige of a product. Take for example, the difference between Godiva® chocolate and a Hershey® bar. Both products have similar nutrient content, but they are perceived to be quite different in quality due to differences in processing, packaging, and cost. In some cases, the prestige of a product may be reflected onto the consumer. For example, the terms *functional foods, organic,* or *natural* also carry a variety of hidden social messages such as "healthfulness" and "self-healing" to individuals who purchase them. Purchasers of these foods may perceive their choices as more healthful than the conventional products, although they may not be so (Saher et al. 2003).

TECHNOLOGICAL INFLUENCES ON FOOD CHOICES

Evolution and culture have allowed humans to adapt to the environment by exchanging primitive methods of foraging for more reliable technological methods of food cultivation. Modern food technology has increased the variety of food choices and the interaction between culturally different societies and has encouraged the sampling of different cuisines. Technology has changed human eating behaviors by allowing food to be easily accessed in industrialized societies. Globally, technology influences food choices by determining what, how much, and where food is available.

Cooking is one technology that has helped to expand human food choices. Early humans used fire to obtain more control over their food environment (Haviland et al. 2004, p. 496). The fire itself may have also protected humans by frightening away predators and decreasing the competition for food. Modern humans continue to use fire as a means of destroying potential pathogenic microorganisms, making foods safer, more edible, and more digestible.

Technological advances have increased exposure to new food choices by allowing food products to be distributed from one continent to another while reducing the risk of spoilage and contamination. Before the nineteenth century, the only methods available for preserving meat were drying, salting, and smoking, none of which were entirely practical since large quantities of food could not be processed or preserved for very long (Clark 2000). The canning process was developed in 1809 and was a product of the Napoleonic wars; the process allowed heat-sterilized food to be stored for longer periods of time without spoiling (Clark 2000). Further methods of processing in the twentieth century involved dehydrating, freezing, and treating with ultrahigh temperatures (UHTs), increasing shelf life, convenience, and variety of food products (Pyke 1972). In addition, refrigeration, vacuum packing, fast freezing, and irradiation ensured that seasonal items would be available year round in economically developed societies (Jones et al. 1983).

The influence of improved transportation and distribution on food choices is also considerable. The manufacturing of steel railroad ties and laying of railroad tracks during the Industrial Revolution and the use of coal and steam for fuel enabled movement of foods across greater distances via railroad and steamship (Clark 2000). The cultivation of tomatoes is an example of advanced technology paired with transportation. Tomatoes were once a seasonal luxury but are now a common international food choice due to applied agricultural science and an extended transportation system. Similarly, tomatoes and tomato-based products such as ketchup, tomato sauce, and tomato paste are available on the worldwide market (Pyke 1972, Scapp and Seitz 1998, p. 299). Food distribution has ultimately supported the mixing of cultural foods, allowing countries thousands of miles apart to sample each other's cuisine without having actual contact with one another (Nabkasorn et al. 2006, pp. 179–184).

Canning, packaging, mechanical refrigeration, and the use of additives and preservatives have increased the shelf life of many food products. Food stability has allowed producers to focus on another aspect of food production: increasing product variety, a factor influencing food choices. Technological advancements in food production are still occurring and are driven by consumer demand for convenient, packaged goods. The number of varieties of food products introduced into the national market is a reflection of this demand. In the United States alone, the number of novel products entering the marketplace has risen annually since the 1990s and hit a record high of 20,031 food and beverage products introduced in 2006 (Martinez 2007).

ECONOMIC INFLUENCES ON FOOD CHOICES

Food security is a global issue that affects millions of people in both impoverished and affluent countries worldwide and ultimately affects their ability to choose appropriate foods for survival. A variety of factors—income, age, gender, education level, cooking ability, and access to food—are responsible for the economic influences on culture and customs. The interrelationship between cultural values and beliefs along with personal skill sets and income all interact to produce variations in food choices (Johns 1990).

Socioeconomic status influences the quality and quantity of food choices. Present-day societies continue to focus on food as one of the motivators for economic development. However, food as an incentive for economic development is more potent in poorer societies, where a higher percentage of income is used to acquire foods (Counihan 1999). In these societies, complex carbohydrates (grains) are relied on more heavily as a dominant source of nutrients as they are more affordable and easily accessible (Drewnowski 1997).

In wealthier countries, economic stability and food abundance have given humans the luxury to choose foods they *wish* to consume. As incomes grow, there is a general transition toward more palatable foods that are high in fat and sugar and low in complex carbohydrates, which are often dense in energy and lacking in nutrients (Drewnowski 1997). This abundance of high-energy food choices has allowed Western societies to easily meet—and exceed—their nutritional needs.

In industrialized societies, the types of foods eaten and the way they are obtained may be used to identify economic status. The value placed on food technology is similar to the value of money that permeates Western societies (Pyke 1972). Wealthier people are of a higher social class and tend to favor the selection of tastier, higher status, and easier-to-prepare foods used to proclaim one's economic standing to others (Atkins and Bowler 2001). Conversely, in times of economic hardship, choices become limited, food portions become meager, and those who must rely on help such as food stamps or food distribution programs are often stigmatized (Atkins and Bowler 2001).

Economic strains may also help to strengthen cultural ties and identity, especially for immigrants (Counihan and Van Esterik 1997, p. 424). For example, some African American communities in the rural South continue to choose traditional foods and cooking methods to reinforce their cultural connections to enduring a formerly brutal society and to ensure that these customs will be passed on to future generations (p. 424).

Limited income may also limit food choices due to lack of knowledge about nutrition (Pelto et al. 1989, p. 217). Instead, families with a low-wage earner may try to imitate the diets of the cultural norm, leading to overexpenditure on choices such as expensive "junk foods" while neglecting the less-expensive—and often more nutritious—alternatives (Atkins and Bowler 2001). Although higher levels of education often allow better access to fresh fruits and vegetables, they do not necessarily lead to healthier food choices (Atkins and Bowler).

POLITICAL INFLUENCES ON FOOD CHOICES

Political influences on food availability may have cultural consequences on food choices. During times of political discord, food can be used as a tool to reward and to punish. Withholding food may be a way in which cultural beliefs are expressed; however, food restriction also limits the variety of food choices. Food policies can also change the way food is acquired, and certain methods of food acquisition may surface that are culturally insensitive.

As food distribution continues to expand across the globe, political decisions that affect interacting food systems will ultimately affect food choices. Government food policies have large impacts on the cost of food insofar as price is a strong determinant of consumption. These policies also have potent influences on food selection. Food strategy designs, price guarantees, and import and export regulations are frequently of political origin (Weinbaum 1982). Since food is required for sustenance, it is a powerful channel for communication to establish connection, create obligations, and exert influence (Counihan 1999).

Regional conflicts and domestic instability often disrupt food supplies and limit food choices and may lead to hunger and malnutrition (Weinbaum 1982, Drèze and Sen 1991). Some recent examples include a famine in North Korea, the food shortages in Darfur, and the conflicts in the Middle East. In regions of conflict, receiving daily food rations may be dependent on safe passage of food convoys. These situations may promote unorthodox food selection methods that may not be culturally acceptable, such as theft, begging for money, and begging for food items (Booth 2006).

Purposefully withholding food from one's own family is an intolerable act in most cultures. However, restricting food choices has universally been used as a political weapon (Lappe and Collins 1978). For example, Stalin's political decisions included the system of collectivization, resulting in the seizure of all privately owned farms, and excessive export of goods, all of which intentionally caused the Ukraine famine (Library of Congress 2004). Another example involved agricultural policies made during Mao Zedong's Great Leap Forward campaign, which placed too much dependence on capital inputs for agricultural progress and led to food shortages and starvation for millions of people in China (Library of Congress 2005). Individuals may also participate in the deliberate avoidance of foods to take a political stance. For example, British suffragists in the early twentieth century refused foods while in prison, and Irish nationals went on hunger strikes later in the twentieth century (Library of Congress 2007).

Embargo and restriction of food imports can also be used to influence food choices. For example, meat from cows with bovine spongiform encephalopathy and fresh spinach contaminated with *Escherichia coli* are usually kept from entering the food system. Food embargoes may also prevent the food trade between countries. By issuing a food embargo, one nation's agricultural abundance is used as leverage to protest against another country's practices (Drèze and Sen 1991). Even if the country is capable of producing the food domestically, higher costs of domestic production may lead to economic difficulties and may force alternative food choices to be made.

Government and other political institutions can influence food choices in other ways as well. For instance, New York City has banned restaurants from using trans fatty acids—fats that contribute to heart disease—when preparing food on site. On a national level, health promotion guidelines such as the Dietary Guidelines for Americans, the National School Lunch Program, food safety regulations, and rules for food labeling and content disclosure are also influencing food choices (Atkins and Bowler 2001). Labeling of food products determines what information producers must provide to inform the consumers about the foods they purchase. Governments may also control food advertising. Food laws, tariffs, and trade agreements affect what is available within and between countries. Government subsidies given to individual farmers, partnerships, and corporations to balance out the "good years from the bad" also affect food production and take some of the risk out of agriculture, but they may also encourage overproduction.

PSYCHOLOGICAL INFLUENCES ON FOOD CHOICES

Culturally related psychological influences such as what foods are appropriate for consumption are reflected in human food choices. Manipulating the human mind is a powerful means by which to influence human food acquisition and eating behaviors.

Human food choices comprise a small fraction of what humans are physiologically capable of eating. For a food to make its way to the mouth, it must first be accepted by the individual, which is often culturally determined (Shepherd and Raats 2006). As a result, new flavors that are introduced into the mainstream culture may be modified by the process of acculturation to meet the cultural and customary tastes of the dominant culture (Lieberman 2003). Still, some members of the dominant culture refuse the new foods (refer to Figure 4.2b).

Some anthropologists suggest that foods must first be removed from "nature" and then transformed through "culture" to be psychologically acceptable and presentable at the table (Levi-Strauss 1983). For example, one seldom chooses meat in a supermarket labeled as cow, calf, pig, or sheep. Some cultures object to choosing whole fish. Instead, one is more likely to choose a T-bone steak, veal cutlet, pork tenderloin, mutton, or salmon steak. Foods are removed from their "natural" forms of "dead animal flesh" and are then trimmed, packaged, and renamed to appear less vulgar (Levi-Strauss 1983). Table 4.2 illustrates the conversion of foods from nature through culture and ultimately into foods that form the dishes and cuisines that are selected by individuals.

Certain foods are accepted or rejected based on whether they are perceived as "natural." For example, an underlying psychological fear of "unnatural" or "unhealthful" substances, such as pesticides, may cause individuals to avoid foods sprayed with pesticides since they are perceived as harmful. However, foods in their natural and unprocessed forms may still provide a breeding ground for microbial contamination and be unsafe. For example, rye can be contaminated with the fungus ergot, which causes ergotism and behavioral disorders (Bennett and Klich 2003, pp. 497–516). Peanuts can be contaminated by aflatoxin-B_1, a known liver carcinogen (pp. 497–516). A wide variety of natural and synthetic substances is added to foods to make them safe for storing and consuming over longer periods of time. Some of these chemicals repel insects; stabilize emulsions to prevent separation; add flavors and colors to increase palatability and visual acceptance; and add antioxidants to minimize oxidation of fats to protect flavor, all of which influence food choices (Williams 1975).

Some cultures categorize foods as hot or cold, not in *temperature*, but in the way the foods supposedly affect the body in different philosophical systems. "Hot" and "cold" foods also affect social relations because they represent nutrition and medicine in different ways (Pelto et al. 1989). Those who consume too much hot or cold foods may fall ill (Pelto et al. 1989). According to the hot-cold system, a balance of hot and cold foods is vital if one is to remain healthy.

While some humans rely on a meat-based diet as the principal source of sustenance, others rely on plants. Some believe human beings are not meant to eat animals. Vegetarian eating styles cover a broad spectrum (Figure 4.3). The bridging characteristic between vegan eating habits, lacto-ovo-vegetarian, lacto-vegetarian, and semivegetarian eating patterns is the complete or partial avoidance of meat. Vegetarian eating and vegetarianism are two different entities, although they often appear together. *Vegetarianism* is based on the belief that particular food and lifestyle behaviors may affect health and nutritional status (Twigg 1983, Dwyer 1995). Religious beliefs may also be expressed in food customs and practices (see Chapter 3).

In fact, there is a biological need for both animal and plant foods. If selected from animal products, protein is more readily available, and of higher biological value, than it is from plants. Some individuals may prefer one form of protein to the other; yet, many combine animal products with plant products. Rather than being a dichotomous choice, there is growing consensus that a spectrum of vegetarian and nonvegetarian eating patterns exists (Dwyer 1995), as illustrated in Figure 4.3. The only absolute taboo is that most humans would agree that choosing to eat other humans is not acceptable (Arens 1979).

For humans to adopt and accept new foods, there are certain psychological and cultural criteria that must be met (Fieldhouse 1996). There must be an advantage to adopting the new foods into their existing diets, and the new foods must be

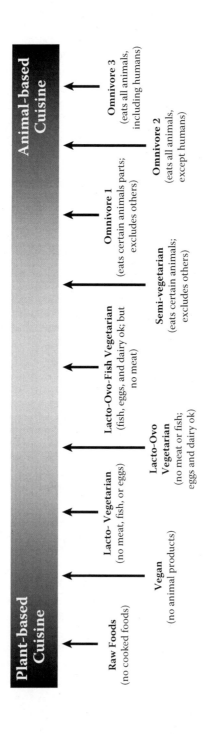

FIGURE 4.3   Plant-versus animal-based food spectrum. (Adapted from Harris and Ross 1987; and Beardsworth and Keil 1997.)

compatible with the already existing foods (Fieldhouse 1996). Furthermore, the new food must also provide minimal risk and have a favorable cost benefit (Fieldhouse 1996). Finally, the adoption of a new food is more likely to occur if another individual of the same culture has already accepted it (Fieldhouse 1996).

Food *neophobia*, or the fear of trying new foods, is also a culturally and psychologically influenced determinant of food choice. Foods that are not a regular part of one's eating habits are likely to be feared because they are unfamiliar. Depending on age and stage of development, children may be more—or less—likely to try new foods than adults because their food habits have not yet been established. Adults are also capable of accepting new foods, but certain conditions to foster acceptance are often needed.

Anxieties about food are additional psychological factors that affect cultural acceptance of foods. Anxiety pertaining to technology is one of the key players. Some fear that humans are straying too far away from nature, forgetting that many preventable diseases and food-borne pathogens also stem from natural causes. In fact, human beings are exposed to more carcinogens through naturally occurring chemicals that are produced by plants as a means of defending themselves against the same harsh environment in which humans adapted. Many human cancers are caused by smoking, lack of sufficient intake of fruits and vegetables, chronic infections, and hormonal factors and not by natural or synthetic pesticides (Abbott 1992, Ames and Gold 1997). Still, these individuals have anxieties not only over technological practices pertaining to food production—pesticides, genetic modification, hormones, and additives—but they also tend to fear technology in general (Devcich et al. 2007, pp. 333–337).

Mass communication and the media also exert powerful psychological influences on food choices. Through technologies such as the radio, television, and the Internet, producers are capable of establishing social norms to society in three ways. First, they provide information about food and sometimes educate. They also project images of social norms and perceived ideal body image. Finally, they encourage a lifestyle that lacks adequate physical activity (Fieldhouse 1996). Advertisements also generate feelings around foods that make people feel good about certain products that are high in excess calories and low in overall nutrition. Advertisers also target children in order to establish food choices that they will continue to follow into adulthood (Fieldhouse 1996).

CONCLUSIONS

"One generation plants the trees under whose shade future generations rest."

—Chinese proverb

Human food choices are influenced by many factors: politics, economics, psychology, biology, technology, social, religion, and the senses. Historically, the natural environment has limited all of these factors. Humans continue to face environmental constraints, increasing global population pressures, and the need to continue to find ways to ensure a safe, reliable, and *sustainable* food supply to meet the biological and cultural needs of future generations.

1. Explain how culture and customs influence the transformation of plants and animals from nature into distinctive cuisines.
2. Compare and contrast the various forms of plant- and animal-based diets and describe the factors influencing the selection of such diets.
3. *Case Study*: Anna is a Russian American female who was not raised in the United States. After moving to the United States, she experienced weight loss. She was referred to a registered dietitian (RD), who discovered that Anna had not been eating well because she was not used to the tastes of American foods. How might the RD help Anna integrate American foods into her diet while suggesting culturally appropriate cuisines?
4. *Project*: You are the owner of a restaurant in a large cosmopolitan city. Your clientele come from a variety of countries and you are trying to find a cost-effective way for your staff to adapt basic ingredients into dishes that appeal to your customers. Using the principles outlined in Table 4.2 and external resources, develop a culturally diverse menu, incorporating at least one staple food item that is common in several regions of the world.

REFERENCES

Abbott, P.J. Carcinogenic chemicals in food: evaluating the health risk. *Food and Chemical Toxicology* 30, no. 4, 1992. pp. 327–332.

Ames, B.N. and L.S. Gold. Environmental pollution, pesticides, and the prevention of cancer: misconceptions. *Federation of American Societies for Experimental Biology* 11, no. 14, 1997. pp. 1041–1052.

Angel, J.L. Health as a crucial factor in the changes from hunting to developed farming in the eastern Mediterranean. In *Paleopathology at the Origins of Agriculture*, Cohen, M.N. and Armelagos, G.J., Eds. New York: Academic Press, 1984.

Arens, W. *The Man-Eating Myth: Anthropology and Anthropophagy*. New York: Oxford University Press, 1979.

Atkins, P. and I. Bowler. *Food in Society: Economy, Culture, Geography*. London: Arnold, 2001.

Beardsworth, A. and Keil, T. *Sociology on the Menu: An Invitation to the Study of Food and Society*. London: Routledge, 1997.

Bennett, J.W. and M. Klich. Mycotoxins. *Clinical Microbiology Reviews* 16, no. 3, Jul 2003. 497–516.

Booth, S. Eating rough: food sources and acquisition practices of homeless young people in Adelaide, South Australia. *Public Health Nutrition* 9, no. 2, 2006. pp. 212–218.

Clark, R. *Global Life Systems: Population, Food, and Disease in the Process of Globalization*. Lanham, MD: Rowman and Littlefield, 2000.

Cohen, M.N. *Health and the Rise of Civilization*. New Haven, CT: Yale University Press, 1989.

Counihan, C. *The Anthropology of Food and Body: Gender, Meaning, and Power*. New York: Routledge, 1999.

Counihan, C. and P. Van Esterik, Eds. *Food and Culture: A Reader*. New York: Routledge, 1997.

de Castro, J.M. Social facilitation of duration and size but not rate of the spontaneous meal intake of humans. *Physiology and Behavior* no. 47, 1990. pp. 1129–1135.

de Castro, J.M. Family and friends produce greater social facilitation of food intake than other companions. *Physiology and Behavior* no. 56, 1994. pp. 445–455.

de Castro, J.M. What are the major correlates of macronutrient selection in Western populations? *Proceedings of the Nutrition Society* no. 58, 1999. pp. 755–763.

DeMan, J.M. *Principles of Food Chemistry*, 3rd ed. Gaithersburg, MD.: Aspen, 1999.

Devcich, D.A., I.K. Pedersen, and K.J. Petrie. You eat what you are: modern health worries and the acceptance of natural and synthetic additives in functional foods. *Appetite* 48, no. 3, May, 2007: 333–337.

Douglas, M. *Culture*. New York: Russell Sage Foundation, 1978.

Drewnowski, A. Taste preferences and food intake. *Annual Review of Nutrition* 17, 1997.

Drewnowski, A. and M.R.C. Greenwood. Cream and sugar: human preferences for high fat food. *Physiology and Behavior* no. 30, 1983.

Drèze, J. and A. Sen, Eds. *The Political Economy of Hunger*. New York: Oxford University Press, 1991.

Dwyer, J.T., Ed. *Nutrition in Women's Health*, Krummel, D.A. and P.M. Kris-Etherton, Eds. Gaithersburg, MD: Aspen, 1995.

Egan, T. *The Worst Hard Time: The Untold Story of Those Who Survived the Great American Dust Bowl*. Boston: Houghton Mifflin, 2006.

Ferraro, G.B. *Cultural Anthropology: An Applied Perspective,* 6th ed. Stamford, CT: Thomson Wadsworth, 2006.

Fieldhouse, P. *Food and Nutrition: Customs and Culture*, 2nd ed. Cheltenham, U.K.: Stanley Thornes, 1996.

Food and Agriculture Organization of the United Nations. *Globalization of Food Systems in Developing Countries: Impact on Food Security and Nutrition*. Rome: Food and Agriculture Organization of the United Nations, 2004.

Harris, M. and E.B. Ross, Eds. *Food and Evolution: Toward a Theory of Human Food Habits*. Philadelphia: Temple University Press, 1987.

Haviland, W.A., H.E.L. Prins, D. Walrath, and B. McBride. *Cultural Anthropology: The Human Challenge*, 11th ed. Belmont, CA: Wadsworth, 2004.

Hillman, G.C., S.M. Colledge, and D.R. Harris. Plant food economy during the Epipalaeothic Period at Tell Abu Hureyra, Syria: dietary diversity, seasonality, and modes of exploitation. In *Foraging and Farming. The Evolution of Plant Exploitation*, Harris, D.R. and G.C. Hillman, Eds. London: Unwin Hyman, 1989.

Hladik, C.M. Seasonal variations in food supply for wild primates. In *Coping with Uncertainty in Food Supply*, de Garine, I. and G.A. Harrison, Eds. Oxford, U.K.: Oxford University Press, 1988. pp. 1–25.

International Food Information Council. *Functional Foods. Backgrounder.* International Food Information Council Foundation Media Guide on Food Safety and Nutrition. Available at http://www.ific.org/nutrition/functional/index.cfm.

Johns, T. *The Origins of Human Diet and Medicine: Chemical Ecology*. Tucson: University of Arizona Press, 1990.

Jones, M.O., B. Giuliano, and R. Krell, Eds. *Foodways and Eating Habits: Directions for Research*. Los Angeles: California Folklore Society, 1983.

Kittler, P.G. and K.P. Sucher. *Food and Culture*, 4th ed. Belmont, CA: Thomson Wadsworth, 2004.

Lang, J.H., Ed. *Larousse Gastronomique: The New American Edition of the World's Greatest Culinary Encyclopedia*. New York: Crown, 1995.

Lappe, F.M. and J. Collins. *Food First: Beyond the Myth of Scarcity*. New York: Ballantine, 1978.

Levi-Strauss, C. *The Raw and the Cooked: Mythologiques* [Le Cru et le cuit], Vol. 1. Translated by Weightman, J. and D. Chicago: University of Chicago Press, 1983.

Library of Congress. Revelations from the Russian Archives: Ukrainian famine. 2004. Available at http://www.loc.gov/exhibits/archives/ukra.html. Accessed January 30, 2008.

Library of Congress. The Great Leap Forward 1958–60. 2005. Available at http://lcweb2.loc.gov/cgi-bin/query/r?frd/cstdy:@field(DOCID+cn0039). Accessed January 30, 2008.

Library of Congress. From abolition to equal rights (John Bull and Uncle Sam). 2007. Available at http://www.loc.gov/exhibits/british/brit-4.html. Accessed January 10, 2008.

Lieberman, L.S. Dietary, evolutionary, and modernizing influences on the prevalence of type 2 diabetes. *Annual Review of Nutrition* no. 23, 2003. pp. 345–377.

Martinez, S. Food product introductions continue to set records. 2007. Available at http://www.ers.usda.gov/amberwaves/november07/findings/foodproduct.htm.

Massachusetts Institute of Technology. Australian recipes. 2005. Available at http://web.mit.edu/~mitsos/www/WCA_cook2/Australian%20cooking%20class.pdf. Accessed October 5, 2008.

Mela, D. Eating behavior, food preferences and dietary intake in relation to obesity and bodyweight status. *Proceedings of the Nutrition Society* no. 55, 1996.

Mela, D.J. and S. Catt. Ontogeny of human taste and smell preferences and their implications for food selection. In *Longterm Consequences of Early Environment, Growth, Development, and the Lifespan: A Developmental Perspective*, Henry, C.J.K. and S.J. Ulijaszek. Oxford, U.K.: Oxford University Press, 1996.

Nabkasorn, C., N. Miyai, A. Sootmongkol, S. Junprasert, H. Yamamoto, M. Arita, and K. Miyashita. Effects of physical exercise on depression, neuroendocrine stress hormones and physiological fitness in adolescent females with depressive symptoms. *European Journal of Public Health* 16, no. 2, Apr 2006: 179–184.

Oxford University Press. *Oxford English Dictionary: The Definitive Record of the English Language*. Available at http://dictionary.oed.com.ezproxy.library.tufts.edu/. Accessed October 12, 2007.

Pelto, G.H., P.J. Pelto, and E. Messer, Eds. *Research Methods in Nutritional Anthropology*. Hong Kong: United Nations University Press, 1989.

Pyke, M. *Technological Eating*. London: Camelot Press, 1972.

Rozin, E. and P. Rozin. Some surprisingly unique characteristics of human food preferences. In *Foods in Perspective. Proceedings of the 34th International Conference on Ethnological Food Research, Cardiff, Wales*, Fenton, A., and T.M. Owens. Edinburgh: John Donal, 1981, pp. 243–252.

Rozin, P. *The Flavour Principle*. New York: Hawthorn, 1973.

Saher, M., A. Arvola, M. Lindemann, and L. Lähteenmäki. Impressions of functional food consumers. *Appetite* no. 42, 2003.

Scapp, R. and B. Seitz, Eds. *Eating Culture*. Albany: State University of New York Press, 1998.

Shepherd, R. and M. Raats, Eds. *The Psychology of Food Choice (Frontiers in Nutritional Science)*, Vol. 3. Wallingford, U.K.: CABI, 2006.

Smith, P., O. Bar-Yosef, and A. Sillen. Archaeological and skeletal evidence for dietary change during the late Pleistocene/early Holocene in the Levant. In *Paleopathology at the Origins of Agriculture*, Cohen, M.N. and G.J. Armelagos. New York: Academic Press, 1984.

Strickland, S.S. and P.S. Shetty, Eds. *Human Biology and Social Inequality*. Cambridge, U.K.: Cambridge University Press, 1998.

Twigg, J. *The Sociology of Food and Eating by Anne Murcott*. Aldershot, U.K.: Gower, 1983.

Ulijaszek, S.J. Human eating behavior in an evolutionary ecological context. *Proceedings of the Nutrition Society* no. 61, 2002.

Ulijaszek, S.J., G. Hillman, J.L. Boldsen, and C.J. Henry. Human dietary changes [and discussion]. *Philosophical Transactions: Biological Sciences* 334, no. 1270, 1991.

United States Department of Agriculture. Global climate change: questions and answers. Economic Research Service. 2004. Available at http://www.ers.usda.gov/Briefing/GlobalClimate/Qa.htm#impacts. Accessed January 31, 2008.

Weinbaum, M. *Food, Development, and Politics in the Middle East*. Boulder, CO: Westview Press, 1982.

Williams, R. *Physician's Handbook of Nutritional Science*. Springfield, IL: Charles C Thomas, 1975.

Section III

The Required Nutrients

5 Digestion and Absorption of Carbohydrates

Roberto Quezada-Calvillo and Buford L. Nichols

CONTENTS

ABSTRACT

Carbohydrates are the major dietary sources of energy for humans. While most dietary carbohydrates are derived from multiple botanical sources, lactose and trehalose are the only animal-derived carbohydrates of relevance for human diet. The nutritional value of all carbohydrates depends on their digestion into monosaccharides by glucosidases of the gastrointestinal tract. Digestion of starch, the carbohydrate most abundantly consumed by humans, depends on the concerted activity of the six enzyme activities: salivary and pancreatic amylases, sucrase-isomaltase, and maltase-glucoamylase. Simpler carbohydrates such as sucrose, lactose, or trehalose only require the activities of sucrase, lactase, and trehalase, respectively, for their digestion. Glucose, galactose, and fructose are the main monosaccharides produced by digestion of dietary carbohydrates that are metabolized by humans. Depending on the requirements of individuals, these monosaccharides may follow metabolic pathways of either energy generation or energy storage. The imbalance between energy expenditure and dietary intake of energy-rich foods has a direct relationship to the development of diseases involving energy metabolism, such as obesity or diabetes. These effects have led to attempting the classification of carbohydrates based on their potential to induce physiologic effects in humans. However, to understand and manipulate the physiologic responses of the human organism to carbohydrate feeding, detailed studies to identify the multiple enzyme/substrate interactions among the large variety of available carbohydrates and the eight glucosidic activities of the human gastrointestinal tract are necessary.

BACKGROUND

CARBOHYDRATES IN FOOD

Carbohydrates constitute the main source of energy supplied by foods and may account for up to 100% of the total dietary caloric intake. However, in an average Western diet carbohydrates contribute around 50% of the ingested calories. These macronutrients can be ingested in a large variety of forms and chemical structures, including monomers or polymers derived mainly from the monosaccharides glucose, fructose, and galactose. Starch is the most abundant digestible carbohydrate present in human foods; however, the consumption of simpler carbohydrates, such as sucrose or natural and synthetic syrups, has increased substantially with the development of industrialized foods. In addition, lactose is the characteristic sugar present in milk of most mammals. Other nondigestible carbohydrates, such as cellulose or inulins, are also present in plant foods and are classified as dietary fiber.

Without exception, the energetic nutritional value of carbohydrates depends on the ability of the human gastrointestinal tract to digest them, breaking them down and transforming them into monomers of glucose, fructose, or galactose. These monosaccharides are susceptible to absorption by specific transporters located in the epithelial cell layer lining the small intestine. Once absorbed, the monosaccharides are deposited into the blood circulation for their subsequent transport to the liver and other tissues throughout the body. In the liver, most monosaccharides different from glucose are enzymatically transformed into glucose for their subsequent processing by the metabolic machinery of the hepatic cells.

Ingestion of carbohydrates in the human diet has received particular attention in recent years due to the duality of their nutritional value. Carbohydrates constitute the main energy source for adult humans and have particular nutritional importance in children due to the high energy requirements during this developmental stage. The child's central nervous system (CNS) has much higher energetic requirement than the adult CNS. It is well known that the main energy source for the CNS is glucose; therefore, some of the most common food supplements in infant formulas are based on maltodextrins produced by partial chemical or enzymatic hydrolysis of starch (Holliday 1971; Elia 1992; Chugani 1998; Peters et al. 2004). In addition, carbohydrates are indispensable sources of energy for high-performance athletes; thus, maltodextrins and fructose are also normal components of energetic or "reconstituting" beverages. Ingestion of nondigestible carbohydrates, generically called dietary fiber, can cause qualitative changes in the intestinal flora, changes in intestinal motility, and changes in the chemical composition of feces. Some of the changes induced by the dietary fiber can be beneficial for gastrointestinal function and constitute the basis for the development of "prebiotic" and "probiotic" foods (Rolfe 2000; Freitas et al. 2003). In contrast with the beneficial nutritional effects of carbohydrates, their ingestion in large quantities has been associated with chronic and degenerative diseases such as obesity, diabetes, coronary disease, and metabolic syndrome (Holliday 1971; McCall 2004; Peters et al. 2004).

Depending on their biochemical and physicochemical properties, carbohydrates can therefore have diverse nutritional and physiologic effects. However, from the point of view of nutritional intake, carbohydrates have been classified as glucogenic and nonglucogenic, which describes their ability to produce absorbable free glucose monomers generating a glycemic response, that is, an increment in the blood glucose concentration (Azad and Lebenthal 1990; Wursch and Pi-Sunyer 1997). In general, glucogenic carbohydrates can be degraded in the gastrointestinal tract into the easily absorbable monosaccharides glucose, galactose, or fructose, which constitute most of the caloric intake in human diets. In contrast, nonglucogenic carbohydrates such as cellulose, inulin, and gum arabic cannot be degraded to monosaccharides due to the absence of enzymatic mechanisms for their digestion in the human gastrointestinal tract. Part of these nonglucogenic carbohydrates can be fermented by colonic bacterial flora with production of short-chain fatty acids like butyrate, propionate, and others but reduced energy production for the host. Nonglucogenic carbohydrates that are not fermented are excreted in the feces, contributing to their moisture and consistency (Anderson 1985; Correa-Matos et al. 2003; Englyst and Englyst 2005; Livesey 2005).

HISTORY OF MAJOR FOOD CARBOHYDRATES

The historic patterns of carbohydrate consumption by humans have experienced important changes. Carbohydrate consumption by primitive human populations most probably was limited, dependent on environmental or geographic conditions, and influenced by their collection patterns. Gathering of fruits, leaves, seeds, and tubers provided the main source of carbohydrates in a seasonal manner. Since the geographic distribution of carbohydrate-rich foods did not overlap with all human populations, some people consumed almost no food carbohydrate. Examples of this are the arctic human populations of Inuits, whose carbohydrate consumption before the arrival of Western civilization was almost nonexistent (McClellan and DuBois 1930; Gudmand-Hoyer and Jarnum 1969).

The development of agriculture marked a change in the type and quantity of carbohydrates available to humans (Dobrovolskaya 2005). This activity made available for human consumption large quantities of grains with a high content of carbohydrates, particularly in the form of starch, which provided and exceeded their caloric requirement. It is well accepted that the domestic crops of wheat, corn, rice, and barley were the foundations on which diverse civilizations developed throughout the world. Thus, starch derived from grain crops was the most abundant source of carbohydrates during the preindustrialized centuries of human civilization.

Although apparently cane sugar or sucrose was known in Europe centuries before Christ as part of imports from oriental countries, it was not until the colonization of America that crystallized sucrose became an important source of sweet carbohydrates for Western civilizations. Crops of cane sugar were among the first established in the new world by the European settlers, and they remain as one of the most important agricultural products for several countries of America and other developing countries (Tangley 1986; Japto 2000). However, the use of sucrose has remained mostly as a food sweetener and candy component. The Industrial Age also had an impact on the patterns of carbohydrate consumption by humans. Starch industrialization with procedures for its refinement and transformation into syrups by chemical or enzymatic methods has become one of the most important sources of products for production of a wide variety of foods. Simple carbohydrates, such as glucose and fructose, are important components in commercial beverages and baking. Presently, carbohydrates are available for human consumption in a multiplicity of forms and products from a large variety of raw materials of diverse botanical origin.

DISEASES ASSOCIATED WITH FOOD CARBOHYDRATE METABOLISM

The rapid change in patterns of carbohydrate consumption by humans, from sparse and occasional to daily large quantities, has not occurred without side effects on the human physiology. Human physiology appears to be well adapted for consumption of small quantities of carbohydrates distributed during the day, resembling the hypothetical carbohydrate consumption of early humans (Dobrovolskaya 2005). When an excess of carbohydrates was available for early humans, their storage in the form of fats was important to provide energy requirements during times of low food availability. However, present feeding patterns for some humans involve the repeated ingestion of relatively large quantities of carbohydrates in short periods of time. In many cases, the energy intake exceeds the requirement, leading to the continuous transformation and storage of carbohydrates in the form of body fat and causing eventual development of obesity. In addition, the frequent glycemic overload resulting from the digestion and absorption of large quantities of carbohydrates imposes a stress on the glycemic homeostatic mechanisms based on the pancreatic response of insulin/glucagon secretion. This chronic stress has been implicated as a damaging factor for pancreatic beta cells and cellular response mechanisms that contribute to the development of type 2 diabetes (Livesey 2005).

INTESTINAL BRUSH BORDER GLUCOSIDASES AND THEIR SUBSTRATES

LACTOSE/LACTASE-PHLORIZIN HYDROLASE

Lactose [β-d-galactopyranosyl(1,4')d-glucopyranoside] is the sugar abundant in the milk of mammalian species (milk sugar, from the Latin *lac* or milk) and is the first carbohydrate consumed during the lifetime of humans. Breast-fed infants consume approximately 50 g/day of lactose, but the nutritional relevance of this sugar is limited to the lactation period, when it provides most of the caloric requirements (Marquis et al. 2003; Foda et al. 2004). After weaning, other nutrients in foods can provide these nutritional requirements. Thus, consumption of dairy products by adult humans is unnecessary from a nutritional point of view. However, dairy products mainly from bovine origin are in fact important foods for a large number of individuals.

Digestion of lactose is dependent on the enzyme called lactase-phlorizinase (LPH), a protein bound to the apical cell membrane of the epithelial cells lining the small intestine. LPH is formed by the two subunits lactase and phlorizinase. The lactase subunit is the only known β-galactosidase of the intestinal lumen with the ability to degrade lactose into its two conforming monosaccharides: glucose and galactose (Buller et al. 1989; Montgomery et al. 1991). The defective expression of this enzyme may cause impaired ability to digest lactose. In most of the human population, the cessation of breast milk consumption after weaning is associated with a decrease in the levels of LPH expression and its corresponding activity in the small intestine, hampering the digestion of ingested lactose. This state, sometimes called *adult hypolactasia*, is therefore a normal developmental process of humans, with results evident during childhood or adolescence. However, in populations of north European origin, the levels of lactase activity may remain unchanged during the entire life. This state, sometimes referred to as *persistent lactasia*, is a distinctive feature of a minority human population that acquired the genetic ability to continue the expression of the enzyme after weaning, probably resulting as an adaptation to the high consumption of dairy products derived primarily from intensive cattle farming.

STARCH, MALTOSACCHARIDES, AND MALTOSE/MALTASE-GLUCOAMYLASE AND SUCRASE-ISOMALTASE

Starch and its derived products are by far the polysaccharides most consumed by the human population. Natural starches are produced in reproductive tissues of plants as a form of energy storage inside of granules in a semicrystalline form (French 1975; Lee 1983). In these granules, starch is present in two main molecular structures: amylose, which consists of long linear chains of glucose associated by α-1,4 glucosidic linkages and occasional branching with α-1,6 glucosidic linkages, and amylopectin, which consists of relatively short α-1,4-bound glucose chains of variable length with a relatively high content of α-1,6 branching chains. The linear structure of amylose molecules allows its aggregation in crystalline structures inside the granules termed A domains, while the branched structure of amylopectin produces its disorganized aggregation, forming the noncrystalline structure of B domains of the granules. Lipids and proteins are also found in small but variable proportion in the starch granule; however, these components have a relatively limited contribution to the physicochemical and organoleptic properties of starch. The proportion of amylose versus amylopectin, the average length of α-1,4 linear chains, and frequency of α-1,6 branching are variable between starches derived from different species or varieties of plants, affecting the degree of crystallization of the respective starch granules (Lee 1983; Robertson 1988; Gray 1992; Englyst et al. 1999). Highly crystalline granules with high amylose content require more severe conditions to attain their hydration and enzymatic hydrolysis. Corn, wheat, potato, tapioca, and rice are the major grain sources of commercial starch. However, hundreds of different botanical species have been used as sources of starch that differ in their physicochemical, organoleptic, or nutritional properties.

Effective total amounts of glucose and rate of release produced during starch digestion are therefore affected by its botanical origin, industrial processing, and form of preparation and presentation in foods. Thus, diverse procedures have been employed to measure starch product digestibility based on their rates and potential amounts of glucose release (α-glucogenesis). The most widely used procedure is the Englyst test based on the rates of glucose release during the enzymatic hydrolysis by preparations of fungal and animal glucosidases and the measurement of the proportions of starch that are rapidly (RD) or slowly (SD) digested, as well as the undigested or resistant starch (RS) fraction after 2 hours of enzymatic treatment (Englyst and Cummings 1990; Englyst et al. 1999; Englyst and Englyst 2005). Alternative methods attempting to measure the potential physiologic effects of starches, the glycemic index (Ludwig 2000; Brand-Miller et al. 2002; Leeds 2002; Englyst et al. 2003; Han et al. 2006) and insulinemic index (timed blood glucose or insulin concentrations, respectively) are measured in human subjects after ingestion of a standardized amount of starchy foods. However, these procedures are limited by the type and number of products that can be administered, the number of human subjects who can be studied, and the inherent physiologic variations among individuals, making difficult the acquisition of reproducible or comparable data sets. More recently, the use of recombinant human and mouse intestinal α-glucosidic enzymes has been introduced, but the convenience of these procedures awaits standardization with a wider collection of starch products and the formal comparison with more common procedures.

In humans, the digestion of starch occurs by enzymatic hydrolysis during its transit through the gastrointestinal tract and requires the participation of six different α-glucosidase enzyme activities (Corring 1980; Jones et al. 1983; Lee 1983; Fujita and Fuwa 1984; Gray 1992). Given the structural complexity and variability of starches as well as the diversity of enzymes involved in their hydrolysis, the detailed mechanisms for the process have not been determined. The most accepted models describe the process only superficially, without taking into account the complexity in structure and composition of starches, their intermediary products of degradation, or the different catalytic properties of each enzyme molecule interacting with specific substrate molecules. In this model, two luminal α-1,4 endoglucosidases, namely salivary and pancreatic amylases, respectively, hydrolyze linear unbranched segments with five or more glucose residues present in amylose and amylopectin molecules, releasing oligomers that have from two (maltose: O-α-d-glucopyrinosyl-(1,4)-α-d-glucopyranoside; the simplest glucose oligomer) to ten glucose residues but minimal production of free glucose (Gray 1975; Brayer et al. 2000; Horvathova et al. 2001). The segments containing α-1,6 branches are resistant to these amylase activities. To attain the effective release of free glucose, linear glucose oligomers resulting from the amylase digestion have to be further hydrolyzed by four α-1,4 exoglucosidic activities present in the mucosal epithelial cells of the small intestine. These activities are constituted by the enzymes sucrase-isomaltase (SI) and maltase-glucoamylase (MGAM), each composed of two subunits containing catalytic sites that act on the nonreducing ends of linear glucose oligomers with substantial release of free glucose monomers (Gray 1975; Corring 1980; Jones et al. 1983; Fujita and Fuwa 1984; Gray 1992). In addition, the isomaltase subunit of SI also shows α-1,6 glucosidic activity that cleaves the linkages present in the branching points of intermediate branched oligomers and isomaltose, releasing free glucose and the corresponding unbranched glucose oligomer.

More recent studies of the details of the catalytic mechanisms for starch digestion have demonstrated the oversimplification of that model (Quezada-Calvillo et al. 2007, 2008). For instance, amylase digestion of starches derived from different botanical sources or subject to different processing yields end products with characteristic composition of glucose oligomers that depended primarily on the nature of the initial substrate. The α-glucogenic abilities of MGAM and SI are not homogeneous for all glucose oligomers since differences of up to two orders of magnitude can be observed in α-glucogenesis from different glucose oligomers, which suggests that the efficiency of α-glucogenesis from starch-derived oligomers depends substantially on their relative abundance. Together, the α-glucogenic activities of luminal pancreatic amylase and of the two mucosal enzymes MGAM and SI are higher than the sum of each of the individual activities, indicating that they have a synergistic interaction enhancing the efficiency of the starch digestion. Despite the high specific activity displayed by MGAM in the hydrolysis of glucose oligomers, in humans this enzyme contributes only 20 to 30% of the total small intestinal mucosa α-glucogenic capacity due in part to the marked predominance of SI molecules in the apical membrane of epithelial cells. In addition, while MGAM experiences substrate inhibition by maltotriose and maltotetrose (three and four glucose residues, respectively), no inhibition of SI has been reported by glucose oligomers. This suggests that MGAM displays full activity only under low concentrations of these starch digestion products, while SI maintains its slow hydrolytic activity even at high concentrations of oligomers (Quezada-Calvillo et al. 2007, 2008). The effect may constitute a built-in physiologic mechanism to constrain the glucose overload that could occur during the digestion of large starchy meals, with MGAM working at full α-glucogenic capacity.

Sucrose and Isomaltose/Sucrase-Isomaltase

Sucrose, which is also known as cane or table sugar, is the disaccharide most abundantly consumed in the human diet as natural sweetener. Table sugar is refined from sugar cane or sugar beets, and various refined forms of sucrose are used in production of foods. Sucrose is also found in high concentration in fruits, such as ripe oranges and apples, where it serves as an easily available reserve of energy for the fast growth of a new plant. Intestinal digestion of sucrose [α-d-glucopyranosyl(1,2′) β′-d-fructofuranoside] requires hydrolysis of the α(1,2′)β′ linkage, which involves the anomeric carbon of both hexoses, to yield glucose and fructose. In mammalian species, the hydrolysis of sucrose is performed by the sucrase activity of the SI enzyme complex of the epithelial cells lining the intestine. The resulting monosaccharides are then absorbed by the intestinal epithelial cells into blood circulation.

In contrast to sucrose, isomaltose [O-α-d-glucopyranosyl-(1,6)-α-d-glucopyranoside] is a disaccharide found in very low concentrations in natural sources. However, this disaccharide is assumed to be a limiting product of the enzymatic hydrolysis of starches. The α-1,6 glucosidic linkage of isomaltose is hydrolyzed by the isomaltase subunit of the SI enzymatic complex present in the small intestinal brush border membrane. The reaction releases two molecules of glucose per molecule of isomaltose. Although isomaltose sometimes is considered as a "slow-digesting" product of amylase starch digestion, SI displays about 50% as much isomaltase activity as its maltase activity, which ensures enough activity to cope with the total isomaltose that may be produced in the starch digestive process. Although there is no experimental evidence, probably other α-1,6 branched oligosaccharides, such as isomaltotriose, isomaltotetrose, and so on, may be responsible for the slow digestion fraction observed in the hydrolysis products of starch.

Trehalose/Trehalase

Trehalose [O-α-d-glucopyranosyl-(1,1)-α-d-glucopyranoside] is a disaccharide of minor importance for the contemporary human diet since it is found in appreciable amounts only in some mushrooms and yeasts as well as in most insects' hemolymph. In insects, this disaccharide serves as a form of energy mobilization and appears to contribute to the preservation of the organism's viability after desiccation or exposure to freezing temperatures (Lillford and Holt 2002; Richards et al. 2002; Elbein et al. 2003). The enzyme responsible for the breakdown of trehalose is trehalase, which, like SI, LPH, and MGAM, is found in the brush border membrane of the intestinal epithelial cells; however, instead of the transmembrane peptide domain that anchors the other three mucosal glucosidic enzymes, trehalase is anchored to the apical membrane of enterocytes through a phophatidyl-inositol anchor (Semenza 1986). The presence of this enzyme in the human intestinal tract seems odd in view of the low consumption of foods containing trehalose; in fact, trehalase deficiency is the least frequently detected anomaly of carbohydrate digestion in humans, probably due to the low or absent presence of trehalose in modern human diets. However, this enzyme may have had an important role in early humans, for whom insect and fungal species may have constituted significant components of their diet. Each molecule of trehalose hydrolyzed renders two molecules of free glucose.

PRIMARY ROUTES OF CARBOHYDRATE METABOLISM IN MAMMALS

Glucose, galactose, and fructose are the three main carbohydrates that can be metabolized by the human body. Most other carbohydrates, if absorbed, have to be transformed into glucose before being metabolized; otherwise, they are excreted into the urine.

Depending on energetic requirements, glucose can follow different metabolic pathways: With high-energy expenditure, glucose follows predominantly a pathway leading to its oxidative transformation into CO_2, H_2O, and energy in the form of adenosine triphosphate (ATP). With low-energy expenditure, glucose follows pathways of storage in liver and muscle that consists of the formation of glycogen (glycogenesis). Storage in adipose tissue consists of glucose transformation into fatty acid precursors of lipids (Mathews et al. 1999).

GLUCOSE METABOLISM

Liver is the main organ controlling the metabolic route followed by glucose after absorption. The hepatic dominance in glucose metabolism is primarily defined by the enterohepatic circulation that carries most absorbed components of diet into the liver via portal circulation. Insulin and glucagon are in turn the main hormonal regulators for the rate of absorption of glucose into hepatic as well as other body tissues. Brain and muscle are the two extrahepatic tissues with the highest oxidative metabolism of glucose. The incorporation and "fixation" of glucose in the tissues depends on the glucose transporter GLUT2 and the enzyme hexokinase, which catalyzes the phosphorylation of glucose into glucose-6P. Hexokinase is activated by insulin, causing the fast absorption and retention of glucose by the body cells, while glucagon has the opposite effects on the enzyme. Once in the cytoplasm, glucose can follow either the oxidative pathways leading to the production of energy and involving glycolysis, citric acid pathway, and mitochondrial oxidative phosphorylation/respiratory chain or the storage pathways leading to glycogen and fat.

For the glycolytic pathway, glucose-6P is isomerized to fructose-6P and further phosphorylated to fructose-1,6-biphosphate. This step is the main regulator for the entrance of glucose into the glycolytic pathway with adenosine monophosphate (AMP), a product from consumption of ATP, as an allosteric activator of the enzyme 6-phosphofructokinase that catalyzes this phosphorylation reaction. Depending on the intracellular redox state, the glycolytic pathway can produce pyruvate or lactate as the main final products: A high concentration of H^+ in the form of NADH or NADPH plus H^+, products of the citric acid pathway, causes accumulation of lactate and a decreased pyruvate concentration, which limits its availability for the citric acid pathway; low concentrations of NADH or NADPH plus H^+ increase pyruvate concentration, which is the main substrate of the citric acid pathway. The entrance of pyruvate into the citric acid pathway, also known as the tricarboxylic acid cycle, involves its irreversible transformation into acetate by the enzyme pyruvate dehydrogenase, with removal of one carbon in the form of CO_2 and association to a cofactor known as coenzyme A (CoA). Pyruvate dehydrogenase is a multisubunit enzyme complex susceptible to allosteric regulation by a multiplicity of metabolites and thus is the main control point for the citric acid pathway; however, the most relevant regulators are the relative concentrations of adenosine diphosphate (ADP)/ATP and NAD/NADH plus H^+: ADP and NAD activate it, while the other two decrease its activity. Although in plant species the production of new glucose molecules (gluconeogenesis) is possible from acetate groups, in most mammals this step is completely irreversible; therefore, gluconoegenesis occurs through completely different pathways that depend on the transformation of amino acids into pyruvate. This is the main reason for the incapacity of mammals to produce glucose from the acetate groups generated from the β-oxidation of lipids. The acetate resulting from the pyruvate decarboxylation is then linked to a molecule of oxaloacetate with production of citric acid, from which the pathway takes its name. In the citric acid pathway, the two-carbon acetate molecule is broken down into CO_2 with the production of large quantities of the reduced intermediaries NADH plus H^+ and $FADH_2$ as well as some guanosine triphosphate (GTP) molecules. NADH plus H^+ and $FADH_2$ are then used as proton donors to generate a transmembrane electrochemical gradient that drives the production of ATP by the mitochondrial respiratory chain, which is the end product of glucose oxidative pathways.

In the storage pathways, when glucose-6P accumulates in the cytoplasm of liver and muscle cells due to low concentrations of AMP and low activity of 6-phosphofructoinase, glucose-6P can be isomerized to glucose-1P and then transformed into uridine diphosphate (UDP)-glucose to follow the pathway of glycogenesis, with production of glycogen as a form of energy storage in animal tissues. In addition, in adipose tissue the absorption of glucose causes the diversion of the glucolysis toward the production of glyceraldehyde-3P, which is used in the synthesis of fatty acids and triacylglycerides, without production of pyruvate, lactate, or energy (Mathews et al. 1999).

FRUCTOSE METABOLISM

Fructose can be metabolized directly by the glycolysis pathway through production of fructose-1P catalyzed by the enzyme fructokinase. By effects of the enzyme aldolase of the glycolysis pathway, fructose-1P is cleaved into dihydroxyacetone-P and glyceraldehyde; the last is further phosphorylated to produce glyceraldehyde-3P. These two products are identical to the two intermediate three-carbon molecules produced during the glycolysis from glucose and afterward follow the same metabolic processing. This metabolic route involves a shortcut from the classical glycolysis followed by glucose and may be the cause for the observed faster rate of fructose metabolism (Mathews et al. 1999).

GALACTOSE METABOLISM

For the catabolism of galactose, its conversion into glucose is indispensable by a complex pathway. First, galactose is phosphorylated by liver galactokinase; then, galactose-1P is exchanged for the glucose group present in UDP-glucose by the enzyme galactose-1P uridine transferase, releasing UDP-galactose and glucose-1P. UDP-galactose is then transformed into UDP-glucose by UDP-galactose 4-epimeraze. Deficiency of galactose-1P uridine transferase is an important form of galactose intolerance, causing jaundice, cirrhosis, cataracts, and CNS malfunction, probably as a consequence of intracellular accumulation of galactose. Once transformed into UDP-glucose, the monosaccharide can follow the glucogenic pathway or can be transformed into glucose-1P to follow any of the regular catabolic pathways of glucose (Mathews et al. 1999).

REGULATION OF ENZYME AND TRANSPORT PROTEIN SYNTHESIS AND PROCESSING

AMYLASE

Salivary and pancreatic amylases are coded by the genes AMY1 and AMY2, respectively, located in the human chromosome 1. In the typical human haploid genome, two copies of AMY2 (AMY2 A and B) are present (Zabel et al. 1983). However, the presence of multiple copies of the segment containing both AMY1 and AMY2 genes is common among human normal populations. In fact, an association between the amounts of starch ingested by selected human ethnic groups and the frequency of multiple copies of the genes has been observed, suggesting that multiple copies of amylase genes may be an adaptive feature for efficient digestion of high-starch diets (Iafrate et al. 2004).

Amylase secretion during early postnatal development is usually low, probably reflecting the still-immature pancreas (Montgomery et al. 1991). After birth, full production of amylases occurs as a weaning adaptation to carbohydrate feeding, and this stimulus remains during life as the primary stimulus leading to the synthesis, processing, and secretion of pancreatic amylase and other digestive enzymes. Thus, signaling for production and release of amylases seems to be mediated by neuronal pathways. Several neuroendocrine mediators, such as cholecystokinin or acetylcholine, have been observed to induce the secretion of zymogen granules from exocrine pancreatic cells via a Ca^{2+} influx-mediated mechanism (Williams 2001; Wasle and Edwardson 2002; Williams et al. 2002). Although the levels of expression of the messenger RNA (mRNA) coding for pancreatic amylase can be modified, the protein synthesis seems to be the actual limiting step controlling the production of pancreatic amylase. Salivary and pancreatic amylases are synthesized as proteins that are 511 amino acids long with molecular weight near 78 kDa. After cleavage of a secretory peptide 15 amino acids long, the mature form is transported and stored into secretory granules. The enzymes require Ca^{2+} and Cl^- ions to display their full activity (Brayer et al. 2000; Maurus et al. 2005).

LACTASE-PHLORIZINASE

The gene coding for human lactase-phlorizin hydrolase is located on chromosome 2 (Kruse et al. 1988). Lactase-phlorizin hydrolase is synthesized as a polypeptide of about 220 kDa, which after intracellular proteolytic processing produces a mature enzyme with a molecular mass of approximately 160 kDa. The exact mechanism of regulation of lactase-phlorizin hydrolase processing is obscure. The rate of processing of lactase-phlorizin hydrolase is accelerated by thyroid hormone, insulin, cortisol, and insulin-like growth factors (Shulman et al. 1992; Dudley et al. 1996, 1998; Burrin et al. 2001). However, its expression has been seen in cultured epithelial cells in the absence of hormonal stimuli (Kendall, Jumawan, and Koldovsky 1979; Simon-Assmann et al. 1986). After weaning, in most humans the lactase-phlorizinase activity decreases to values of less than 10% of those observed during infancy (Montgomery et al. 1991; Wang et al. 1998).

Genetic *cis*-regulation of lactase-phlorizin hydrolase transcription is mediated by its promoter region (Boudreau et al. 2001; Troelsen et al. 2003) as well as by diverse upstream sequences that bind transcription factors such as NF-LPH1 (Troelsen et al. 1992; Boudreau et al. 2002), GATA factors, hepatocyte nuclear factor HNF-1a, and homeodomain protein Cdx-2 (Spodsberg et al. 1999; Krasinski et al. 2001). In addition, single-nucleotide polymorphisms located at positions −13910 (C/T) and −22018 (G/A) from the start codon have been associated with the phenotype of persistent lactasia. These locations seem to be transcriptional enhancer sequences that activate the LPH promoter.

SUCRASE-ISOMALTASE

The SI gene is located on human chromosome 3 (Chantret et al. 1992). SI is synthesized as a single membrane protein 1827 amino acids long with a molecular size close to 210 kDa; it is inserted in the membrane of the enterocytes through its N-terminus (Sigrist et al. 1975; Hunziker et al. 1986) and subjected to intracellular glycosylation (Danielsen 1982). Once transported to the apical membrane, it is subjected to extracellular processing by pancreatic proteolytic enzymes generating a free C-terminal sucrase subunit and a membrane-bound N-terminal isomaltase subunit, which remain associated through noncovalent interactions (Cowell et al. 1986; Hu, Spiess, and Semenza 1987; Saphiro et al. 1991).

The SI activity is tightly linked to enterocyte differentiation and responds to similar transcription factors as the lactase-phlorizin hydrolase promoter (Cdx-2, HNF-1a, and GATA factors) (Wu et al. 1994; Taylor et al. 1997; Lynch et al. 2000; Silberg et al. 2000; Boudreau et al. 2002). Several major regulatory target sequences for the binding of these and other factors have been located upstream of the SI promoter, directing the cell-line-specific transcription of the gene. SI footprints (i.e., SIF1, SIF2, and SIF3) are regulatory elements of the SI gene promoter that bind the transcription factors Cdx-1 and Cdx-2 caudal-related proteins, HNF-1a, GATA-4, and GATA-6, playing important roles in the regulation of the tissue-specific gene expressions (Wu et al. 1994; Taylor et al. 1997; Lynch et al. 2000; Silberg et al. 2000; Boudreau et al. 2002).

Transport of SI molecules to the apical membrane requires its correct intracellular processing and, apparently, recognition by transporter systems. Diverse mutations affecting signals required for processing and transport have been described and appear to be the main causative factor of SI genetic deficiencies in humans

(Ouwendijk et al. 1996; Ouwendijk et al. 1998; Jacob et al. 2000; Spodsberg et al. 2001a, 2001b; Propsting et al. 2003; Ritz et al. 2003).

Maltase-Glucoamylase

The human MGAM gene is located on chromosome 7 (Nichols et al. 2003) and codes for a protein at least 1857 amino acids long. The studies of the regulation of MGAM activities have been hampered by the overlapping of substrates and activities of SI and MGAM. Some studies have suggested that MGAM expression and activity parallel those of SI (McDonald and Henning 1992). In addition, although the expression of MGAM mRNA has been observed in diverse tissues, its extraintestinal functions have not been described (Pereira and Sivakami 1991; Ben Ali et al. 1994). Elucidation of MGAM expression patterns has been complicated even more by the evident amplification of the gene segment coding for the C-terminal subunit of the protein (Naumoff 2007). In humans, at least four different tandem C-terminal segment amplifications can be found in the genome, and there is some evidence indicating the existence of alternative splicing within the primary transcript from these amplified segments. Studies of the mechanisms for gene regulation of this protein await to be performed.

Similar to SI, MGAM is inserted in the plasma membrane by its N-terminal end, but in contrast to SI, no proteolytic processing has been documented to occur for human MGAM (Naim et al. 1988). Although single-nucleotide gene polymorphisms have been described involving changes in the amino acid sequence of the protein, no association of these changes with deficiencies have been observed.

Monosaccharide Transporters

The transport of monosaccharides through cellular membranes, either at the intestinal epithelial layer or in other body tissues, is mediated by integral membrane proteins known as glucose transporters (GLTs). At the apical membrane of intestinal epithelial cells, the transport of glucose and galactose occurs against a concentration gradient that requires a special transporter able to overcome this energetic barrier (Wright et al. 1980; Wright et al. 1997, 2003, 2004). This active transporter is called sodium-dependent glucose transporter 1 (SGLT1) on chromosome 22, which is able to use the energy produced by the sodium concentration gradient existing across the cellular membranes to drive the transport of glucose or galactose into the epithelial cells. In contrast, fructose utilizes GLT5 on chromosome 1, which is a passive transporter that depends on the simple diffusion of this monosaccharide to cross the apical membrane of epithelial cells (Blakemore et al. 1995). Once inside the epithelial cells, all monosaccharides reach high enough concentration to exit the cell assisted by a facilitated diffusion transporter called GLT2 located on chromosome 3.

Synthesis of the GLT proteins seems to be directly dependent on transcription of the respective genes (Corpe and Burant 1996); however, transport of the synthesized proteins into the apical membrane seems to be dependent on hormonal and neuronal signals (Shu et al. 1997). While cyclic AMP (cAMP) and protein kinases play important roles in the transcription of the respective mRNA coding for the transporters

(Hirsch et al. 1996; Wright et al. 2004), circadian cycle and luminal carbohydrates have been observed to trigger the delivery of intracellular vesicles containing GLTs into the apical membrane. Some selectivity of these regulatory mechanisms can be observed since GLT5 expression at the cellular membranes depends largely on the availability of its respective mRNA, while expression of SGLT1 and GLT2 transporters depends primarily on their fast transport from intracellular storage into the plasma membrane.

CONCLUSIONS AND STUDY TOPICS

- Carbohydrates are the main source of human dietary energy intake.
- Most carbohydrates are botanical products for energy storage and transport.
- The monosaccharides glucose, galactose, and fructose are the basic building blocks of most polysaccharides, but their multimeric association gives rise to an almost infinite number of molecular structures with one, two, or up to millions of monosaccharide residues.
- Starch is the polysaccharide with the highest consumption by the human population.
- Salivary amylase, pancreatic amylase, SI, MGAM, and trehalase are the hydrolytic enzymes of the human gastrointestinal tract involved in hydrolysis of α-glucosidic bonds of polysaccharides for releasing their conforming monomers. In contrast, lactase-phlorizinase is the only β-glucosidase of the human gastrointestinal tract involved in the hydrolysis of lactose.
- Given the multiplicity of chemical structures of polysaccharides, particularly starch, their physiologic effect in terms of rate and total amount of monosaccharide release is variable for each preparation obtained from specific botanical sources.
- The absorption of monosaccharides for their subsequent metabolism requires membrane proteins called glucose or hexose transporters that passively or actively contribute to the mobilization of monosaccharides across the plasmatic membrane of cells.
- Depending on the individual energetic expenditure, glucose, galactose, and fructose may follow a metabolic pathway leading to the production of energy or storage of energy in the form of glycogen and fat.

REFERENCES

Anderson, J.W. (1985) Physiological and metabolic effects of dietary fiber. *Fed. Proc.* 44(14):2902–2906.

Azad, M.A. and Lebenthal, E. (1990) Role of rat intestinal glucoamylase in glucose polymer hydrolysis and absorption. *Pediatr. Res.* 28(2):166–170.

Ben Ali, H., et al. (1994) Relationship between semen characteristics, alpha-glucosidase and the capacity of spermatozoa to bind to the human zona pellucida. *Int. J. Androl.* 17(3):121–126.

Blakemore, S.J., et al. (1995) The GLUT5 hexose transporter is also localized to the basolateral membrane of the human jejunum. *Biochem. J.* 309(Pt 1):7–12.

Boudreau, F., et al. (2002) Hepatocyte nuclear factor-1 alpha, GATA-4, and caudal related homeodomain protein Cdx2 interact functionally to modulate intestinal gene transcription. Implication for the developmental regulation of the sucrase-isomaltase gene. *J. Biol. Chem.* 277(35):31909–31917.

Boudreau, F., Zhu, Y., and Traber, P.G. (2001) Sucrase-isomaltase gene transcription requires the hepatocyte nuclear factor-1 (HNF-1) regulatory element and is regulated by the ratio of HNF-1 alpha to HNF-1 beta. *J. Biol. Chem.* 276(34):32122–32128.

Brand-Miller, J.C., et al. (2002) Glycemic index and obesity. *Am. J. Clin. Nutr.* 76(Suppl.):S281–S285.

Brayer, G.D., et al (2000) Subsite mapping of the human pancreatic alpha-amylase active site through structural, kinetic, and mutagenesis techniques. *Biochemistry* 39(16):4778–4791.

Buller, H.A., et al. (1989) New insights into lactase and glycosylceramidase activities of rat lactase-phlorizin hydrolase. *Am. J. Physiol.* 257(4 Pt 1):G616–G623.

Burrin, D.G., et al. (2001) Oral IGF-I alters the posttranslational processing but not the activity of lactase-phlorizin hydrolase in formula-fed neonatal pigs. *J. Nutr.* 131(9):2235–2241.

Chantret, I., et al. (1992) Sequence of the complete cDNA and the 5' structure of the human sucrase-isomaltase gene. Possible homology with a yeast glucoamylase. *Biochem. J.* 285:915–923.

Chugani, H.T. (1998) A critical period of brain development: studies of cerebral glucose utilization with PET. *Prev. Med.* 27(2):184–188.

Corpe, C.P. and Burant, C.F. (1996) Hexose transporter expression in rat small intestine: effect of diet on diurnal variations. *Am. J. Physiol.* 271(1 Pt 1):G211–G216.

Correa-Matos, N.J., et al. (2003) Fermentable fiber reduces recovery time and improves intestinal function in piglets following *Salmonella typhimurium* infection. *J. Nutr.* 133(6):1845–1852.

Corring, T. (1980) The adaptation of digestive enzymes to the diet: its physiological significance. *Reprod. Nutr. Dev.* 20(4B):1217–1235.

Cowell, G.M., et al. (1986) Topology and quaternary structure of pro-sucrase/isomaltase and final-form sucrase/isomaltase. *Biochem. J.* 237:455–461.

Danielsen, E.M. (1982) Biosynthesis of intestinal microvillar proteins. Pulse-chase labelling studies on aminopeptidase N and sucrase-isomaltase. *Biochem. J.* 204(3):639–645.

Dobrovolskaya, M.V. (2005) Upper paleolithic and late stone age human diet. *J. Physiol. Anthropol. Appl. Hum. Sci.* 24(4):433–438.

Dudley, M.A., et al. (1996) Lactase phlorhizin hydrolase turnover in vivo in water-fed and colostrum-fed newborn pigs. *Biochem. J.* 320(Pt 3):735–743.

Dudley, M.A., et al. (1998) Protein kinetics determined in vivo with a multiple-tracer, single-sample protocol: application to lactase synthesis. *Am. J. Physiol.* 274 (3 Pt 1):G591–G598.

Elbein, A.D., et al. (2003) New insights on trehalose: a multifunctional molecule. *Glycobiology* 13(4):17R–27R.

Elia, M. (1992) Organ and tissue contribution to metabolic rate. In: Kenney, J.M. and Tucker, H.N. (Eds.), *Energy Metabolism: Tissue Determinants and Cellular Corollaries.* New York: Raven Press, pp. 61–79.

Englyst, H.N. and Cummings, J.H. (1990) Non-starch polysaccharides (dietary fiber) and resistant starch. *Adv. Exp. Med. Biol.* 270:205–225.

Englyst, K.N. and Englyst, H.N. (2005) Carbohydrate bioavailability. *Br. J. Nutr.* 94(1):1–11.

Englyst, K.N., et al. (1999) Rapidly available glucose in foods: an in vitro measurement that reflects the glycemic response. *Am. J. Clin. Nutr.* 69(3):448–454.

Englyst, K.N., et al. (2003) Glycaemic index of cereal products explained by their content of rapidly and slowly available glucose. *Br. J. Nutr.* 89(3):329–340.

Foda, M.I., et al. (2004) Composition of milk obtained from unmassaged versus massaged breasts of lactating mothers. *J. Pediatr. Gastroenterol. Nutr.* 38(5):484–487.

Freitas, M., et al. (2003) Host-pathogens cross-talk. Indigenous bacteria and probiotics also play the game. *Biol. Cell* 95(8):503–506.

French, D. (1975) *Chemistry and Biochemistry of Starch.* Baltimore: Butterworth.

Fujita, S. and Fuwa, H. (1984) The influence of hardly digestive starch granules on sucrase and isomaltase in small intestinal mucosa of rats. *J. Nutr. Sci. Vitaminol. (Tokyo)* 30(2):135–142.

Gray, G.M. (1975) Carbohydrate digestion and absorption. Role of the small intestine. *N. Engl. J. Med.* 292(23):1225–1230.

Gray, G.M. (1992) Starch digestion and absorption in nonruminants. *J. Nutr.* 122(1):172–177.

Gudmand-Hoyer, E., and Jarnum, S. (1969) Lactose malabsorption in Greenland Eskimos. *Acta Med. Scand.* 186:235–237.

Han, X.Z., et al (2006) Development of a low glycemic maize starch: preparation and characterization. *Biomacromolecules* 7(4):1162–1168.

Hirsch, J.R., Loo, D.D., and Wright, E.M. (1996) Regulation of Na+/glucose cotransporter expression by protein kinases in *Xenopus laevis* oocytes. *J. Biol. Chem.* 271(25):14740–14746.

Holliday, M.A. (1971) Metabolic rate and organ size during growth from infancy to maturity and during late gestation and early infancy. *Pediatrics* 47(Suppl)(1):169–179.

Horvathova, V., Janecek, S., and Sturdik, E. (2001) Amylolytic enzymes: molecular aspects of their properties. *Gen. Physiol. Biophys.* 20(1):7–32.

Hu, C.B., Spiess, M., and Semenza, G. (1987) The mode of anchoring and precursor forms of sucrase-isomaltase and maltase-glucoamylase in chicken intestinal brush-border membrane. Phylogenetic implications. *Biochim. Biophys. Acta* 896(2):275–286.

Hunziker, W., et al (1986) The sucrase-isomaltase complex: primary structure, membrane-orientation, and evolution of a stalked, intrinsic brush border protein. *Cell* 46(2):227–234.

Iafrate, A.J., Feuk, L., Rivera, M.N., Listewnik, M.L., Donahoe, P.K., Qi, Y., Scherer, S.W., and Lee, C. (2004) Detection of large-scale variation in the human genome. *Nat. Genet.* 36(9):949–951.

Jacob, R., et al. (2000) Congenital sucrase-isomaltase deficiency arising from cleavage and secretion of a mutant form of the enzyme. *J. Clin. Invest.* 106(2):281–287.

Japto, M. (2000) Sugarcane as history in Paule Marshall's "To Da-Duh, in memoriam." *African American Rev.* 34(3):475–482.

Jones, B.J., et al. (1983) Glucose absorption from starch hydrolysates in the human jejunum. *Gut* 24(12):1152–1160.

Kendall, K., Jumawan, J., and Koldovsky, O. (1979) Development of jejunoileal differences of activity of lactase, sucrase and acid beta-galactosidase in isografts of fetal rat intestine. *Biol. Neonate* 36(3–4):206–214.

Krasinski, S.D., et al. (2001) Differential activation of intestinal gene promoters: functional interactions between GATA-5 and HNF-1 alpha. *Am. J. Physiol. Gastrointest. Liver Physiol.* 281(1):G69–G84.

Kruse, T.A., et al. (1988) The human lactase-phlorizin hydrolase gene is located on chromosome 2. *FEBS Lett.* 240:123–126.

Lee, P.C. (1983) Digestibility of starches and modified food starches. *J. Pediatr. Gastroenterol. Nutr.* 2(Suppl. 1):S227–S232.

Leeds, A.R. (2002) Glycemic index and heart disease. *Am. J. Clin. Nutr.* 76 (Suppl.):S286–S289.

Lillford, P.J. and Holt, C.B. (2002) In vitro uses of biological cryoprotectants. *Philos. Trans. R. Soc. Lond. B Biol. Sci.* 357(1423):945–951.

Livesey, G. (2005) Low-glycaemic diets and health: implications for obesity. *Proc. Nutr. Soc.* 64(1):105–113.

Ludwig, D.S. (2000) Dietary glycemic index and obesity. *J. Nutr.* 130(Suppl. 2):S280–S283.

Lynch, J., et al. (2000) The caudal-related homeodomain protein Cdx1 inhibits proliferation of intestinal epithelial cells by down-regulation of D-type cyclins. *J. Biol. Chem.* 275(6):4499–4506.

Marquis, G.S., et al. (2003) An overlap of breastfeeding during late pregnancy is associated with subsequent changes in colostrum composition and morbidity rates among Peruvian infants and their mothers. *J. Nutr.* 133(8):2585–2591.

Mathews, C.K., van Holde, K.E., and Ahern, K.G. (1999) 13. *Carbohydrate Metabolism I: Anaerobic Processes in Generating Metabolic Energy. Biochemistry.* New York: Prentice Hall.

Maurus, R., et al. (2005) Structural and mechanistic studies of chloride induced activation of human pancreatic alpha-amylase. *Protein Sci.* 14(3):743–755.

McCall, A.L., et al. (2004) Cerebral glucose metabolism in diabetes mellitus. *Eur. J. Pharmacol.* 490(1–3):147–158.

McClellan, W.S. and DuBois, E.F. (1930) Clinical calorimetry XLV. Prolonged meat diets with a study of kidney function and ketosis. *J. Biol. Chem.* 87:651–668.

McDonald, M.C. and Henning, S.J. (1992) Synergistic effects of thyroxine and dexamethasone on enzyme ontogeny in rat small intestine. *Pediatr. Res.* 32(3):306–311.

Montgomery, R.K., et al. (1991) Lactose intolerance and the genetic regulation of intestinal lactase-phlorizin hydrolase. *FASEB J.* 5:2824–2832.

Naim, H.Y., Sterchi, E.E., and Lentze, M.J. (1988) Structure, biosynthesis, and glycosylation of human small intestinal maltase-glucoamylase. *J. Biol. Chem.* 263:19709–19717.

Naumoff, D.G. (2007) Structure and evolution of the mammalian maltase-glucoamylase and sucrase-isomaltase genes. *Mol. Biol. (Mulekulyarnaya Biologiya)* 41(6):962–973.

Nichols, B.L., et al. (2003) The maltase-glucoamylase gene: common ancestry to sucrase-isomaltase with complementary starch digestion activities. *Proc. Natl. Acad. Sci. U. S. A.* 100(3):1432–1437.

Ouwendijk, J., et al. (1996) Congenital sucrase-isomaltase deficiency. Identification of a glutamine to proline substitution that leads to a transport block of sucrase-isomaltase in a pre-Golgi compartment. *J. Clin. Invest.* 97(3):633–641.

Ouwendijk, J., et al. (1998) Analysis of a naturally occurring mutation in sucrase-isomaltase: glutamine 1098 is not essential for transport to the surface of COS-1 cells. *Biochim. Biophys. Acta* 1406(3):299–306.

Pereira, B. and Sivakami, S. (1991) A comparison of the active site of maltase-glucoamylase from the brush border of rabbit small intestine and kidney by chemical modification studies. *Biochem. J.* 274 (Pt. 2):349–354.

Peters, A., et al. (2004) The selfish brain: competition for energy resources. *Neurosci. Biobehav. Rev.* 28(2):143–180.

Propsting, M.J., Jacob, R., and Naim, H.Y. (2003) A glutamine to proline exchange at amino acid residue 1098 in sucrase causes a temperature-sensitive arrest of sucrase-isomaltase in the endoplasmic reticulum and cis-Golgi. *J. Biol. Chem.* 278(18):16310–16314.

Quezada-Calvillo, R., et al. (2007) Luminal substrate "brake" on mucosal maltase-glucoamylase activity regulates total rate of starch digestion to glucose. *J. Pediatr. Gastroenterol. Nutr.* 45(1):32–43.

Quezada-Calvillo, R., et al. (2008) Luminal starch substrate "brake" on maltase-glucoamylase activity is located within the glucoamylase subunit. *J. Nutr.* 138:685–692.

Richards, A.B., et al. (2002) Trehalose: a review of properties, history of use and human tolerance, and results of multiple safety studies. *Food Chem. Toxicol.* 40(7):871–898.

Ritz, V., et al. (2003) Congenital sucrase-isomaltase deficiency because of an accumulation of the mutant enzyme in the endoplasmic reticulum. *Gastroenterology* 125(6):1678–1685.

Robertson, J.A. (1988) Physicochemical characteristics of food and the digestion of starch and dietary fibre during gut transit. *Proc. Nutr. Soc.* 47:143–152.

Rolfe, R.D. (2000) The role of probiotic cultures in the control of gastrointestinal health. *J. Nutr.* 130(2S Suppl.):396S–402S.

Saphiro, G.L., et al. (1991) Postinsertional processing of sucrase-alpha-dextrinase precursor to authentic subunits: multiple steep cleavage by trypsin. *Am. J. Physiol.* 24:G847–G857.

Semenza, G. (1986) Anchoring and biosynthesis of stalked brush border membrane proteins: glycosidases and peptidases of enterocytes and renal tubuli. *Annu. Rev. Cell Biol.* 2:255–313.

Shu, R., David, E.S., and Ferraris, R.P. (1997) Dietary fructose enhances intestinal fructose transport and GLUT5 expression in weaning rats. *Am. J. Physiol.* 272(3 Pt. 1):G446–G453.

Shulman, R.J., et al. (1992) Effect of oral insulin on lactase activity, mRNA, and posttranscriptional processing in the newborn pig. *J. Pediatr. Gastroenterol. Nutr.* 14(2):166–172.

Sigrist, H., Ronner, P., and Semenza, G. (1975) A hydrophobic form of the small-intestinal sucrase-isomaltase complex. *Biochim. Biophys. Acta* 406(3):433–446.

Silberg, D.G., et al. (2000) Cdx1 and cdx2 expression during intestinal development. *Gastroenterology* 119(4):961–971.

Simon-Assmann, P., et al. (1986) Maturation of brush border hydrolases in human fetal intestine maintained in organ culture. *Early Hum. Dev.* 13(1):65–74.

Spodsberg, N., et al. (1999) Transcriptional regulation of pig lactase-phlorizin hydrolase: involvement of HNF-1 and FREACs. *Gastroenterology* 116(4):842–854.

Spodsberg, N., Alfalah, M., and Naim, H.Y. (2001a) Characteristics and structural requirements of apical sorting of the rat growth hormone through the O-glycosylated stalk region of intestinal sucrase-isomaltase. *J. Biol. Chem.* 276(49):46597–46604.

Spodsberg, N., Jacob, R., Alfalah, M., Zimmer, K.P., and Naim, H.Y. (2001b) Molecular basis of aberrant apical protein transport in an intestinal enzyme disorder. *J. Biol. Chem.* 276(26):23506–23510.

Tangley, L. (1986) Sugarcane: Cuba's "noble crop." *BioScience* 3(7):414–420.

Taylor, J.K., et al. (1997) Activation of enhancer elements by the homeobox gene Cdx2 is cell line specific. *Nucleic Acids Res.* 25(12):2293–2300.

Troelsen, J.T., et al. (1992) A novel intestinal trans-factor (NF-LPH1) interacts with the lactase-phlorizin hydrolase promoter and co-varies with the enzymatic activity. *J. Biol. Chem.* 267:20407–20411.

Troelsen, J.T., et al. (2003) An upstream polymorphism associated with lactase persistence has increased enhancer activity. *Gastroenterology* 125(6):1686–1694.

Wang, Y., et al. (1998) The genetically programmed down-regulation of lactase in children. *Gastroenterology* 114(6):1230–1236.

Wasle, B. and Edwardson, J.M. (2002) The regulation of exocytosis in the pancreatic acinar cell. *Cell Signal.* 14(3):191–197.

Williams, J.A. (2001) Intracellular signaling mechanisms activated by cholecystokinin-regulating synthesis and secretion of digestive enzymes in pancreatic acinar cells. *Annu. Rev. Physiol.* 63:77–97.

Williams, J.A., et al. (2002) Cholecystokinin activates a variety of intracellular signal transduction mechanisms in rodent pancreatic acinar cells. *Pharmacol. Toxicol.* 91(6):297–303.

Wright, E.M., et al. (1997) Regulation of Na+/glucose cotransporters. *J. Exp. Biol.* 200 (Pt. 2):287–293.

Wright, E.M., et al. (2004) Surprising versatility of Na+-glucose cotransporters: SLC5. *Physiology (Bethesda)* 19:370–376.

Wright, E.M., Martin, M.G., and Turk, E. (2003) Intestinal absorption in health and disease— sugars. *Best. Pract. Res. Clin. Gastroenterol.* 17(6):943–956.

Wright, E.M., van Os, C.H., and Mircheff, A.K. (1980) Sugar uptake by intestinal basolateral membrane vesicles. *Biochim. Biophys. Acta* 597(1):112–124.

Wu, G.D., et al. (1994) Hepatocyte nuclear factor-1 alpha (HNF-1 alpha) and HNF-1 beta regulate transcription via two elements in an intestine-specific promoter. *J. Biol. Chem.* 269(25):17080–17085.

Wursch, P., and Pi-Sunyer, F.X. (1997) The role of viscous soluble fiber in the metabolic control of diabetes. A review with special emphasis on cereals rich in beta-glucan. *Diabetes Care* 20(11):1774–1780.

Zabel, B.U., et al. (1983) High-resolution chromosomal localization of human genes for amylase, proopiomelanocortin, somatostatin, and a DNA fragment (D3S1) by in situ hybridization. *Proc. Natl. Acad. Sci. U. S. A.* 80(22):6932–6936.

6 Lipids[1]

Duane E. Ullrey

CONTENTS

ABSTRACT

A brief review of the nature of lipids, their historical uses, and early studies of their chemistry and essentiality is followed by a discussion of their structural and functional roles and their digestion and absorption. Fatty acids are classified as saturated, monounsaturated, or polyunsaturated, and *trans* fatty acids are characterized. Essential fatty acids are described, and potential food sources are listed, along with estimated human adequate intakes of lipids and essential fatty acids.

ADIPOSITY TO ESSENTIALITY

Lipids are substances that are greasy to the touch, insoluble in water, and soluble in ether and include fats, oils, waxes, fatty acids, cholesterol, and related compounds. An understanding of the physiological significance of dietary lipids was late arriving, but early observers noted that well-fed animals or humans accumulated body fat deposits that disappeared rapidly as a consequence of starvation or hard work. In ancient Egypt, oils were used as cosmetics and as medicines, and one report stated that a mixture of fats from gazelle, serpents, crocodiles, and hippopotami was good for growing hair. Soap-making by boiling tallow (beef fat) with alkali was an ancient art, and boiling cattle hooves in water produced a floating liquid, called neat's-foot oil, that was a useful lubricant because it did not solidify, even at freezing temperatures.

In Paris, F. C. H. Vogel conducted studies with lard (hog fat) and human fat and noted in 1805 that when they were broken down by distillation, the product was acidic. Fatty acids were described by M. E. Chevreul, a professor at Lycée Charlemagne, in an 1823 treatise on the chemistry of fats; also noted was that glycerol separates from fats when they are converted into soaps. Glycerol was originally discovered in 1783 by C. W. Scheele, a Swedish chemist, and named glycerin (after the Greek word *glykeros*, for sweet). Marcellin Berthelot established the chemical nature of a solid fat (comparable to that found in tallow) in 1854 by heating glycerol with stearic acid (a fatty acid) in a closed vessel to synthesize tristearin (three stearic acid molecules combined with glycerol, known generically as a triacylglycerol).

The late eighteenth and the nineteenth century was an extremely active period for lipid chemists. Oxidative rancidity was described, and the chemical nature of waxes, phospholipids, cholesterol, and some of the saturated and unsaturated (containing one or more double bond) fatty acids was explored. One of the more interesting controversies during this period concerned whether body fat originated only from fat in the diet or whether dietary carbohydrate could be converted to body fat. Ultimately, it was established that body fat can have its origins in dietary fat, carbohydrate, or protein.

It was not until 1930 that George and Mildred Burr, at the University of Minnesota, discovered the essential nature of a fatty acid. When linoleic acid was missing from the diet, specific deficiency signs appeared, indicating that, for this lipid, carbohydrate was no substitute. For the cat (and perhaps most carnivores), it was shown in 1975 that arachidonic acid is also a dietary essential because it cannot be synthesized from linoleic acid in the cat's tissues, even though such synthesis is possible in humans. Evidence for the essentiality of other fatty acids was equivocal until 1967, when α-linolenic acid was shown to be required by rainbow trout and in 1982 was shown to be required by humans.

LIPID FUNCTIONS

ENERGY RESERVE

As humans grow, mature, and age, characteristic changes in body composition occur. These changes are influenced by gender, heredity, stature, and of course, food supply in relation to need. One of these changes involves deposition of fat in adipocytes (fat cells), producing substantial deposits of body fat that can insulate against body heat loss and serve as an energy reserve. Various procedures have been used to measure the size of these adipose deposits, and about 59 million adult Americans have fat stores so large as to be classified as obese. In these individuals, consumption of energy far exceeds their energy expenditure. Fat is an efficient energy storage form because every gram of fat can provide more than twice as much energy in support of body metabolism as can carbohydrate or protein. However, excess body fat is often accompanied by health problems—a serious consequence of *too much* food and too little exercise. Unfortunately, *too little* food characterizes the lives of much of the world's population, and the benefits of a substantial body fat store to buffer short periods of food scarcity are not available to them.

THERMOGENESIS

Specialized body fat deposits, with an unusually rich blood vessel supply, have been designated brown adipose tissue. They are characterized by the ability to generate heat in response to food intake or to prolonged cold exposure by a process called *nonshivering thermogenesis*. Although present in newborns, particularly between the shoulder blades (interscapular area) and useful in elevating body temperature after birth, brown adipose tissue appears to play little role in maintaining the body temperature of human adults.

CELL MEMBRANE STRUCTURE AND FUNCTION

Particular fatty acids are found in phospholipids that are critical for cell membrane structure and function. They provide physical support to the membranes, serve as a source of physiologically active compounds, and modulate cross-membrane movement of metabolically active substances.

NUTRIENT TRANSPORT

In addition to serving as a source of essential fatty acids, lipids are involved in transport of the lipid-soluble vitamins A, D, E, and K and provitamin A carotenoids.

HORMONAL ACTIVITY

No longer is body fat considered just a source of stored energy. In recent years, white adipose tissue has been shown to secrete over ten peptide hormones—compounds that contain two or more amino acids. By means of these hormones (such as leptin, which affects appetite and the immune system), fat serves also as an endocrine organ, sustaining energy homeostasis through regulation of glucose and lipid metabolism, modulating the immune response, and influencing reproduction.

DIGESTION AND ABSORPTION

Fats are hydrophobic (have a limited affinity for water). However, the enzymes that digest them normally function in an aqueous environment. Thus, fat digestion is enhanced through emulsification by bile salts and fat breakdown products, producing small fat droplets with greater surface area and an increased accessibility to fat-splitting enzymes (lipases). Fat digestion may begin in the stomach through the action of lingual lipase, secreted by glands lying underneath the tongue. However, most fat digestion occurs in the lumen of the upper small intestine, where pancreatic lipase splits fatty acids from triacylglycerol to produce diacylglycerol, then monoacylglycerol. Only a small percentage of triacylglycerol is completely hydrolyzed to glycerol and free fatty acids.

The products of digestion combine with bile salts to form small aggregates called *micelles* (which also may contain fat-soluble vitamins and cholesterol). These micelles interact with the luminal membrane of the enterocyte (intestinal mucosal cell), releasing their contents to the enterocyte interior. Short-chain fatty acids (fewer than 10 to 12 carbon atoms) may pass through the cell and the basolateral membrane into the portal blood, where they combine with albumin and are transported to the liver. Longer-chain fatty acids and monoacylglycerols are reassembled into triacylglycerols within the interior of the enterocyte, are coated with protein to which carbohydrate is attached, and are formed into chylomicrons (which also carry fat-soluble vitamins, cholesterol, and cholesterol esters [cholesterol combined with a fatty acid]), which pass through the basolateral membrane into the lymphatics and ultimately to the blood.

FATTY ACIDS

CLASSIFICATION

Saturated
Fatty acids with carbon chain lengths of 4 to 24 carbon atoms are found in foods and body tissues. When connected by single bonds and with hydrogen atoms occupying all available positions, the fatty acids are designated as saturated. Examples include myristic acid (with a 14-carbon chain and no double bonds, 14:0), palmitic acid (16:0), stearic acid (18:0), arachidic acid (20:0), and lignoceric acid (24:0).

Monounsaturated
Fatty acids with one unsaturated bond include palmitoleic acid (a 16-carbon chain with 1 double bond, 16:1) and oleic acid (18:1).

Polyunsaturated
Fatty acids with two or more double bonds include linoleic acid (18:2), α-linolenic acid (18:3), arachidonic acid (20:4), eicosapentaenoic acid (20:5), and docosahexaenoic acid (22:6).

Trans Isomers
Fatty acids with double bonds can exist in either a *cis* (folded) or a *trans* (linear) configuration. Natural fats contain mostly *cis* fatty acids, but when vegetable oils are partially hydrogenated (by catalytically adding hydrogen at double bonds) to make them solid at room temperature (as in production of margarines), electronic shifts

occur in some of the remaining, unhydrogenated double bonds, resulting in *trans* configurations that are energetically more stable. However, these *trans* fatty acids tend to increase blood cholesterol, decrease high-density (good) lipoprotein (HDL) cholesterol, and raise low-density (bad) lipoprotein (LDL) cholesterol. Conjugated *trans* fatty acids (e.g., vaccenic acid and conjugated linoleic acid) are natural components of foods derived from ruminants, such as cattle, that may have antiatherogenic and anti-inflammatory effects and may protect against diabetes and certain cancers.

ESSENTIAL FATTY ACIDS

Linoleic Acid

Fatty acids in dietary fat that cannot be synthesized in the body from other precursors and that have specific metabolic roles are considered *essential*. Linoleic acid is a polyunsaturated fatty acid required for integrity of the skin, water retention in tissues, normal growth, and optimal reproduction. As noted, this fatty acid is 18 carbon atoms long with two double bonds (18:2). It is also a member of the ω-6 fatty acid series based on the position of the first double bond, counting from the methyl, or omega (ω), end of the carbon chain. Sometimes, the letter *n* is substituted for the omega symbol, so linoleic acid may be variously designated as a member of the ω-6 or *n*-6 fatty acid series.

α-Linolenic Acid

α-Linolenic acid (18:3) is a polyunsaturated fatty acid required for normal brain development in the fetus and that may influence postnatal vision, behavior, and cognition. Essentiality is derived largely from its role as a starting point for synthesis of the metabolically irreplaceable eicosapentaenoic (20:5) and docosahexaenoic (22:6) fatty acids. It may be important that at least some of the last two forms be present preformed in the diet of preterm infants. To the extent that these fatty acids are available preformed, the need for α-linolenic acid is diminished. These fatty acids are members of the ω-3 (or *n*-3) fatty acid series. Eicosapentaenoic and docosahexaenoic acids are also reported to reduce the risk of cardiovascular disease by reducing blood cholesterol and promoting favorable shifts in ratios of LDL cholesterol and HDL cholesterol in blood plasma.

FOOD SOURCES

Foods vary widely in fat concentration and proportions of fatty acids within that fat. Animal fats tend to have higher proportions of saturated or monounsaturated fatty acids than do vegetable fats, although polyunsaturated concentrations in animal fats can be increased by diet. Corn, safflower, soybean, and sunflower oils are high in linoleic acid (50% or more of total fat), but only soybean oil has a significant amount (7%) of α-linolenic acid. Flaxseed and canola oils have significant concentrations of α-linolenic acid but are relatively low in linoleic acid. Fish (particularly fatty fish) and shellfish are generally good sources of eicosapentaenoic and docosahexaenoic fatty acids.

ESTIMATED ADEQUATE INTAKE

Total Fat

Estimated adequate intakes of fat have been established only for infants from birth to 6 months (31 g/day) and from 6 months to 1 year (30 g/day).

Essential Fatty Acids

Estimated adequate intakes of linoleic acid range from 7 g/day for 1- to 3-year-old children to 17 g/day for 14- to 30-year-old males.

Estimated adequate intakes of α-linolenic acid range from 0.7 g/day for 1- to 3-year-old children to 1.6 g/day for 14- to 30-year-old males.

Because ω-3 and ω-6 fatty acids appear to evoke opposing biological activity in certain physiological systems, both their absolute concentrations and relative proportions in the diet require consideration.

CONCLUSIONS AND STUDY TOPICS

Food lipids are concentrated sources of energy and essential fatty acids (which cannot be synthesized in body tissues). Diseases of dietary excess and deficiency of lipids are addressed in Chapters 9 and 10, respectively. Topics deserving study include the influence of economic status, culture, and political forces on the availability, use, and misuse of lipid-containing foods.

NOTE

1. To save space, most of the information on lipids was derived from reviews in the references listed in the bibliography; see them for more detail. Specific citations are not included in the text.

BIBLIOGRAPHY

Carpenter, K.J. 2003. A short history of nutritional science: part 1 (1785–1885). *J. Nutr.* 133:638–645.

Carpenter, K.J. 2003. A short history of nutritional science: part 2 (1885–1912). *J. Nutr.* 133:975–984.

Carpenter, K.J. 2003. A short history of nutritional science: part 3 (1912–1944). *J. Nutr.* 133:3023–3032.

Carpenter, K.J. 2003. A short history of nutritional science: part 4 (1945–1985). *J. Nutr.* 133:3331–3342.

Gropper, S.S., J.L. Smith, and J.L. Groff. 2005. *Advanced Nutrition and Human Metabolism.* Belmont, CA: Thomson Wadsworth.

Guerre-Millo, M. 2002. Adipose tissue hormones. *J. Endocrinol. Invest.* 25:855–861.

Jones, P.J.H. and A.A. Papamandjaris. 2006. Lipid: cellular metabolism. In *Present Knowledge in Nutrition,* 9th ed., Vol. 1, Bowman, B.A. and R.M. Russell, Eds. Washington, DC: International Life Sciences Institute, pp. 125–137.

Lichtenstein, A.H. and P.J.H. Jones. 2006. Lipids: absorption and transport. In *Present Knowledge in Nutrition,* 9th ed., Vol. 1, Bowman, B.A. and R.M. Russell, Eds. Washington, DC: International Life Sciences Institute, pp. 111–124.

McCollum, E.V. 1957. *The History of Nutrition.* Boston: Houghton Mifflin.

7 Protein and Amino Acids[1]

Duane E. Ullrey

CONTENTS

ABSTRACT

The historical discovery of nitrogen in animal tissues and its association with protein and amino acids is described. Important roles of proteins in body structure, as enzymes and hormones, and in immunity and nutrient transport are characterized. Details of protein digestion and absorption are briefly presented. Those dietary amino acids found to be indispensable, conditionally indispensable, and dispensable are identified. A source of information on concentrations of protein and amino acids in food is given, and systems for evaluating protein quality are described. Recommended dietary allowances for protein and indispensable amino acids are summarized.

FIRST CAME NITROGEN

Claude Berthollet's report to the French Academy of Sciences in 1785 that a gas emitted by decomposing animal tissue (ammonia) contained 83% nitrogen and 17% hydrogen (by weight) provided an early clue to the presence of nitrogen in protein. Research by others found no nitrogen in sugars, starches, or fats, but modest amounts were present in wheat flour, which was thus credited with containing "animal matter." Since it was known that air contained nitrogen, there was speculation whether the high amounts in animal tissue were derived from this source. François Magendie fed single foods, like sugar or olive oil (thought to be nutritious but containing no nitrogen) to dogs, and the dogs died. When dogs were fed bread (containing some nitrogen), they lived longer but died nevertheless. Likewise, he reported in 1841 that gelatin obtained by boiling bones was not a complete food for dogs, even though it was high in nitrogen. As further research would reveal, Magendie's diets had several nutrient deficiencies, but the quantity and quality (identity and proportions of indispensable amino acids) of dietary protein were certainly limited, and there was no evidence that nitrogen in air helped at all.

Coincident with and relevant to the studies of Magendie, Jean Baptiste Boussingault reported in 1836 that cereal crops were unable to use atmospheric nitrogen to support their own growth but legumes could, and he proposed that the relative nutrient value of plants could be assessed from their nitrogen content. He also reported in 1839 on nitrogen balance trials with a horse and a dairy cow in which he measured nitrogen intake from their food and nitrogen excreted in feces, urine, and (in the case of the dairy cow) milk. J. B. Dumas, a contemporary of Boussingault, concluded in 1841 that only plants could synthesize the nitrogenous substances found in animals, a statement found to be false when it was discovered that dispensable amino acids can be synthesized by animal tissues and incorporated into their nitrogenous substances. However, clarification of this issue required further research. These nitrogenous substances were ultimately named proteins, from the Greek word *prōte*, for primary, referring to their role as primary components of the animal body. Several of these animal substances were characterized and given names, such as albumin, fibrin, and casein. Although they differed in physical properties, they were found to contain about 16% nitrogen. Liebig analyzed animal muscles and reportedly found no carbohydrate or fat—only protein. Thus, he stated in his 1842 book on application of chemistry to physiology that protein is the only true nutrient because it serves as the machinery of the body and provides the fuel for its work.

Carl Voit, Liebig's protégé, supervised a laboratory in Munich where nitrogen balance procedures were used to estimate protein requirements. Wilbur Atwater of Wesleyan University in Connecticut studied in Voit's laboratory and set a protein standard for American workmen of 125 g per day. Russell Chittenden of Yale University challenged this high value in 1895 when he found that his own needs did not exceed 40 g per day. Soon, the significance of amino acids became apparent with the discovery of tyrosine and leucine—released when proteins were boiled with strong acids. Further, their connection to the physiology of digestion was established when these same amino acids were found when pepsin-treated proteins were exposed to pancreatic trypsin. In 1902, F. G. Hopkins and S. W. Cole at Cambridge isolated the amino acid tryptophan by enzymic digestion of protein. T. Osborne and L. B. Mendel reported in 1916 that rats fed the corn protein zein failed to grow unless supplemented with tryptophan and lysine. This and further work established that certain amino acids were dietary essentials (indispensable), and the last of these to be discovered, threonine, was identified in 1935 by W. C. Rose and associates at the University of Illinois.

PROTEIN FUNCTIONS

STRUCTURAL ROLES

Those proteins that serve structural roles in the body are commonly divided into contractile proteins and fibrous proteins. The principal contractile proteins are actin and myosin and are found in muscle, where they, when activated, power movement of body parts, such as the limbs, head and neck, heart, lungs, and intestinal tract. The fibrous proteins are characteristic of connective tissue and include collagen, elastin, and keratin. They are found in mucus, cartilage, tendons, blood vessels, skin, matrices of bones and teeth, hair, and nails.

ENZYMES

Enzymes are proteins that act as catalysts to change the rate of metabolic reactions. They may be found inside or outside body cells, depending on their function. They combine selectively with substrate molecules to split them; to transfer atoms, electrons, or functional groups; or to join molecules. Some enzymes require cofactors, such as iron, copper, or zinc, or coenzyme forms of certain B vitamins to fulfill their functions. Physiological processes that require enzymes include digestion of food, generation of energy by tissues, generation and propagation of neurological impulses, clotting of blood, and contraction of muscle.

HORMONES

Hormones are chemical messengers that may regulate metabolic processes by promoting enzyme synthesis or affecting enzyme activity. They are synthesized and secreted by endocrine tissue, transported in the blood, and produce specific responses in target organs. Some hormones are steroids derived from cholesterol. Others are derived from amino acids that may be linked together or metabolically altered. These include thyroid and parathyroid hormones, melatonin, insulin, growth hormone, adrenocorticotropic hormone, antidiuretic hormone, glucagon, and calcitonin.

IMMUNITY

Proteins called *immunoglobulins*, or antibodies, are produced by plasma cells, derived from specialized white blood cells called B-lymphocytes. They function by binding and inactivating foreign objects (antigens) such as bacteria or viruses that enter the body with potential to cause disease. The immunoglobulin-antigen complexes that are formed may then be destroyed by complement proteins produced mainly in the liver or by cytokines produced by white blood cells, such as T-helper cells or macrophages. Prenatal or childhood protein-energy malnutrition may have permanent adverse affects on immune function.

NUTRIENT TRANSPORT

Proteins that combine with substances requiring transport in the blood, within cells, or across cell membranes include albumin, transthyretin (formerly prealbumin), heme proteins (hemoglobin, myoglobin), transferrin, and ceruloplasmin. Albumin transports several nutrients, including calcium, zinc, and vitamin B_6. Transthyretin complexes with retinol-binding protein to transport retinol (vitamin A). Hemoglobin and myoglobin are iron-containing proteins that bind or transport oxygen. Transferrin transports iron, and ceruloplasmin transports copper.

PROTEIN DIGESTION AND ABSORPTION

Digestion of dietary protein begins in the stomach through the action of hydrochloric acid and pepsin. End products of gastric digestion include primarily large polypeptides, some oligopeptides (polymers of just a few amino acids), and free amino acids. These end products are emptied from the stomach into the duodenum, stimulating release of alkaline pancreatic juice containing water, bicarbonate, various electrolytes, and proenzymes that undergo activation to trypsin, chymotrypsin, and carboxypeptidases. Peptidases are also produced by the brush border of enterocytes (intestinal mucosal cells), extending from the duodenum into the distal small intestine (ileum). The main end products of digestion in the small intestine are dipeptides, tripeptides, and free amino acids.

Absorption of amino acids from the intestinal lumen into the enterocyte involves multiple energy-dependent transport systems through the brush border, some of which also require sodium cotransport. Dipeptide and tripeptide transport through the brush border involves carrier systems different from those for amino acids, and peptide absorption tends to be more rapid. The inward movement of hydrogen ions provides an absorptive driving force, followed by outward pumping of hydrogen ions into the lumen in exchange for inward movement of sodium ions. Evidence suggests that about two thirds of the amino acids are absorbed as small peptides, whereas the rest are absorbed in free form. Peptides inside the enterocytes are hydrolyzed by cytoplasmic peptidases, generating free amino acids.

A significant proportion of absorbed amino acids is used by the intestinal cell for synthesis of new digestive enzymes, hormones, and other nitrogen-containing compounds and for energy. Transport through the enterocyte basolateral membrane into the interstitial fluid (and ultimately into the capillaries and portal vein) appears to be largely by diffusion and sodium-independent systems, although sodium-dependent pathways may be important when luminal amino acid concentrations are low.

AMINO ACIDS

Protein is required in the diet to supply the indispensable amino acids that cannot be synthesized in the body and to supply nitrogen for synthesis of the dispensable amino acids. Dietary protein quality is largely dependent on how closely the quantity and proportions of biologically available indispensable amino acids match the needs of the consumer. Whole egg and milk proteins are particularly high in quality, whereas grain proteins tend to be low in lysine, with corn also low in tryptophan. By combining two or more protein sources that have complementary amino acid concentrations, a high-quality food may be created from items that are individually poor. Such a combination can result by using appropriate proportions of grain and legume proteins.

INDISPENSABLE AMINO ACIDS

Those (indispensable) amino acids that cannot be synthesized by normal human tissues at a rate sufficient to meet needs are histidine, isoleucine, leucine, lysine, methionine, phenylalanine, threonine, tryptophan, and valine. Thus, required amounts must be present in the diet. Preformed cystine (or cysteine) can meet some of the need for methionine, and preformed tyrosine can meet some of the need for phenylalanine.

CONDITIONALLY INDISPENSABLE AMINO ACIDS

Premature infants or humans with disease-associated organ malfunction may have dietary requirements for amino acids that would otherwise be dispensable. Their rates of synthesis appear to be inadequate under certain physiological or pathological circumstances to meet cellular needs. As a consequence, cysteine, tyrosine, proline, arginine, or glutamine may be conditionally indispensable. For example, metabolic functions of the organs of premature infants may not be fully developed, and the ability to synthesize dispensable amino acids, such as cysteine and proline, is often limited. There is evidence that activity of the liver enzyme cystathianase, required for conversion of methionine to cysteine, does not reach adult levels even in full-term infants until they are at least 4 months old. Cirrhosis of the liver in adults, from excessive long-term alcohol consumption, may impair the synthesis of tyrosine from phenylalanine and of cysteine from methionine and, as a consequence, impair synthesis of metabolically important proteins. The genetic disorder phenylketonuria is associated with limited phenylalanine hydroxylase activity required for conversion of phenylalanine to tyrosine.

DISPENSABLE AMINO ACIDS

Amino acids that can be synthesized from tissue or dietary precursors at a rate sufficient to meet needs (if protein-nitrogen supplies are adequate) are considered dispensable. These normally include alanine, arginine, asparagine, cysteine, cystine, glutamic acid, glutamine, glycine, proline, serine, and tyrosine.

FOOD SOURCES

Muscle and organ meats, poultry, fish, cheese, and legumes (beans, peas) generally have high concentrations of protein on a dry matter basis. Grains, many vegetables, and fruits usually have less. Typical concentrations may be found in the periodically updated Nutrient Database for Standard Reference (SR) available from the U.S. Department of Agriculture, Agricultural Research Service, Nutrient Data Laboratory (www.ars.usda.gov/nutrientdata).

EVALUATION OF PROTEIN QUALITY

As noted, dietary protein is important as a source both of nitrogen and of indispensable amino acids. A high-quality protein contains indispensable amino acids in an available form in the approximate proportions needed by the body. Protein quality has been estimated in a variety of ways. The *amino acid score* is an expression of the concentration of indispensable amino acids in a test protein as a fraction of those amino acids in a reference protein, such as that from whole eggs. This score is sometimes corrected for true digestibility. The *protein efficiency ratio* is a measure of the grams of weight gained during growth on a test protein divided by the grams of protein consumed. *Nitrogen balance* is a measure of food nitrogen retained by the body after accounting for losses in feces, urine, and shed skin cells. *Biological value* is a measure of the nitrogen in a test protein that is retained as a fraction of nitrogen absorbed. *Net protein utilization* is a measure of the nitrogen in a test protein that is retained as a fraction of nitrogen consumed.

RECOMMENDED DIETARY ALLOWANCES

Protein

Recommended dietary allowances for protein range from 13 g/day for 1- to 3-year-old children to 56 g/day for men 19 years of age or older.

Indispensable Amino Acids

Estimates of indispensable amino acid requirements for humans vary with methods used in their determination. Current estimates are appreciably higher than those proposed in 1985 by the Food and Agriculture Organization of the United Nations World Health Organization (FAO/UN/WHO). Presently, recommended dietary allowances (mg/kg body weight/day) for adults are histidine, 14; isoleucine, 19; leucine, 42; lysine, 38; methionine plus cysteine, 19; phenylalanine plus tyrosine, 33; threonine, 20; tryptophan, 5; valine, 24.

CONCLUSIONS AND STUDY TOPICS

Food proteins are potential sources of amino acids that are metabolically indispensable because they cannot be synthesized in body tissues. Dispensable amino acids in foods are used in consort with those that are indispensable to produce proteins found in muscle and bone, enzymes, hormones, immune globulins, and nutrient transport systems. Diseases of dietary excess and deficiency of protein and amino acids are addressed in Chapters 9 and 10, respectively. Topics worthy of study include the influence of economic status, regional availability, and culture on the ability to meet amino acid needs.

NOTE

1. To save space, most of the information on proteins and amino acids was derived from reviews in the references listed in the bibliography; see them for more detail. Specific citations are not included in the text.

BIBLIOGRAPHY

Carpenter, K.J. 2003. A short history of nutritional science: part 1 (1785–1885). *J. Nutr.* 133:638–645.

Carpenter, K.J. 2003. A short history of nutritional science: part 2 (1885–1912). *J. Nutr.* 133:975–984.

Carpenter, K.J. 2003. A short history of nutritional science: part 3 (1912–1944). *J. Nutr.* 133:3023–3032.

Carpenter, K.J. 2003. A short history of nutritional science: part 4 (1945–1985). *J. Nutr.* 133:3331–3342.

Gropper, S.S., J.L. Smith, and J.L. Groff. 2005. *Advanced nutrition and human metabolism.* Belmont, CA: Thomson Wadsworth.

Holden, J.M., J.M. Hamly, and G.R. Beecher. 2006. Food composition. In *Present knowledge in nutrition,* 9th ed., Vol. 2, Bowman, B.A. and R.M. Russell, Eds. Washington, DC: International Life Sciences Institute, pp. 781–794.

McCollum, E.V. 1957. *The history of nutrition.* Boston: Houghton Mifflin.

Murphy, S.P. 2006. Dietary standards in the United States. In *Present knowledge in nutrition,* 9th ed., Vol. 2, Bowman, B.A. and R.M. Russell, Eds. Washington, DC: International Life Sciences Institute, pp. 859–875.

Pencharz, P.B. and V.R. Young. 2006. Protein and amino acids. In *Present knowledge in nutrition,* 9th ed., Vol. 1, Bowman, B.A. and R.M. Russell, Eds. Washington, DC: International Life Sciences Institute, pp. 59–77.

Shankar, A. 2006. Nutritional modulation of immune function and infectious disease. In *Present knowledge in nutrition,* 9th ed., Vol. 2, Bowman, B.A. and R.M. Russell, Eds. Washington, DC: International Life Sciences Institute, pp. 604–624.

8 Vitamins and Mineral Elements[1]

Duane E. Ullrey

CONTENTS

ABSTRACT

The history of the discovery of vitamins and essential minerals has been reviewed, including particularly significant findings from the seventeenth to the late twentieth century. Current knowledge of functions, absorption, metabolism, and dietary sources of 4 fat-soluble vitamins, 9 water-soluble vitamins, and 15 mineral elements required by humans and many animals is included. Quantitative requirements for these nutrients at various ages, physiological states, levels of physical activity, and environmental circumstances are regularly revised based on new data. Ranges of currently recommended allowances for these nutrients in human diets are presented.

VITAMIN BEGINNINGS

Identification of biologically vital elements in the late 1700s, during the "chemical revolution" in France, provided a foundation for subsequent discoveries of the essential nutrients. In 1842, Justus von Liebig, a German organic chemist, stated that protein was the only true nutrient, providing both the structure of muscle and the energy for its contraction. However, an 1847 account of scurvy in Scottish prisoners consuming ample protein (but no potatoes) and its prevention by lemon juice (containing negligible nitrogen) belied Liebig's conclusion. In fact, James Lind showed in 1746, in a controlled experiment, that citrus fruit, but not sulfuric acid or vinegar, cured scurvy. Thus, the effectiveness of oranges, lemons, and limes was known for almost 200 years before the active agent was finally identified in 1932 as ascorbic acid, 4 years after its isolation from adrenal glands by Albert Szent-Györgi.

Rickets was common in young children in large industrialized cities in Western Europe during the late 1800s, even when calcium intakes appeared adequate. Those affected were often fed breast milk substitutes (which tended to be low in fat) and had limited sun exposure due to airborne pollution. Walter Cheadle in 1888 concluded that rickets could be prevented by cod liver oil, and Theobald Palm noted in 1890 that rickets was rare in regions with lots of sunlight. In 1924, it was established that ultraviolet (UV) irradiation of rats or of their diet would prevent the disease. The activated dietary factor was found to be lipid soluble, was named vitamin D (because ascorbic acid had previously been designated vitamin C), and was crystallized in 1931.

Other observations relevant to the vitamin story were made by the microbiologist Pekelharing in 1888 concerning beriberi in the army of the Dutch East India colony. Although he thought an unusual bacterial infection might be responsible for this condition, an infectious origin could not be confirmed. Knapp in 1909 observed eye lesions (xerophthalmia and keratomalacia) that were responsive to cod liver oil in rats fed a purified diet. Ultimately, the curative factor was identified and named vitamin A.

Despite this early history and studies from 1900 to 1911 by Gerritt Grijns, who concluded that a polyneuritis in chickens fed white rice was caused by a deficiency of an unstable, water-soluble, organic compound, the vitamin era is commonly stated to have begun in 1912 with the studies of beriberi in humans by Casimir Funk. He isolated a water-soluble organic compound containing an amine group from rice polishings that was effective in preventing or curing this disease. Funk believed this was a vital amine and coined the term "vitamines" to include this and other, yet-to-be-identified, vital factors. The "e" was dropped when it was established that not all vitamins contained amine groups. E. V. McCollum and associates concluded in 1916 from their studies with purified diets (begun at least 3 years before) and from studies of others that there was an unidentified fat-soluble A (needed for growth and prevention of xerophthalmia) and an unidentified water-soluble B, which eventually proved to be the antiberiberi factor found by Funk. Thiamin (B_1) was isolated in 1926 and its structure established in 1936.

Pellagra, a disease characterized by dermatitis, gastrointestinal problems, and mental disturbances, was commonly observed in the early 1900s, particularly in the southern United States where corn was a dietary staple, and meat and milk intakes were limited. Although some proposed that it was infectious or caused by a pathogenic mold, Joseph Goldberger and associates, using a dog model, showed in 1928 that pellagra could be cured by yeast. Following isolation of a known chemical, nicotinic acid, from yeast, it was established in 1937 that either nicotinic acid or nicotinamide (jointly designated niacin) were effective antipellagrins. Later, it was found that corn is low in tryptophan, an amino acid that can serve as a precursor for niacin synthesis in animal tissues (if the tryptophan supply is adequate). In addition, an appreciable amount of the niacin in corn is bound and unavailable for absorption unless treated with alkali, as in production of tortillas.

Some patients with pellagra still showed lesions about the mouth (cheilosis) even after treatment with niacin. Yeast that had been autoclaved lost its thiamin activity but still was effective against these lesions. The effective agent was termed B_2 but was soon found to be a complex of factors. The agent effective against cheilosis was identical with a greenish-yellow fluorescent pigment isolated from whey in 1935 and was subsequently named riboflavin.

Again, yeast proved useful when in 1931 it was found to cure a macrocytic anemia of pregnancy commonly seen in Mohammedan women in Bombay, India. Lucy Wills induced this anemia, and a leukopenia, in rhesus monkeys fed a poor Bombay diet. These conditions responded to extracts from yeast or from liver but did not respond to any of the then-known vitamins. The effective factor was temporarily designated vitamin M for monkey. By 1944, it was established that a compound isolated from spinach supported growth in bacteria and in chicks and prevented macrocytic anemia in the latter. It was named folic acid from its origin in foliage.

The 1930s and 1940s were a particularly active period in vitamin research, and additions to the water-soluble list by 1937 included pantothenic acid, pyridoxine (B_6), and biotin. A fat-soluble dietary factor shown in 1922 to be required for reproduction by rats, and eventually to prevent certain muscular dystrophies, was named vitamin E and was isolated in 1935. Another fat-soluble vitamin, found by Henrik Dam in 1935 to prevent hemorrhages in chicks, was named vitamin K (for "koagulation" in Danish). The last on the list of generally accepted vitamins, isolated in 1948, was water soluble, contained cobalt, and was named vitamin B_{12}. It was found in animal tissues, soil, and certain bacteria but not in higher plants. If the diet contained adequate cobalt, microorganisms in the rumen of cattle or sheep, or in the cecum and colon of horses, were found to synthesize sufficient vitamin B_{12} to meet their needs. The structure of vitamin B_{12} was established in 1955.

VITAMIN A AND PROVITAMIN A CAROTENOIDS

The term *vitamin A* refers to a group of compounds possessing the biological activity of all-*trans*-retinol. They are important for vision, cell growth, communication between cells, and differentiation of cells into specific functional types. They are variably soluble in lipids and organic solvents but insoluble in water. Following digestion and release from the dietary matrix, their absorption is enhanced by bile salts and pancreatic lipase, which promote formation of small lipid droplets (micelles) in the intestinal lumen, within which retinyl esters, such as retinyl palmitate (a combination of retinol and palmitic acid), are dissolved and hydrolyzed. Retinol is taken up by the mucosal cells lining the intestine (enterocytes) and incorporated into chylomicrons (lipoprotein particles) that are transported into the lymph and then into the general blood circulation. The chylomicrons deliver retinyl esters, some unesterified retinol, and carotenoids to the liver and other (extrahepatic) tissues. Retinyl esters that have been stored in the liver are hydrolyzed before transport to the peripheral tissues in association with a complex of retinol-binding protein and transthyretin.

Over 600 carotenoids have been identified, but fewer than 60 have provitamin A activity. Those that do undergo conversion in the body to retinol. One of the most abundant and most active is β-carotene, which can be split into two retinol molecules; other common dietary carotenoids, α-carotene and β-cryptoxanthin, potentially yield one. This bioconversion may occur during digestion, followed by absorption of retinol as described. In some species, including humans, provitamin A carotenoids also may be absorbed intact and are subject to postabsorption bioconversion. If not converted to retinol, some carotenoids may function as antioxidants.

Historically, vitamin A activity has been expressed in international units (IU), with 1 IU equivalent to 0.3 μg of retinol. IUs are still used in the literature, but the Food and Agriculture Organization of the United Nations/World Health Organization (FAO/WHO) has adopted retinol equivalents (REs), and the U.S. Institute of Medicine (IOM) has adopted retinol activity equivalents (RAEs), with 1 RE or 1 RAE equivalent to 1 μg of retinol. Although these two systems are equivalent in expressing vitamin A activity of preformed vitamin A compounds, they differ in their assignment of activity to carotenoids. Furthermore, the vitamin A activity of provitamin A carotenoids varies with the species fed, its vitamin A status, and the nature of the diet. Thus, it is often preferable to use mass as the basis for estimating contributions of provitamin A carotenoids to the requirement for vitamin A. The IOM has set 1 RAE equal to 12 μg of β-carotene or 24 μg of α-carotene or β-cryptoxanthin in food matrices consumed by humans.

WHO/FAO and IOM estimates of requirements and recommended intakes of vitamin A for humans are not identical but tend to be similar for comparable age classes. Recommended daily dietary allowances range from 300 RAE for 1- to 3-year-old children to 1300 RAE for lactating females. Whole milk, cheese, butter, eggs, organ meats (particularly liver), and fish are important sources of vitamin A in U.S. diets. The main natural form of vitamin A in these foods is esters (e.g., retinyl palmitate).

Carrots, spinach, broccoli, peas, sweet potatoes, and squash are good sources of provitamin A carotenoids. Ready-to-eat cereals may be fortified with vitamin A. Carotenoids may be added to margarine and to diets for poultry or salmon—usually to contribute desired color to edible products. Other foods have been fortified, and in the Philippines, additions of vitamin A to coconut oil significantly improved human retinol status. In addition, it has been shown that provitamin A concentrations in several vegetables (e.g., carrots, cauliflower, yams, cassava, and rice) can be increased by genetic selection or biotechnical amplification. Termed *biofortification*, it may be a useful public health measure for control of vitamin A deficiency in economically poor cultures when applied to staple food crops. Pharmaceutical vitamin preparations commonly contain retinyl acetate or retinyl palmitate, and those consumed orally may contain carotenoids as well.

Vitamin D

Ultraviolet (UV) irradiation of ergosterol in plants or yeast produces ergocalciferol (vitamin D_2), whereas UV irradiation of 7-dehydrocholesterol in skin produces previtamin D_3, followed by thermal (body heat) conversion to cholecalciferol (vitamin D_3). Effective UV wavelengths are in the UVB range, 280 to 315 nm, although solar wavelengths below 290 seldom reach the skin because they are usually screened out by atmospheric ozone and molecular oxygen. Both vitamin D compounds are moderately soluble in lipids and insoluble in water. They are best known for the ability of their metabolically active forms to promote intestinal absorption of calcium and its incorporation into bone, thus preventing rickets in growing young and osteomalacia in adults. However, these active forms appear to have broader metabolic functions and may inhibit the proliferation and growth of certain types of cancer, particularly breast, colon, and prostate. In company with the hormones parathormone and calcitonin (as well as others), vitamin D maintains normal plasma Ca^{2+} and phosphate concentrations and thus has an impact on a variety of soft tissue events such as neuromuscular activity, reproduction, and immune function. Vitamin D_2 appears to be only about one third as effective as vitamin D_3 in elevating human serum 25-hydroxy vitamin D levels (used to assess vitamin D status).

Vitamin D_2 or D_3 from the diet is absorbed by dissolution in micelles within the intestinal lumen, passive diffusion into enterocytes, and incorporation into chylomicrons, which enter the general circulation via the lymph. Chylomicrons may deliver vitamin D to the liver and extrahepatic tissues, or some may be transferred to and transported by D-binding protein (DBP). Vitamin D_3, formed following irradiation of the skin, slowly diffuses into the blood and is bound to DBP for transport. To become metabolically active, ergocalciferol and cholecalciferol undergo conversion in the liver to 25-hydroxy ergocaliferol and 25-hydroxy cholecalciferol, respectively, followed by conversion in the kidneys to 1,25-dihydroxy ergocalciferol and 1,25-dihydroxy cholecalciferol (calcitriol). This last renal conversion also results in a 24,25-dihydroxy cholecalciferol, which appears to have metabolic functions alone or in combination with calcitriol.

Vitamin D activity is expressed in IU, with 1 IU equivalent to 0.025 µg cholecalciferol. Dietary vitamin D comes primarily from animal products, such as liver, fish, beef, eggs, and fortified milk (400 IU/quart) in the United States. Some other foods, such as cereals, bread, and margarine, also may be fortified. Estimated adequate dietary intakes range from 200 IU per day for children to 600 IU per day for adults over the age of 70. It has been estimated that 10 minutes of summer sun exposure of the hands and face will supply about 400 IU, but latitude and time of day (as well as season) may greatly affect this estimate.

Vitamin E

The term *vitamin E* may be applied to eight related natural compounds: α-, β-, γ-, and δ-tocopherol and α-, β-, γ-, and δ-tocotrienol. Unlike most vitamins that have specific metabolic roles or that function as cofactors, vitamin E serves as a chain-breaking antioxidant to protect unsaturated fatty acids in cell membrane phospholipids from oxidative damage by scavenging peroxyl radicals that originate from metabolic reactions. Most effective in humans is *RRR*-α-tocopherol (the natural form, also known as d-α-tocopherol), which has a stereoisomeric *R*-configuration at 2, 4′, and 8′ positions in the tocopherol molecule. Synthetic α-tocopherol (*all racemic* or dl) has eight potential stereoisomers, with less biological activity than the natural form and with activity dependent on the *R*-configuration in position 2. Interaction of α-tocopherol with peroxyl radicals results in its oxidation, but it can be restored to its reduced, and functional, form by vitamin C or glutathione. Selenium-dependent glutathione peroxidase also has antioxidant functions in the body, so the quantitative supply of selenium influences the quantitative need for vitamin E and vice versa.

Steps in the digestion and absorption of vitamin E compounds differ somewhat, depending on chemical form. Tocopherols are found as free alcohols in food, but tocotrienols are commonly esterified. Thus, ester bonds must be hydrolyzed by pancreatic esterase or duodenal mucosal esterase before absorption can take place. This is also true for synthetic ester forms of tocopherols, such as *all rac-* or dl-α-tocopheryl acetate.

Absorption of vitamin E alcohols from micelles in the intestinal lumen occurs primarily in the jejunum by passive diffusion. Absorbed vitamin E alcohols are incorporated into chylomicrons within the enterocyte and reach the liver via the lymph and general circulation. Equilibration with or transfer to plasma lipoproteins may occur during transport. Delivery of vitamin E to extrahepatic tissues appears restricted to *RRR*-α-tocopherol, which is incorporated into very low density lipoproteins (VLDLs) and bound to a very specific protein made in the liver called α-tocopherol transfer protein (αTPP). Other vitamin E forms are poorly recognized by this transfer protein. A genetic defect in the ability of the liver to synthesize αTPP may result in vitamin E deficiency.

Plants, particularly green leaves and seed oils, are good sources of vitamin E, with α-tocopherol predominating in green leaves and in canola, cottonseed, olive, safflower, and sunflower oils. Corn and soybean oils contain some α-tocopherol, but γ-tocopherol predominates. Vitamin E levels in animal products tend to be low and are concentrated in fatty tissues, with most as α-tocopherol. Recommended dietary allowances range from 6 mg per day for 1- to 3-year-old children to 19 mg per day for lactating women.

VITAMIN K

Naturally occurring forms of vitamin K include phylloquinone, produced by green plants, and at least eight menaquinones, produced by anaerobic bacteria in the lower digestive tract. Menadione is a synthetic form of vitamin K that must be alkylated in the liver to become active. Dietary phylloquinone is absorbed from the small intestine via dissolution in micelles and passive diffusion into the enterocyte. Menaquinones synthesized by bacteria in the lower tract are absorbed by passive diffusion from the ileum and colon. In the enterocyte, vitamin K is incorporated into chylomicrons and carried via lymph into the general circulation. Vitamin K is involved in blood clotting and in bone mineralization by promoting carboxylation of glutamyl residues in specific proteins required for these processes. There also is evidence of activity of vitamin K in promoting nerve growth and neuronal survival. Phylloquinone concentrations are particularly high (>200 µg/100 g) in broccoli, collards, kale, spinach, Swiss chard, and watercress. Recommended adequate intakes range from 30 µg per day for 1- to 3-year-old children to 120 µg per day for adult men.

THIAMIN

Thiamin (vitamin B_1), when phosphorylated to thiamin diphosphate (TDP, also known as thiamin pyrophosphate, TPP), plays an essential role as a coenzyme in energy metabolism. It promotes conversion of pyruvate (a three-carbon acid) to acetate (a two-carbon acid), which then enters the citric acid cycle, a key cycle in interconversions of energy metabolites in the body. TDP/TPP is also involved in synthesis of five-carbon sugars (pentoses) and nicotinamide adenine dinucleotide phosphate (NADP), in metabolism of branched-chain amino acids, and in oxidation of certain branched-chain fatty acids. Thiamin, as thiamin triphosphate (TTP), appears to activate ion transport in nerve membranes and may be involved in nerve impulse transmission. Thiamin is absorbed from the intestine by an energy- and sodium-dependent transport mechanism at low dietary concentrations, but when intakes are high, absorption is mostly by passive transport (diffusion). Thiamin in the blood occurs either in the free form, bound to the protein albumin, or as thiamin monophosphate (TMP). However, most of blood thiamin is in the red cells, not the plasma. Free thiamin is taken up by the liver and phosphorylated. Most of the thiamin in the liver and other (extrahepatic) tissues is in the form of TDP/TPP. Thiamin is widely distributed in foods of animal and plant origin, including liver, muscle meats, legumes, whole grains, and fortified breads and cereals. Thiamin supplements are usually in the form of thiamin hydrochloride or thiamin mononitrate. Recommended dietary allowances range from 0.5 mg per day for 1- to 3-year-old children to 1.4 mg per day for pregnant and lactating women.

RIBOFLAVIN

Riboflavin (vitamin B_2) exists in free form or as part of two coenzymes, flavin mononucleotide (FMN) or flavin adenine dinucleotide (FAD). These coenzymes function as prosthetic groups (nonprotein constituents) for flavoprotein enzymes involved in oxidation-reduction reactions within intermediary metabolic cycles that are central to energy production. They function in these reactions as oxidizing agents through their ability to accept a pair of hydrogen atoms. Absorption of riboflavin requires that it be freed from any bound forms (protein bound or phosphorylated) before transport into the intestinal mucosa via a saturable, energy-dependent carrier. If food riboflavin concentrations are high, some may be absorbed by diffusion. In the enterocyte, riboflavin is phosphorylated to FMN and then dephosphorylated at the basolateral membrane before entering the blood. Free riboflavin or riboflavin bound to plasma proteins (albumin and globulins) is carried via the portal vein to the liver, where it is phosphorylated again to FMN or FAD. Riboflavin is found in many foods, but animal products (such as milk, cheese, eggs, and meat) and legumes are particularly good sources. Green vegetables supply some, and grains and fruits are generally low, although cereals and breads may be enriched with riboflavin supplements. Recommended dietary allowances range from 0.5 mg per day for 1- to 3-year-old children to 1.6 mg per day for lactating women.

NIACIN

Niacin is a generic term for nicotinic acid and nicotinamide, both of which have vitamin activity as a consequence of their interconvertability. As nicotinamide adenine dinucleotide (NAD) or nicotinamide adenine dinucleotide phosphate (NADP), niacin serves coenzyme functions as a hydrogen donor or electron acceptor in oxidation-reduction reactions. NAD is reduced to NADH during catabolism of carbohydrates (glycolysis), β-oxidation of fatty acids, oxidation of ethanol, oxidative decarboxylation of pyruvate, and oxidation of acetyl coenzyme A (CoA) via the Krebs cycle. NAD also catalyzes the conversion of vitamin B_6, as pyridoxal, to its excretory product pyridoxic acid. NAD can serve as a donor of adenosine diphosphate ribose (ADP-ribose) in nonredox reactions involving proteins that function in repair, replication, and differentiation of nuclear DNA. NADP is reduced to NADPH during synthesis of fatty acids, cholesterol, steroid hormones, and DNA precursors (deoxyribonucleotides); oxidation of glutamate; and regeneration of glutathione, vitamin C, and thioredoxin. NADPH is required in several reactions involving folate metabolism.

Absorption of nicotinamide or nicotinic acid from the stomach has been demonstrated, but most absorption takes place in the small intestine. When dietary concentrations are low, niacin is absorbed by sodium-dependent, carrier-mediated diffusion. At high concentrations, most is absorbed by passive diffusion. If NAD or NADP is present in the diet, hydrolysis in the intestinal lumen or enterocyte is required to release free nicotinamide for further transport. Nicotinamide is the predominant niacin form in plasma, but nicotinic acid is also present—up to a third bound to plasma proteins. These free forms move across cell membranes by simple diffusion, but once they are converted to NAD or NADP within cells, they are trapped until hydrolyzed.

Foods of animal origin, such as fish and organ and muscle meats, are good sources of niacin. Legumes and enriched breads and cereals also provide appreciable amounts. Nicotinamide is the usual supplemental form. Niacin is complexed with carbohydrates (niacytin) or small peptides (niacinogens) in some foods and is poorly available unless treated with alkali, such as soaking corn in lime water. The need for niacin may be partially or totally met by hepatic synthesis of NAD from tryptophan in some species if dietary tryptophan supplies are adequate (i.e., if tryptophan supplies exceed needs for protein synthesis). Recommended dietary allowances range from 6 mg per day for 1- to 3-year-old children to 18 mg per day for pregnant women.

PANTOTHENIC ACID

Pantothenic acid is a water-soluble, yellow, viscous oil that is relatively unstable unless combined with calcium to form white, crystalline calcium pantothenate. It functions in the body as a component of 4'-phosphopantetheine and CoA. Synthesis of these compounds requires pantothenic acid, Mg^{2+}, the amino acid cysteine, and adenosine triphosphate (ATP). They are important in liberation of energy from carbohydrates, lipids, and proteins and in synthesis of heme (for hemoglobin), cholesterol, steroid hormones, bile salts, fatty acids, and ketone bodies. CoA is involved in acetylations of certain proteins, thus affecting their activity, location, and function. CoA also acetylates amino sugars that function structurally to provide molecular recognition sites on cell surfaces. Choline is acetylated by CoA to form the neurotransmitter acetylcholine.

Much of pantothenic acid in food is present as CoA. During digestion, CoA is hydrolyzed first to pantetheine and then to pantothenic acid. The latter appears to be absorbed principally from the jejunum—at low concentrations by a sodium-dependent, active process and at high concentrations by passive diffusion. Pantothenic acid is then transported via plasma in free form to the tissues, where the majority is converted to 4'-phosphopantetheine and CoA. This vitamin is widely distributed in foods and is found in quite good concentrations in meats (particularly liver), eggs, legumes, broccoli, potatoes, and whole-grain cereals. Vitamin supplements usually contain calcium pantothenate or panthenol. Estimated adequate intakes range from 2 mg per day for 1- to 3-year-old children to 7 mg per day for lactating women.

VITAMIN B_6

Vitamin B_6 occurs in three interchangeable forms, pyridoxine, pyridoxal, and pyridoxamine, each having a 5'-phosphate derivative. Pyridoxal phosphate (PLP) is the principal coenzyme involved in amino acid metabolism (transamination, decarboxylation, transulfhydration, desulfhydration, cleavage, racemization, and synthesis), although pyridoxamine phosphate (PMP) also functions as a coenzyme in transaminations. In addition, PLP is involved in synthesis of heme, sphingolipids, niacin, carnitine, and taurine and participates in glycogen catabolism and modulation of the activity of some steroid hormones.

Pyridoxine and pyridoxine phosphate (PNP) are the forms found mostly in plants, although pyridoxine β-glucoside is sometimes present. PLP and PMP are the forms found primarily in animals. To be absorbed, phosphorylated forms must be dephosphorylated by intestinal phosphatases, one of which (alkaline phosphatase) is zinc dependent. Absorption of the free forms into the enterocyte occurs by passive diffusion, primarily in the jejunum, followed by release into the portal blood plasma. When taken up by the liver, most of the free forms are converted to PLP and stored or bound to albumin and delivered to peripheral tissues by the systemic circulation.

Good food sources of vitamin B_6 include meats, vegetables, bananas, whole-grain breads and cereals, and nuts. Pyridoxine hydrochloride is commonly used in vitamin supplements. Recommended dietary allowances range from 0.5 mg per day for 1- to 3-year-old children to 2.0 mg per day for pregnant women.

BIOTIN

Biotin is a white, water-soluble vitamin that functions as a coenzyme for four important carboxylases. Pyruvate carboxylase replenishes oxaloacetate for the Krebs cycle and is necessary for carbohydrate synthesis from noncarbohydrate precursors (gluconeogenesis). Acetyl CoA carboxylase furnishes acetate for synthesis of fatty acids. Propionyl CoA carboxylase is involved in metabolism of some amino acids and fatty acids with an odd-numbered chain. β-Methylcrotonyl CoA carboxylase is involved in catabolism of leucine and certain isoprenoid compounds. Biotin also may be involved in gene expression.

Biotin in foods is often bound to protein or to lysine (biocytin). These bound forms must be hydrolyzed by proteases in the jejunum and ileum for absorption to occur. Some biocytin may be absorbed, but it is not functional and is excreted in the urine. Biotin transport into the enterocyte at low concentrations is sodium dependent and carrier mediated. Transport across the basolateral membrane is carrier mediated but does not require sodium. Biotin synthesized by colonic bacteria can be absorbed in the proximal and midtransverse regions. In plasma, biotin is mostly in the free form, although some may be bound to albumin, α- and β-globulins, and biotinidase. Biotin is taken up by tissues via active, sodium-dependent carriers. Good food sources include liver, legumes, egg yolk, nuts, and some cereals. Raw egg white contains a glycoprotein, called avidin, that binds biotin. However, avidin is heat labile, and cooking frees biotin for absorption. The amount of biotin synthesized by bacteria in the gut of cattle, sheep, and horses is considerable, but the amounts available from colonic bacterial synthesis in humans are unlikely to meet needs. Estimated adequate intakes range from 8 μg per day for 1- to 3-year-old children to 35 μg per day for lactating women.

FOLACIN

Folacin (or *folate*) is a generic term for compounds with the activity of folic acid (pteroylmonoglutamate). The basic structure of folic acid consists of a pteridine nucleus conjugated with para-aminobenzoic acid, forming pteroic acid, which in turn is bound to glutamic acid. In food, folacin may contain up to nine glutamate residues. Before absorption, the polyglutamate forms must be hydrolyzed to the monoglutamate form. This hydrolysis is performed by glutamate carboxypeptidases in pancreatic juice, bile, and the intestinal cell brush border. This last enzyme is zinc dependent, and zinc deficiency will inhibit folic acid absorption. Transport into the enterocyte is energy and sodium dependent. Within the enterocyte, folic acid is reduced to dihydrofolate

(DHF) and then to tetrahydrofolate (THF) by a NADPH-dependent reductase. THF is then converted to either 5-methyl THF or 10-formyl THF, and these compounds enter the portal blood plasma. Hepatic uptake of folic acid is carrier mediated, and THF, 5-methyl THF, 10-formyl THF, and 5-formyl THF may be found in the liver. In systemic blood plasma, about two thirds of the folacin compounds are bound to protein, whereas about one third are free. THF derivatives function as coenzymes by accepting one-carbon groups from amino acid metabolism and interact with vitamin B_6 and vitamin B_{12} in conversion of homocysteine to cystathionine. They also are involved in synthesis of purines and pyrimidines and are thus essential for cell division (particularly important for production of red cells).

Good food sources of folacin include green vegetables (such as spinach, broccoli, asparagus, and turnip greens), legumes, citrus fruits, liver, and fortified breads and cereals. Recommended dietary allowances range from 150 µg per day for 1- to 3-year-old children to 600 µg per day for pregnant women.

VITAMIN B_{12} (COBALAMIN)

Cobalamin is a generic term for compounds having a corrin nucleus with a central cobalt atom to which 5,6-dimethylbenzimidazole and a cyanide, hydroxyl, water, nitrite, 5′-deoxyadenosyl, or methyl group is attached. Their respective names are cyanocobalamin, hydroxocobalamin, aquocobalamin, nitritocobalamin, 5′-deoxyadenosylcobalamin (adenosylcobalamin), and methylcobalamin. Only the last two are active as coenzymes. Methylcobalamin is required for the conversion of homocysteine to methionine. The methyl group for this conversion is acquired from 5-methyl tetrahydrofolate. Adenosylcobalamin is required for the conversion of l-methylmalonyl CoA to succinyl CoA in the Krebs cycle (l-methylmalonyl CoA is made from d-methylmalonyl CoA, which is generated from propionyl CoA in a biotin-dependent reaction). Digestion and absorption of cobalamin involves a cobalamin-binding R protein in saliva and gastric juice that may bind to the vitamin before release from food proteins. Cobalamin bound to R protein enters the duodenum, where pancreatic proteases release the cobalamin. Then, intrinsic factor (IF), a glycoprotein produced by parietal cells in the stomach, binds the cobalamin and travels to the ileum, where cobalamin receptors (cubilins) form a cubilin-IF-cobalamin complex that may be absorbed by receptor-mediated endocytosis (transport inward by a cell vesicle that surrounds the complex). However, if cobalamin concentrations are high, passive diffusion may account for most of the absorption. Following absorption and transport to the liver, enterohepatic circulation is significant, and cobalamin secreted in bile can be reabsorbed in the ileum. Cobalamin in the plasma is bound to one of three transcobalamins: TCI, TCII, or TCIII. Uptake of cobalamin by tissues appears to involve receptors for TCII, and the TCII-cobalamin complex is taken into cells by endocytosis, followed by lysosomal degradation of TCII and release of cobalamin within the cytosol (cell fluid).

Food sources of cobalamin are animal products. However, microbes are the ultimate source since the tissues of higher plants or animals are incapable of synthesizing cobalamin. Meat, meat products, poultry, fish, shellfish, and eggs are good sources and contain predominantly adenosyl- and hydroxocobalamin. Milk, cheese, and yogurt contain somewhat less, mostly as methyl- and hydroxocobalamin. Vitamin supplements usually contain cyanocobalamin or hydroxocobalamin (dark red in pure form, pink when diluted). Recommended dietary allowances range from 0.9 µg per day for 1- to 3-year-old children to 2.8 µg per day for lactating women.

Vitamin C

Vitamin C (ascorbic acid or ascorbate) is not manufactured by tissues of invertebrate and fish species that have been studied. It is likewise a dietary essential for humans, guinea pigs, fruit bats, some birds, and most nonhuman primates because of a missing tissue enzyme, l-gulanolactone oxidase, required for its synthesis. It can be synthesized in the liver or kidneys of other terrestrial animal species that have been investigated, including some amphibians, reptiles, and other mammals. Ascorbic acid functions as an antioxidant and as a cofactor for hydroxylating enzymes that participate in synthesis of collagen, carnitine, and neurotransmitters (e.g., norepinephrine and serotonin) and in synthesis and catabolism of tyrosine. Its role in these reactions is to maintain iron and copper atoms in their reduced state within metalloenzymes.

Absorption of lower dietary concentrations in the small intestine involves a sodium-dependent active transport system in the brush border of the enterocyte; higher intakes may be absorbed by simple diffusion. Prior to absorption, ascorbate may be oxidized to dehydroascorbate, followed by reduction (requiring glutathione) within the enterocyte back to ascorbate. Transport across the basolateral membrane into the portal blood is carrier mediated. Free ascorbic acid is the predominant form in the plasma.

Good food sources include citrus fruits and juices, asparagus, broccoli, kale, papaya, and strawberries. Vitamin supplements may contain ascorbic acid, calcium ascorbate, sodium ascorbate, ascorbyl palmitate, or ascorbyl polyphosphate. Recommended dietary allowances range from 15 mg per day for 1- to 3-year-old children to 120 mg per day for lactating women.

MINERALS ARE NOT JUST ROCKS

Recognition that certain mineral elements are required nutrients was associated with discovery of their regular presence in body tissues. In 1874, J. Forster concluded that elements consistently found by analysis in animal tissues must be essential for life and thus should be present in the diet. However, connections between some elements and human or animal health were apparent even earlier. The English physician Thomas Sydenham noted in 1664 that oral iron salts restored normal skin color in anemic humans, and V. Menghini reported in 1747 that blood contains iron. J. G. Gahn, of Sweden, discovered in 1748 that bones contained phosphorus as phosphate of lime. The "calcareous" nature of bone had been noted for some time, but previously bone was considered some peculiar kind of earth. Calcium, the element, was not discovered until 1808 (simultaneously) by H. Davy and J. Berzelius. An association of calcium with the health of bone was made in 1842 by Charles Chossat, a Swiss physiologist and physician, who observed that the addition of calcium carbonate to a diet for pigeons prevented bone fragility. The German chemist, Justus von

Liebig, discovered in 1847 that of the two elements sodium and potassium, concentrations of the former were relatively higher in blood and lymph, whereas concentrations of the latter were relatively higher in soft tissues. J. B. Boussingalt reported in 1849 that low-sodium diets (apparently sodium deficient) fed to oxen resulted in signs of illness. In 1881, Sydney Ringer, an English physician, found that a solution containing the chlorides of sodium, potassium, and calcium was useful in sustaining the function of tissues in vitro, and Ringer's solution is still in use today. Hans Huebner reported in 1909 that a low-phosphorus diet produced rickets in dogs, and T. B. Osborne and L. B. Mendel reported in 1918 that dietary phosphorus restriction retarded growth in rats. The importance of the dietary calcium:phosphorus ratio for bone formation was demonstrated in rats by H. C. Sherman in 1921. The essentiality of magnesium was discovered in 1931 by E. V. McCollum, who found that rats fed a low-magnesium diet exhibited dilation of blood vessels and extreme hyperirritability. Low-chloride diets were reported by E. R. Orent-Keiles in 1937 to retard growth and to cause hypersensitivity in rats.

Discovery of the essentiality of several trace elements awaited development of analytical techniques that could detect them in low concentrations. Iodine was found in 1908 to prevent goiter in rat pups when it was given to their mothers. In 1928, E. B. Hart reported that copper as well as iron is required to prevent anemia in rats fed a milk-based diet. In 1931, Hart reported that mice failed to grow and ovulate on a low-manganese diet, whereas McCollum found that rat pups failed to suckle and survive (due to alterations in normal behavior). Essentiality of zinc was established in 1934 by Hart, who observed growth retardation and hair loss in rats fed a low-zinc diet. Cobalt was found in 1935 by E. J. Underwood and H. R. Marston to prevent anorexia (loss of appetite), anemia, and lethargy in sheep and was ultimately proven to be essential for synthesis of vitamin B_{12} by rumen bacteria. Fluoride, at appropriate concentrations in drinking water, was reported by H. T. Dean in 1938 to inhibit development of dental caries in children. Molybdenum was found in 1953 by D. A. Richert and W. W. Westerfield to be a component of the tissue enzyme xanthine oxidase. Selenium was found essential by Klaus Schwarz in 1957 to prevent liver necrosis in rats. Walter Mertz reported in 1959 that chromium was a factor in maintenance of glucose tolerance. Other elements, such as silicon (1972), nickel (1975), arsenic (1976), lithium (1981), lead (1981), and boron (1981), have been assigned essential status under special circumstances.

Calcium

Calcium is the fifth most abundant element on Earth and the most abundant mineral element in the body. Over 99% is present in the skeleton and teeth, mainly as hydroxyapatite $[Ca_{10}(PO_4)_6(OH)_2]$. Thus, the skeleton serves not only as the body's structural framework but also as a ready calcium reserve. Bone remodeling proceeds throughout life, with mineral resorption performed by osteoclasts and mineral deposition by osteoblasts. This remodeling is an essential feature of bone growth, repair of structural damage to bone, and homeostatic maintenance of blood calcium levels. Mineral deposition exceeds mineral resorption during skeletal growth, whereas mineral resorption may predominate during age-related bone loss.

Absorption of calcium from food requires that it be released in ionized form (Ca^{2+}). Solubilization is promoted by HCl in the stomach. Active absorption occurs principally in the duodenum and proximal jejunum and requires energy and a calcium-binding protein called calbindin. This process is regulated by calcitriol, the active form of vitamin D_3. Calbindin not only facilitates absorption into the enterocyte but also serves as a transport protein to deliver calcium to the basolateral membrane for extrusion. As calcium is extruded into the blood plasma, magnesium moves into the enterocyte in a process described as a calcium-magnesium pump. If calcium levels in the intestinal lumen are high, absorption in the jejunum and ileum may proceed by passive diffusion. Calcium in blood plasma is found in ionized form (~50%); bound to proteins (~40%), mainly albumin and prealbumin; and complexed (~10%) with sulfate, phosphate, or citrate.

Calcium concentrations, both within and outside cells, are tightly controlled by parathormone (from the parathyroid), calcitonin (from thyroidal parafollicular C cells), and cacitriol. Although most of the body's calcium is associated with integrity of bone, calcium's nonosseous functions are extremely important and include membrane permeability, nerve conduction, muscle contraction, blood clotting, and enzyme regulation. Interactions with other nutrients, such as protein, phosphorus, magnesium, sodium, potassium, iron, zinc, and boron, may variably affect absorption, metabolism, or excretion.

Good food sources include milk and dairy products, such as cheese and yogurt; leafy vegetables, such as kale, turnip and mustard greens; sardines (with bones); legumes and legume products, such as tofu; and calcium-fortified orange juice. Calcium supplements include calcium carbonate, calcium acetate, calcium citrate, calcium citrate-malate, calcium gluconate, calcium lactate, and calcium monophosphate. Estimated adequate calcium intakes range from 500 mg per day for 1- to 3-year-old children to 1300 mg per day for teenagers and young pregnant or lactating women. Estimated adequate intakes for males or females 51 years of age or older are 1200 mg per day, assuming intakes early in life were adequate.

Phosphorus

Phosphorus is the second most abundant mineral element in the body, with about 85% in the skeleton, 14% in soft tissues, and 1% in blood and other body fluids. In foods and in the body, it is found in combination with inorganic elements or in organic compounds. It is absorbed as inorganic phosphate; thus, organically bound phosphorus in food must be hydrolyzed before absorption can occur. Alkaline phosphatase, a zinc-dependent enzyme, in the duodenum and jejunum frees phosphate from many bound forms (although not from phytate), and phospholipase C, another zinc-dependent enzyme, frees phosphate from phospholipids. Phosphorus absorption occurs primarily in the duodenum and jejunum and appears to be a carrier-mediated, sodium-dependent, active process modulated by calcitriol. At higher luminal phosphorus concentrations, some may be absorbed by passive diffusion. Excessive intakes of calcium, magnesium, or aluminum may inhibit absorption. Phytase (a phosphate esterase) produced by yeast or intestinal bacteria can liberate some of the phosphorus from phytate if it is not complexed with calcium, zinc, or iron. About 70% of the phosphorus in blood plasma is in organic form as phospholipids in lipoproteins. The remaining 30% is inorganic, primarily as phosphate ions (HPO_4^{2-}, $H_2PO_4^{1-}$, or PO_4^{3-}) or associated with minerals such as calcium, magnesium, or sodium.

Phosphorus has many functions in the body, including mineralization of bone, energy storage and transfer in metabolic reactions, formation of nucleic acids (DNA and RNA), maintenance of acid-base balance, and as a structural component (in phospholipids) of cellular membranes.

Animal products (meat, poultry, fish, eggs, milk, and cheese) are food sources of high phosphorus bioavailability. Legumes, nuts, and grains also have considerable phosphorus but much may be bound in phytate (~80% in corn, wheat, and rice) and poorly available. Genetic modification of corn to incorporate a phytase of bacterial origin has shown promise for improving phosphorus availability. Cola drinks contain phosphoric acid and may contribute substantially to the phosphorus intake of habitual consumers. Recommended dietary allowances range from 460 mg per day for 1- to 3-year-old children to 1250 mg per day for teenagers and young pregnant or lactating women.

MAGNESIUM

Magnesium constitutes about 0.05% of the weight of adult humans, with about 55–60% in bone, 40–45% in soft tissues, and about 1% in extracellular fluids. Absorption occurs primarily in the distal jejunum and ileum and involves a carrier-mediated, active transport system at low magnesium intakes and simple diffusion at higher intakes. Absorption may be enhanced by calcitriol, lactose, or fructose and may be inhibited by phytate, nonfermentable fiber, and excessive calcium, phosphorus, or unabsorbed fatty acids. Efflux out of the enterocyte into the blood involves a sodium and energy-dependent carrier and possibly a calcium-dependent carrier. In blood plasma, about 50–55% of magnesium is present as free Mg^{2+}, about 33% is bound to protein (mostly to albumin), and about 13% is complexed with citrate, phosphate, sulfate, or other anions.

About 70% of the magnesium in bone is associated with calcium and phosphorus in the crystal lattice, but the approximately 30% found on bone surfaces is thought to serve as an exchangeable reserve that assists in maintaining normal plasma magnesium levels. This element participates in over 300 enzyme reactions, either as a structural cofactor or as an enzyme activator. A very high proportion of intracellular magnesium is associated with ATP or ADP and the enzyme systems in which they participate. Thus, magnesium is important at numerous metabolic steps in energy production; synthesis of DNA, RNA, and protein; contractility of cardiac and smooth muscle; hydroxylation of vitamin D in the liver; regulation of ion movement across cell membranes; and many other reactions.

Foods high in magnesium include nuts, legumes, whole-grain breads and cereals, green leafy vegetables (chlorophyll contains magnesium), chocolate, coffee, tea, certain spices, and seafood. Supplements include a variety of magnesium salts, such as magnesium sulfate, chloride, acetate, lactate, citrate, gluconate, or oxide. Recommended dietary allowances range from 80 mg per day for 1- to 3-year-old children to 420 mg per day for adult men.

SODIUM

Sodium is the most abundant cation in the body, comprising about 93% of the total cation supply. Approximately 30% is found as a sodium reserve on the surface of bone crystals within the skeleton. The remainder is found in soft tissues and extracellular fluids. Sodium is involved in maintenance of tissue fluid balance, transmission of nerve impulses, and contraction of muscles. An absorptive pathway in the small intestine involves cotransport of sodium ions and glucose; thus, as it is absorbed, sodium also plays a role in providing energy to the tissues. At the brush border of the enterocyte, Na^+ and glucose bind to a carrier that shuttles them from the intestinal lumen to the cell interior. They are released from the carrier, and Na^+ is pumped across the enterocyte basolateral membrane by the Na^+/K^+-ATPase (adenosine triphosphatase) pump, while glucose crosses into the plasma by facilitated diffusion. A second sodium absorptive pathway in the small intestine and proximal colon involves electroneutral cotransport of Na^+ and Cl^-, with Na^+ pumped across the basolateral membrane, as discussed, and Cl^- diffusing into the plasma. In the colon, sodium enters the luminal membrane of the enterocyte through Na^+-conducting channels, diffusing inwardly as a consequence of the downhill concentration gradient. It is accompanied by water and anions and is pumped through the basolateral membrane into the blood plasma by the Na^+/K^+-ATPase pump.

Processed foods, such as soups, canned meats, pickled foods, condiments, and salted snacks, account for nearly 75% of sodium intake in the United States. Naturally occurring sodium in animal and vegetable products accounts for only about 10%. Water generally provides less than 10%, and the remainder comes from salt (sodium chloride) added during cooking or at the table. Estimated adequate sodium intakes range from 1000 mg per day for 1- to 3-year-old children to 1500 mg per day for teenagers and men and women to 50 years of age. Estimated adequate intakes then decline to a low of 1200 mg per day at ages greater than 70. Actual sodium intakes in the United States are estimated to range from 1800 to 5000 mg per day.

POTASSIUM

Potassium constitutes about 0.35% of the body mass of a 70-kg adult human. It is the major intracellular cation, and about 95–98% of the total supply is found within body cells. Potassium is involved in the contractility of muscle, the excitability of nervous tissue, and the maintenance of tissue electrolyte and pH balance. It appears to be absorbed both in the small intestine and colon. Studies of absorption in the colon indicate it enters the enterocyte via a K^+/H^+-ATPase pump. Alternatively, it may enter the enterocyte via a membrane channel. Passage through the basolateral membrane into blood plasma appears to be by diffusion through a K^+ channel down a concentration gradient.

There are many food sources of potassium, including fruits, vegetables, meats, whole grains, legumes, and milk. Salt substitutes may contain potassium instead of sodium. Estimated adequate potassium intakes range from 3000 mg per day for 1- to 3-year-old children to 5100 mg per day for lactating women.

CHLORIDE

Chloride (Cl^-) constitutes about 0.15% of the body mass of a 70-kg adult human, with about 88% in extracellular fluid and 12% inside the cells. As a negative ion, it neutralizes the positive charge of sodium and is important in maintaining electrolyte balance. Other roles include production of hydrochloric acid by the parietal cells of the stomach, chloride release by white blood cells during phagocytosis to aid destruction of microorganisms, and as an exchange ion for bicarbonate (HCO_3^-) during the movement of the latter out of cells as CO_2 is generated during metabolism (the so-called chloride shift). This involves a protein transporter that moves Cl^- and HCO_3^- in opposite directions across the cell membrane so HCO_3^- can enter plasma, be transferred to red blood cells, and be carried to the lungs for exhalation of CO_2.

Most of ingested chloride comes from salt (NaCl). Salt is about 60% chloride, and processed foods and salty snacks supply liberal amounts. Eggs, meat, and seafood are good natural sources. Estimated adequate chloride intakes range from 1500 mg per day for 1- to 3-year-old children to 2300 mg per day for teenagers and men and women to 50 years of age. Estimated adequate intakes then decline to a low of 1800 mg per day at ages greater than 70.

IRON

The adult human body contains about 2 to 4 g of iron, with more than 65% in hemoglobin, about 10% in myoglobin, about 1–5% in enzymes, and the rest in transport form (transferrin) or storage forms (ferritin and hemosiderin). It exists in two oxidation states in the body, ferrous (Fe^{2+}) and ferric (Fe^{3+}). The presence of iron in heme is vital for transport of oxygen in hemoglobin to tissues and for transitional storage of oxygen in myoglobin in muscle. Oxygen is held in a loose coordinate bond with iron in these heme proteins, allowing for ready oxygen transfer to tissues for support of cellular metabolism. There also are heme-containing enzymes, such as catalase, the cytochromes, and myeloperoxidase, and nonheme, iron-containing enzymes, such as the oxygenases and oxidoreductases. These enzymes are variously involved in electron transfer; amino acid metabolism; synthesis of carnitine, collagen, DNA, and vitamin A; and destruction of cytotoxic compounds.

Absorption of iron found in heme, as part of hemoglobin or myoglobin in animal products, requires that heme be hydrolyzed from globin by proteases in the stomach and small intestine. The released heme is soluble and enters the enterocyte intact via an uncharacterized transporter. Within the mucosal cell, heme is hydrolyzed, and ferrous iron is released and used within the cell or transported to the basolateral membrane. Bound nonheme iron in food is freed for absorption by hydrochloric acid and pepsin in gastric secretions and by intestinal proteases. Most of the iron released is present in the ferric form and remains soluble as long as the milieu is acidic. However, intestinal and pancreatic juices tend to be alkaline, and some of the ferric iron may undergo reduction to ferrous iron. Ferrous iron may enter the enterocyte in association with transporter proteins that, to a lesser extent, can also transport copper, zinc, manganese, nickel, and lead. Absorption of ferric iron is facilitated by association with chelators, small organic compounds that form ligands and maintain ferric iron solubility. Iron absorption may be enhanced by ascorbic acid, citric acid, low molecular weight sugars, and mucin (digestive juice glycoproteins). Iron absorption may be inhibited by polyphenols in food and drinks (such as tannins in tea and coffee), oxalic acid in spinach, phytates in soy and some cereal grains, and excessive intakes of calcium, zinc, manganese, and nickel. Research has shown that genetic modification that lowers polyphenol concentrations in common beans (*Phaseolus vulgaris*) may improve both iron and zinc bioavailability. Transport of iron across the interior of the enterocyte appears to involve association with amino acids or proteins. Movement through the basolateral membrane appears to involve a transport protein, coupled with oxidation of any ferrous iron to ferric iron by copper-containing ferroxidases, followed by association with transferrin, the ferric iron transporter in blood plasma.

Iron in foods is found either in heme or nonheme forms. Heme iron is found mainly in hemoglobin and myoglobin in meat, poultry, and fish, comprising about 50–60% of the total iron in these foods; the remainder is nonheme iron. Plant foods, milk, cheese, and eggs contain mostly nonheme iron, although iron concentrations are very low in dairy products. Organ meats, particularly liver, are high in iron, and red meat, clams, oysters, molasses, nuts, legumes, green leafy vegetables, and iron-enriched whole-grain breads and cereals are good sources. Rice is generally considered a poor source, but by selective breeding, iron concentrations in polished rice can be increased two to four times. Iron sources approved for food fortification include elemental iron, ferrous ascorbate, ferrous carbonate, ferrous citrate, ferrous fumarate, ferrous gluconate, ferrous lactate, ferric ammonium citrate, ferric chloride, ferric citrate, ferric pyrophosphate, and ferric sulfate. However, these substances differ in their bioavailability and reactivity with other compounds—issues that must be considered when selecting them for use. Heme iron is highly available, and its bioavailability is generally affected less by other dietary components than is nonheme iron. Recommended dietary iron allowances range from 7 mg per day for 1- to 3-year-old children to 27 mg per day for pregnant women.

Copper

There are about 50 to 150 mg of copper in the adult human body in either cuprous (Cu^{1+}) or cupric (Cu^{2+}) valence states. Copper functions as a structural component of, or a cofactor for, several enzymes. Ceruloplasmin is an α-2 glycoprotein that transports copper in the blood but also has a ferroxidase function as ceruloplasmin-Cu^{2+} oxidizes Fe^{2+} to Fe^{3+} and is reduced to ceruloplasmin-Cu^{1+}. Superoxide dismutase is a copper- (and zinc-) dependent enzyme that protects cell membranes against peroxidative damage by superoxide radicals generated during inflammation. Cytochrome c oxidase is a copper-containing enzyme that participates in electron transfer such that molecular oxygen (O_2) is reduced to form water, and energy is released to permit ATP production. Some other copper-dependent enzymes are involved in amino acid metabolism and influence synthesis of elastin and collagen—and thus the integrity of connective tissue—and melanin, thus affecting skin and hair color.

Copper absorption is prominent in the duodenum but occurs throughout the small intestine. It appears to be absorbed by an active, carrier-mediated transport system at low luminal copper concentrations and by passive diffusion when luminal concentrations are high. Absorption may be enhanced by histidine, methionine, cysteine, and several organic acids, such as acetic, citric, gluconic, lactic, and malic acids. Inhibitors of copper absorption include excessive dietary concentrations of zinc, iron, calcium, phosphorus, vitamin C, and antacids. Transport of copper through the enterocyte has not been well characterized, but because free copper ions damage cells, it is believed to be bound to amino acids, proteins, or peptides, such as glutathione. Active transport across the basolateral membrane is the current presumption. Copper is transported in the portal blood loosely bound to albumin.

Rich food sources of copper include organ meats and shellfish. Nuts, legumes, potatoes, whole grains, and dried fruits are good sources. Copper sulfate is a primary supplemental form. Recommended dietary copper allowances range from 340 µg per day for 1- to 3-year-old children to 1300 µg per day for lactating women.

ZINC

About 1.5 to 2.5 g of zinc are present in the adult human body, with most found in bone, liver, kidneys, muscle, and skin. Different valence states exist, but most zinc in the body is in the divalent (Zn^{2+}) form. At least 70 enzymes are zinc dependent, requiring this element as a structural component or as a cofactor. Functions are diverse but include cellular respiration, conversion of retinol to retinal in the visual cycle, cleavage of amino acids during protein digestion, defense against damage by superoxide radicals, hydrolysis of glutamate residues from dietary folate to permit its absorption, and support of nucleic acid metabolism. As a consequence, zinc is important for cell growth and replication, formation of bone and integrity of skin, immunity, hormone function, glucose tolerance, and many other processes.

Zinc is released from protein and nucleic acids in food by proteases and nucleases in the stomach and small intestine, with absorption occurring mainly in the proximal small intestine. Glutathione, some amino acids, such as histidine and cysteine, and endogenous chelators, such as citric acid and picolinic acid, appear to enhance zinc absorption. Phytate, oxalate, polyphenols, and some soluble fibers appear to inhibit absorption. High levels of divalent cations, like Fe^{2+}, Cu^{2+}, and Ca^{2+}, also may inhibit absorption by competing for binding ligands. Transport into the enterocyte appears to be carrier mediated at low intakes and by passive diffusion at high intakes. Movement through the enterocyte and across the basolateral membrane is presumed to involve a zinc transporter. High zinc intakes tend to induce thionein formation, which when bound with metals is termed *metallothionein*. When bound with zinc, it may serve as a tissue zinc reserve, or when bound with other metals, it may modulate their absorption and toxicity. In the portal blood, zinc is mainly loosely bound to albumin.

Zinc in animal products is typically complexed with amino acids and with nucleic acids, and organ meats, beef, oysters, and mollusks are very good food sources. Poultry, pork, dairy products, whole grains, and leafy and root vegetables also are good sources. Fruits and refined cereals are poor, although selected varieties of rice that contain 20–25 mg of zinc per kilogram have been identified. Many salts of zinc have been used as supplements, but gastric irritation is a common side effect, and zinc acetate seems to be the best tolerated. Recommended dietary zinc allowances range from 3 mg per day for 1- to 3-year-old children to 12–14 mg per day for lactating women.

Manganese

About 10 to 20 mg of manganese is found in the body of a 70-kg adult male human, typically as Mn^{2+} or Mn^{3+}. Manganese functions as a constituent of metalloenzymes or as an enzyme cofactor. Some of the transferases that are activated by manganese can also be activated by magnesium. However, several enzymes concerned with connective tissue metabolism, inhibition of lipid peroxidation, and regulation of calcium-dependent processes specifically require manganese.

Mechanisms of manganese absorption are poorly understood. A quickly saturable, active transport system has been demonstrated in rats, and at high intakes, absorption efficiency declines, possibly to protect against manganese toxicity. Low molecular weight ligands, such as citrate or histidine, may enhance absorption, whereas phytate, oxalate, nonsoluble fiber, and excess iron or copper may inhibit it. Manganese entering the portal blood as Mn^{2+} may remain free or be bound to α-2 macroglobulin. After traversing the liver, Mn^{2+} may be transported to other tissues in free form, bound to α-2 macroglobulin or albumin, or oxidized to Mn^{3+} by ceruloplasmin and then complexed with transferrin.

Foods that are relatively high in manganese include whole-grain breads and cereals, nuts, leafy vegetables, and dried fruits. Milk and cheeses are low. Estimated adequate manganese intakes range from 1.2 mg per day for 1- to 3-year-old children to 2.6 mg per day for lactating women.

Iodine

Iodine is typically found as iodide (I^-) in the body. Its primary function resides in its role as a constituent of the thyroid hormones thyroxine (T_4) and triiodothyronine (T_3) and their regulation of metabolic rate.

Iodine in the diet may exist free as I^- or iodate (IO^{3-}) or bound to amino acids. During digestion, dietary IO^{3-} is reduced to I^- by glutathione. Iodide is readily absorbed from the stomach as well as from the intestinal tract. Iodinated amino acids and T_4 and T_3 may be absorbed intact, thus allowing oral administration of T_4 medication. Iodine in the portal blood may be present as I^- or as T_4 and T_3.

Terrestrial plant food sources are variable in their iodine concentration, reflecting regional differences in the concentration of iodine in soil. Likewise, iodine in drinking water is a reflection of the origin of rock strata from which water is withdrawn. Animals raised for food receive diets supplemented with salts of iodine, and iodized NaCl is commonly used for salting food at the table in the United States. Iodate also may be added as an oxidizing agent to bread dough to improve the physical cross-linking of gluten. Since oceans contain considerable iodine, edible portions of salt-water fish may have 300 to 3000 μg of iodine per kilogram compared to 20 to 40 μg per kilogram in freshwater fish. Iodophors used for cleansing cattle udders and milking equipment may increase iodine concentrations in dairy products. Recommended dietary iodine allowances range from 90 µg per day for 1- to 3-year-old children to 290 µg per day for lactating women.

Fluoride

Fluorine (a gas) bound to inorganic or organic compounds is designated fluoride (F^-), the form that predominates in nature. Inclusion of fluoride among the essential nutrients resides with its effectiveness in reducing the incidence of dental caries and its possible role in maintaining skeletal integrity. During formation of apatite crystals in bone and tooth enamel, fluoride can substitute for some of the hydroxyl ions, producing a fluorohydroxyapatite that is quite acid resistant. Mineralized tissues account for about 99% of fluoride in the body.

Fluoride absorption appears to occur by passive diffusion in the stomach, where it exists primarily as hydrogen fluoride (HF). Both ionic and nonionic fluoride is found in the blood, with some of the latter bound to plasma proteins.

Foods are generally low in fluoride, although marine fish consumed with bones (such as anchovies) and tea (which concentrates fluoride in leaves) may supply considerable amounts. In the United States, many drinking water sources are fluoridated by addition of sodium fluorosilicate. Some may enter the body from toothpaste containing added sodium fluoride or monofluorophosphate. Estimated adequate fluoride intakes range from 0.7 mg per day for 1- to 3-year-old children to 3.0 mg per day for pregnant or lactating women and 4.0 mg per day for adult men.

Molybdenum

Molybdenum is found primarily as Mo^{4+} or Mo^{6+} and is generally bound in biological systems to sulfur or oxygen. It functions as a component of three enzymes that catalyze oxidation-reduction reactions, xanthine dehydrogenase/oxidase, aldehyde oxidase, and sulfite oxidase.

Absorption has not been well studied but is thought to be by passive diffusion. Transport in the blood may be as molybdate (MoO_4^{2-}), or molybdenum may be bound to albumin or α-2 macroglobulin.

Molybdenum is found in many foods, but concentrations tend to vary regionally, coincident with soil levels. Legumes, meat, poultry, and fish are good sources. Fruits and dairy products are low in molybdenum. Recommended dietary molybdenum allowances range from 17 µg per day for 1- to 3-year-old children to 50 µg per day for pregnant or lactating women.

Selenium

About 15 mg of selenium is found in the adult human body. Its chemical characteristics are similar to those of sulfur, and it can replace sulfur in amino acids such as cysteine, cystine, or methionine. Selenium is an essential cofactor for glutathione peroxidases that accept reducing equivalents from glutathione to convert hydrogen peroxide to water or lipid peroxides to hydroxy lipids, thus protecting cellular membranes against damage. Selenium-containing deiodinases catalyze removal of iodine from T_4 to yield T_3, a major active hormone form. Thioredoxin reductase contains selenocysteine and transfers reducing equivalents from NADPH to produce $NADP^+$. Selenophosphate synthetase (containing selenocysteine) catalyzes synthesis of selenophosphate from selenide (H_2Se), a compound needed for synthesis of the mentioned selenium-activated or selenium-containing enzymes. Other selenium-containing enzymes that play antioxidant roles also have been discovered.

Selenium absorption appears to occur mainly in the duodenum, with less in the jejunum and ileum. Organic forms, such as selenocysteine, selenocystine, or selenomethionine, enter the enterocyte via an amino acid transporter, whereas inorganic forms appear to cross the luminal membrane as selenite (H_2SeO_3) or as selenate (H_2SeO_4). On entering the portal blood, selenium binds to sulfhydryl groups in lipoproteins or to selenoprotein P for transport in plasma.

Selenium levels in plant foods vary regionally with available selenium concentrations in soil, and geographical areas of deficiency and toxicity have been identified. Animal food products tend to be less variable because of the need to supplement animal diets to prevent regional deficiencies. Seafoods may be among the better sources due to selenium supplies in the ocean, although elevated mercury levels in some saltwater fish may reduce selenium bioavailability. Forms in food include the selenium analogues of sulfur-containing amino acids and inorganic selenites and selenates. Recommended dietary selenium allowances range from 20 μg per day for 1- to 3-year-old children to 70 μg per day for lactating women.

CHROMIUM

The chromium content of the adult human body has been estimated to be about 4 to 6 mg. In nature, chromium exists in several valence states, but trivalent chromium (Cr^{3+}) is the most stable and is thought to be the most important functional form. It potentiates the action of insulin by mechanisms still under investigation but which may involve Cr_4-chromodulin (four chromium atoms bound to an oligopeptide composed of glycine, cysteine, aspartate, and glutamate).

Absorption of chromium may require formation of complexes with low molecular weight ligands (e.g., methionine or histidine) in the acid environment of the stomach (in which Cr^{3+} is soluble), followed by movement into the enterocyte by either diffusion or a carrier-mediated transporter. Absorption is enhanced by picolinate and ascorbate and inhibited by phytate and antacids. Inorganic chromium (Cr^{3+}) is transported in the blood by transferrin and possibly by albumin or globulins. Some may circulate unbound.

Good food sources include liver, beef, pork, poultry, whole grains, cheese, mushrooms, some spices, tea, and wine. An organic complex of chromium (glucose tolerance factor) is found in brewer's yeast. Estimated adequate chromium intakes range from 11 μg per day for 1- to 3-year-old children to 45 μg per day for lactating women.

CONCLUSIONS AND STUDY TOPICS

Diseases of excess and deficiency of vitamins and mineral elements are addressed in Chapters 9 and 10, respectively. The provision of essential vitamins and minerals in appropriate amounts assumes that foods supplying them are available and afford-able. Because this may not be the case in many parts of the world, topics worthy of study include the influence of culture, poverty, political policy, and geographical fac-tors (such as mineral concentrations in soil and water) on the ability to meet needs.

NOTE

1. To save space, most of the information on vitamins and mineral elements was derived from reviews in the references listed in the bibliography; see them for more detail. Specific citations are not included in the text.

BIBLIOGRAPHY

Bailey, L.B. and J.F. Gregory III. 2006. Folate. In *Present Knowledge in Nutrition*, 9th ed., Vol. 1, Bowman, B.A. and R.M. Russell, Eds. Washington, DC: International Life Sciences Institute. pp. 278–301.

Bates, C.J. 2006. Thiamin. In *Present Knowledge in Nutrition*, 9th ed., Vol. 1, Bowman, B.A. and R.M. Russell, Eds. Washington, DC: International Life Sciences Institute, pp. 242–249.

Camporeale, G. and J. Zempleni. 2006. Biotin. In *Present Knowledge in Nutrition*, 9th ed., Vol. 1, Bowman, B.A. and R.M. Russell, Eds. Washington, DC: International Life Sciences Institute, pp. 314–326.

Carpenter, K.J. 2003. A short history of nutritional science: part 1 (1785–1885). *J. Nutr.* 133:638–645.

Carpenter, K.J. 2003. A short history of nutritional science: part 2 (1885–1912). *J. Nutr.* 133:975–984.

Carpenter, K.J. 2003. A short history of nutritional science: part 3 (1912–1944). *J. Nutr.* 133:3023–3032.

Carpenter, K.J. 2003. A short history of nutritional science: part 4 (1945–1985). *J. Nutr.* 133:3331–3342.

Ferland, G. 2006. Vitamin K. In *Present Knowledge in Nutrition*, 9th ed., Vol. 1, Bowman, B.A. and R.M. Russell, Eds. Washington, DC: International Life Sciences Institute, pp. 220–230.

Gropper, S.S., J.L. Smith, and J.L. Groff. 2005. *Advanced Nutrition and Human Metabolism*. Belmont, CA: Thomson Wadsworth.

HarvestPlus. Breeding crops for better nutrition. Available at: www.HarvestPlus.org.

Jacob, R.A. 2006. Niacin. In *Present Knowledge in Nutrition*, 9th ed., Vol. 1, Bowman, B.A. and R.M. Russell, Eds. Washington, DC: International Life Sciences Institute, pp. 260–268.

Johnson, C.S. 2006. Vitamin C. In *Present Knowledge in Nutrition*, 9th ed., Vol. 1, Bowman, B.A. and R.M. Russell, Eds. Washington, DC: International Life Sciences Institute, pp. 233–241.

Lindshield, B.L. 2006. Carotenoids. In *Present Knowledge in Nutrition*, 9th ed., Vol. 1, Bowman, B.A. and R.M. Russell, Eds. Washington, DC: International Life Sciences Institute, pp. 184–197.

McCollum, E.V. 1957. *The History of Nutrition*. Boston: Houghton Mifflin.

McCormick, D.B. 2006. Vitamin B$_6$. In *Present Knowledge in Nutrition*, 9th ed., Vol. 1, Bowman, B.A. and R.M. Russell, Eds. Washington, DC: International Life Sciences Institute, pp. 269–277.

Miller, J.W., L.M. Rogers, and R.B. Rucker. 2006. Pantothenic Acid. In *Present Knowledge in Nutrition,* 9th ed., Vol. 1, Bowman, B.A. and R.M. Russell, Eds. Washington, DC: International Life Sciences Institute, pp. 327–339.

Nestel, P., H.E. Bouis, J.V. Meenakshi, and W. Pfeiffer. 2006. Biofortification of staple food crops. *J. Nutr.* 136:1064–1067.

Norman, A.W. and H.H. Henry. 2006. Vitamin D. In *Present Knowledge in Nutrition,* 9th ed., Vol. 1, Bowman, B.A. and R.M. Russell, Eds. Washington, DC: International Life Sciences Institute, pp. 198–210.

O'Dell, B.L. and R.A. Sunde, Eds. 1997. *Handbook of Nutritionally Essential Minerals.* New York: Marcel Dekker.

Rivlin, R.S. 2006. Riboflavin. In *Present Knowledge in Nutrition,* 9th ed., Vol. 1, Bowman, B.A. and R.M. Russell, Eds. Washington, DC: International Life Sciences Institute, pp. 250–259.

Solomons, N.W. 2006. Vitamin A. In *Present Knowledge in Nutrition,* 9th ed., Vol. 1, Bowman, B.A. and R.M. Russell, Eds. Washington, DC: International Life Sciences Institute, pp. 157–183.

Stabler, S.P. 2006. Vitamin B_{12}. In *Present Knowledge in Nutrition,* 9th ed., Vol. 1, Bowman, B.A. and R.M. Russell, Eds. Washington, DC: International Life Sciences Institute, pp. 302–313.

Traber, M.G. 2006. Vitamin E. In *Present Knowledge in Nutrition,* 9th ed., Vol. 1, Bowman, B.A. and R.M. Russell, Eds. Washington, DC: International Life Sciences Institute, pp. 211–219.

Section IV

Foods and Health

9 Overweight, Obesity, and Related Diseases

Gail G. Harrison and Summer Hamide

CONTENTS

ABSTRACT

At the same time that problems of undernutrition have not been solved in much of the world, overweight and obesity have emerged in the last quarter century as major public health problems globally. For most developing countries, there is thus now a double burden of malnutrition with associated costs to health and quality of life. More than a billion adults worldwide are now overweight, and a third of these are overweight to a degree that classifies them as obese. The health consequences include premature deaths and chronic diseases that reduce quality of life and threaten to overwhelm the resources of health care systems widely. The causes of the obesity epidemic include changes in the age structure of many populations, with more adults and fewer children as fertility is reduced in most countries; rapid urbanization with consequent decreases in energy expenditure; and globalization of food supplies with rapid change toward diets higher in animal products, oil, and sugars in many countries. It will require committed effort on a large scale to turn the epidemic around, with attention to changes in social and physical environments as well as individual behavior.

THE EMERGENCE OF OBESITY AS A GLOBAL PUBLIC HEALTH PROBLEM

At the same time that problems of undernutrition have not been solved in much of the world, overweight and obesity have emerged in the last quarter century as major public health problems globally. For most developing countries, there is thus now a double burden of malnutrition with associated costs to health and quality of life. More than a billion adults worldwide are now overweight, and a third of these are overweight to a degree that classifies them as obese (World Health Organization [WHO] 2003). The health consequences include premature deaths and chronic diseases that reduce quality of life and threaten to overwhelm the resources of health care systems widely. The impact on morbidity is greater than that on mortality, but the latter is also substantial. The overall result is that, in major part due to rising obesity, for the first time in history life expectancy at birth is expected to level off and even decline in the first half of the twenty-first century (Olshansky et al. 2005). In other words, there is a distinct possibility that the next generation may be the first in recorded history to have shorter life expectancies than did their parents.

DEFINING OVERWEIGHT AND OBESITY

Clinically, *obesity* is an amount of body fat, relative to lean tissue such as muscle and bone, that is elevated to a level at which there are clear adverse effects on health. *Overweight* refers to a lesser degree of excess body fat, in spite of the fact that weight is a unit of mass rather than body composition.

Because of ease of measurement and relatively good correlation with body composition, anthropometric measurements are most commonly used to determine the prevalence of overweight and obesity. International definitions for adults are based on the body mass index (BMI), which is calculated as weight in kilograms divided by height in meters, squared (kg/m^2), and multiplied by 100. This index is an expression of body weight (mass) adjusted for height and is at the population level a good proxy for body fatness. At the individual level, there is variation; for example, a football linebacker may have a high BMI but no excess body fat, and a sedentary elderly woman might have a normal-range BMI but have excess body fat for optimal health. But for populations, BMI is well correlated with body composition across a wide range and provides a simple index that enables epidemiological comparisons across populations. Internationally used diagnostic criteria for adult obesity (BMI > 30) and overweight (BMI 25–29.9) are widely used in nutritional surveillance. These cutoffs are conservative since there is evidence that risk of chronic disease in some populations increases progressively from a BMI of about 21 (Popkin 2004).

For children, definitions are less clear and simple, at least in part because during childhood the distribution of BMI varies with age. In the United States, the Centers for Disease Control and Prevention (CDC) published age- and sex-specific reference data for BMI for children, and individuals with BMI over the 95th percentile are termed overweight, and those between the 85th and 95th percentiles are "at risk for overweight" (2008). An international reference has been proposed for children 2 to 18 years of age, with the reference based on nationally representative data from six countries (Cole et al. 2000), with cutoff points for overweight at the 90th percentile of BMI for age and that for obesity at about the 97th percentile. And, WHO has published reference charts for children from birth to 5 years with the charts based on a multicountry study of normal growth (WHO 2006). All of these definitions carry some degree of arbitrariness but do identify the extreme end of the distribution of relative weight, where there are clear adverse effects on lifetime health.

ETIOLOGY OF OBESITY

Obesity is a complex condition with multiple potential causes, all leading to a long-term imbalance between energy intake and energy expenditure. The fact that many individuals maintain their weight for long periods of time despite major day-to-day fluctuations in both intake and expenditure speaks to exquisite metabolic and physi-ological regulating mechanisms. However, it is clear that such mechanisms are not perfect, and that a large portion of the human species will respond to increases in dietary energy intake or decreases in physical activity with gain in body weight and deposition of fat.

The amount of imbalance required to result in weight gain and eventual obesity is very small; it has been estimated that a systematic error of only 2% in children is sufficient to produce obesity—the amount accounted for by trading 15 minutes of active play for watching television, for example (Goran 2000). For an adult to gain a pound of body fat, the adult must consume 3500 kilocalories more than expended. This means that an increase in intake of 100 kcal/day (about the energy content of a slice of bread) would result in roughly 10 pounds of weight gain over the course of a year. If this continued for 5 years, the individual would gain 50 pounds. Thus, very small changes in diet or physical activity can, over time, result in major changes in the prevalence of obesity in populations.

PREVALENCE AND RECENT TRENDS IN OBESITY AMONG POPULATIONS

Major increases in the prevalence of adult obesity have occurred during the last two decades in many parts of the world, including threefold increases since 1980 in North America, the United Kingdom, Eastern Europe, the Middle East, the Pacific Islands, Australasia, and China (Seidell and Visscher 2004). Notably, sub-Saharan Africa has been exempted from this trend (Martorell 2002), and even there we see some exceptions, particularly in cities. In the United States, currently two thirds of adults are overweight or obese, and almost one third of these are obese. The recent increases in prevalence of obesity affect both men and women (although in most countries the absolute prevalence is higher in women) and all socioeconomic and ethnic groups. Figure 9.1 and Figure 9.2 show graphically the global distribution of adult obesity for men and women.

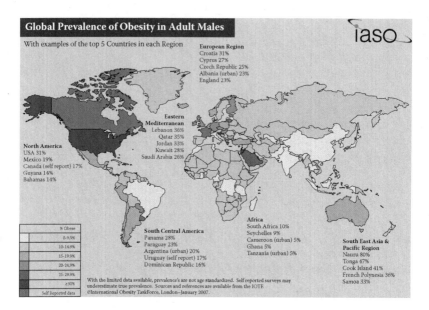

FIGURE 9.1 *A color version of this figure follows page 198.* Global prevalence of obesity in adult males.

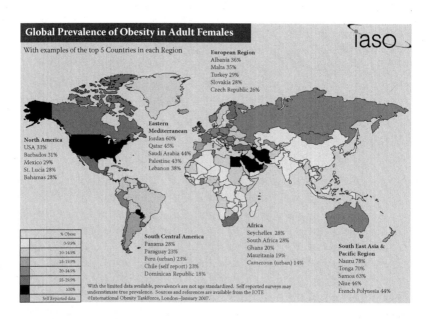

FIGURE 9.2 *A color version of this figure follows page 198.* Global prevalence of obesity in adult females.

Overweight in children has also increased dramatically in the last several years. In the United States, the prevalence of overweight among children 6 to 11 years old has doubled, and the number of overweight adolescents has trebled since the 1960s (Flegal and Trioano 2000, Ogden et al. 2002). Obesity is the most common health problem faced by American children, with African American and Hispanic children and children in low-income households the most severely affected. However, childhood obesity is a global phenomenon and increasingly extends into the developing world; for example, in Thailand the prevalence of obesity in 5- to 12-year-olds rose from 12.2 to 15.6% in just 2 years (World Health Organization 2003). Similar statistics have been documented in many other low- and middle-income countries. Rising rates of obesity in children are particularly problematic, not only because overweight children have a high risk of growing into obese adults, but also because obesity-associated illnesses, including diabetes and high blood pressure, accompany obesity even at young ages.

While the prevalence of obesity has increased across all population segments in the United States, rates are highest among women, middle-aged adults, and those with less education and income. Interestingly, in the United States the association of higher socioeconomic status with lower rates of obesity in women is less marked among Hispanics and African Americans than among non-Hispanic whites (Friel et al. 2007). In developing countries, the long-standing pattern has been that higher socioeconomic status is associated with higher rates of obesity, but that pattern is changing. In Brazil, for example, the obesity/socioeconomic status relationship has already reversed for women, with poorer women now at higher risk than richer ones (Monteiro et al. 2004b). Assuming that this trend will repeat itself in other large populations, the obesity epidemic threatens to widen already existing health disparities in a number of populations (Monteiro et al. 2004a). In other words, the burden will increasingly fall on the poor, as does the burden of undernutrition.

THE REAL CAUSES OF THE OBESITY EPIDEMIC

While body weight has a high degree of heritability and genetics contributes 25–40% to individual variation in weight, the rapidity with which obesity has emerged as a highly prevalent condition argues that profound changes in the environment are responsible for the current global trend. These include urbanization and economic development with attendant decreases in demands for physical activity, as well as globalization of food commodities and trade resulting in widespread dietary change. These quite powerful environmental changes have occurred superimposed on unprecedented changes in the age structure of many of the world's populations. The term *nutrition transition* is being used to describe the shift toward obesity and nutrition-related chronic disease dominating the public health landscape in developing countries (Popkin 1998). To understand it, it is necessary to appreciate the demographic and epidemiologic transitions that underlie it.

CHANGING POPULATION STRUCTURES

The last two decades have seen major decreases in fertility and increases in life expectancy in most parts of the world, including all areas with the single exception of parts of sub-Saharan Africa and perhaps Haiti. The rapid population growth that characterized the period from about 1700 to the present is projected to slow markedly in the next few decades. Population growth is forecast to continue for about the next 50 years, with the world's population stabilizing at somewhere between 8 and 12 billion (Zlotnik 2002). The period of continued growth that will characterize the next few decades will be very uneven across regions. For example, fertility rates in Europe are already below replacement levels, while some regions of the developing world still have very high fertility rates. In most of the developing world, however, fertility is declining dramatically, and life expectancies have increased markedly in the last few years. The consequence is that populations are aging—the proportion comprised of adults is higher and that comprised of children lower. In terms of disease burden and health care needs, this means that health problems of adulthood are becoming more important.

The shift in population structure will be a crucial driver of income growth in many developing countries; with more adults, a larger portion of populations will be economically active, earning incomes that are spent raising a smaller number of children. The implications for the obesity epidemic are already evident. In China, 20 years of a one-child population policy and a growing economy have meant that families invest more resources in a single child. A sharp rise in childhood overweight has been seen already (Jing 2000).

URBANIZATION

Even more dramatic than the slowdown in overall population growth is the shift to a largely urbanized world. Cities are home to almost half the world's population, up from only 30% as recently as 1950 (Zlotnik 2002), and virtually all of the projected population growth globally in the next 20 years will be urban. Life in cities involves major differences in both diet and energy expenditure from life in rural areas. The effect on physical activity patterns is generally profound. Children and young adults spend more time in school as countries invest in education for all; adults are likely to work in service sector, marketing, managerial, or professional occupations rather than agriculture; transportation systems, including buses and automobiles, replace more energy-demanding walking and carrying; high-rise apartment buildings provide little or nothing in the way of space for active play for children; and television sets and computers are increasingly the centers for leisure time. The net effect is a decline in physical activity, fitness, and energy expenditure on a striking scale. Urbanization generally also means more female participation in the workforce and consequent change in food intake patterns within the home, away from time-intensive traditionally prepared meals toward more convenient meals and snacks (Schmidhuber and Shetty 2004). And, compared to rural populations, city dwellers globally consume diets that are higher in fat, animal products, sugar, and processed foods and lower in unrefined grains (Popkin 2002).

GLOBALIZATION OF FOOD SYSTEMS

An acceleration of the nutrition transition is being driven in part by a major change in food distribution and marketing systems that accompanies urbanization in developing countries. Supermarkets are emerging in cities throughout the developing world. This phenomenon occurred earliest in Latin America, where supermarkets' share of retail food sold rose from 15% in 1990 to 60% in 2000 (Reardon and Berdegué 2002). The rise of supermarkets has the potential for major changes in the food environment, making nonperishable foods available widely and vastly increasing the number and variety of food items available to populations of consumers. They also provide a distribution system that can be efficiently used by large corporations that produce a large variety of processed food items. Trade policies have made it possible for multinational corporations to market their products throughout the world, and access to mass media (and thus direct exposure to advertising) has increased very rapidly in the developing world, particularly in cities.

Thus, the combination of changes in population structure, rapid urbanization, and global marketing of food products has produced changes in the energy balance equation for millions of people within a very short period of a few decades. These forces will continue for some time to come, and with them will arrive increasing rates of obesity and related diseases unless committed and targeted efforts are made on a large scale to enable access to high-quality diets and adequate physical activity.

CONSEQUENCES OF THE OBESITY EPIDEMIC

NUTRITION-RELATED NONCOMMUNICABLE DISEASES: WHY DOES OBESITY MATTER?

Rising rates of obesity in populations are of concern because obesity is intricately associated with various diseases that compromise quality of life, predispose to premature death, and place enormous cost burdens on health care systems. These include type 2 diabetes (non-insulin-dependent diabetes mellitus), high blood pressure, dyslipidemias, coronary heart disease, and certain types of cancers (breast, endometrial, prostate, and colon). One analysis for the United States estimates that compared to adults of normal weight, those who are severely obese (BMI > 40) have 7.4 times higher risk of diabetes, 6.4 times higher risk for high blood pressure, 2.7 times higher risk for asthma, almost twice the risk of high cholesterol levels, and overall more than 4 times higher risk for overall fair or poor health (compared to self-reported excellent, very good, or good health) (Mokdad et al. 2004). A recent major analysis indicated that 30% or more of the global cancer burden could be avoided by optimizing the quality of diets (American Institute for Cancer Research/ World Cancer Research Federation [AICR/WCRF] 2007). Deaths in the United States attributable to poor diet and physical inactivity rose by 33% over the last 10 years, while deaths from other causes have declined. Estimates are that if current trends continue, obesity will soon overtake tobacco as the leading preventable cause of death in the United States.

Children who are overweight suffer from increased health problems in several ways. First, many overweight children are developing health problems that in days past were thought of as exclusively adult problems, including type 2 diabetes and high blood pressure. These conditions begin taking their toll in terms of complications earlier in life, producing risk of early mortality. Second, children who are overweight are at risk of becoming severely overweight adults (Ferraro et al. 2003). And third, overweight children suffer from social stigma and emotional ill health in many cultural environments.

The predictable obesity-driven chronic diseases—type 2 diabetes and atherosclerotic heart disease the most visible and consequential—have major implications for quality of life and for life expectancy. The risk of type 2 diabetes and of high blood pressure increase steadily with rising body fatness. The vast majority of persons with diabetes have type 2 diabetes, and of these, 90% are obese or overweight. Globally, the Emerging Market Economy (former Soviet Union) countries had the highest number of diabetics in 1995; by 2025, India and the Middle East will lead if current trends continue (WHO, 2003).

Cardiovascular diseases (CVDs) are now the leading cause of death and disability in most regions of the world (Reddy 2002a,b). Low- and middle-income countries account for more than three quarters of all CVD deaths globally, with the large majority of these attributable to coronary heart disease (CHD). Of great importance is the fact that both type 2 diabetes and CHD are affecting adults at younger ages in developing countries than they do in Europe and North America, with the greatest risk in the age range 45 to 65; thus, the projection is for a major burden from deaths among economically productive adults throughout the developing world (O'Dea and Piers 2002, Reddy 2002a).

IMPACT ON SOCIAL DISPARITIES IN HEALTH

Globally, problems of undernutrition persist in many countries and are actually worsening in sub-Saharan Africa. The Millennium Development Goal to reduce hunger by 50% by the year 2015 seems unattainable. At the same time, rising rates of overweight and obesity add to health care costs and reduced quality of life in the same populations, creating a true double burden. In addition, overweight and obesity are becoming increasingly concentrated among the poor, particularly in urban areas and in the more economically developed countries. Thus, the burden of chronic disease will increasingly fall on society's most disadvantaged, increasing health disparities and compromising public health gains achieved through combating infectious diseases and malnutrition (Hawkes et al. 2004).

IMPACT ON COSTS AND HEALTH CARE SYSTEMS

Health care costs associated with obesity-driven diseases are substantial. In developed countries, estimates are that obesity accounts for 2 to 7% of all medical care costs (WHO 2003). In poorer countries, there has been less research on the topic, but one analysis (Popkin 2006) estimates that the increasing costs of the shift toward nutrition-related chronic disease will overwhelm the health care system of China and slow its economic growth. It is clear that the proportion of the population who are overweight or obese in many developed countries is now so large that there are no longer sufficient health care resources to offer treatment to all, and in developing countries limited resources will quickly be exhausted by the need for expensive treatment for obesity-related disease (WHO 2000).

CONTROLLING THE PROBLEM: THE PREVENTION IMPERATIVE

Reversal of obesity once established at the level of the individual is extremely difficult. Many clinical approaches have been tried, ranging from behavior modification techniques to pharmacological and surgical approaches. All have produced minimal to only moderate successes long term. Thus, prevention is the only productive way forward.

Preventive interventions have until recently also focused on individual behavior change and have produced some short-term success, but there is limited evidence for sustainability. Recently, there has been a rather remarkable recognition that broader, more "upstream" policy and program actions are called for if the obesity epidemic is to be controlled. WHO has called for multisectoral policies that simultaneously address prevention of excess weight gain, promotion of weight maintenance in adulthood, medical management of obesity-driven comorbidities, and promotion of weight loss (WHO 2000). A few national-level programs, notably in Norway and in Mauritius, have successfully reversed the trend toward high-fat, energy-dense diets with a combination of agricultural policies, subsidies and price manipulation, and public education (Friel et al. 2007). Norway, over a period of 20 years beginning in 1975, utilized a combination of consumer and producer price and income subsidies, regulation of food processing and labeling, and public education and saw a reduction in consumption of saturated fat and reduced average blood cholesterol in the population (Norum et al. 1997). Mauritius, through a combination of price policy, mass media, and educational activity in various venues saw a fall in the prevalence of hypertension by 15%, of cigarette consumption by 11% in men and 7% in women, and in mean serum cholesterol by about 10% (Dowse et al. 1995). A number of local and community-based initiatives have had success in increasing physical activity and altering the food environment in key venues such as schools and workplaces.

Turning the obesity epidemic around requires attention at many levels and cannot be managed by health care professionals alone. Just as rapid environmental change has promoted the epidemic, changes in the physical and social environment will be required to control it. Urban planning to promote more walking; school systems that plan curricula to allow for physical activity and the availability of healthy foods and beverages; work site innovations to improve worker health; regulation of marketing and advertising of energy-dense, low-nutrient food products, particularly to children; and health promotion to the public all are required. Health promotion activities need to be directed at more than individual behavior but also at making physical activity and healthy eating societal norms for all parts of the population.

It is clear that committed and targeted actions are required at many levels if the obesity epidemic and its consequences are to be controlled or reversed within the next decades (Schwarz and Brownell 2005, Friel et al. 2007). The fact that overweight may be regarded as culturally desirable or at least acceptable in many cultural contexts means that the health consequences need to be clearly spelled out for policy makers and the public. The need for counteracting the health impacts of sedentary lifestyles and poor diets is too important to be ignored and is a matter of cost, quality of life, and social justice as well as a matter of health.

CONCLUSIONS AND STUDY TOPICS

1. Overweight and obesity have emerged as major public health problems at the same time that basic problems of malnutrition have not been solved. What do you think the priority should be for prevention in developing countries?
2. Most of the predicted diabetes and premature heart disease in the world (and associated health care costs) could be prevented by controlling the obesity epidemic. What argument would you make to policy makers for putting this first on the health care agenda?
3. The obesity epidemic threatens to exaggerate socioeconomic inequities in health. Describe why this is the case and what could be done to avoid it.

REFERENCES

American Institute for Cancer Research/World Cancer Research Federation. *Food, Nutrition and Physical Activity and the Prevention of Cancer: A Global Perspective.* Washington, DC: American Institute for Cancer Research, 2007.

Centers for Disease Control and Prevention. 2000 CDC growth charts. Available at: www.cdc. gov/growthcharts. Accessed March 10, 2008.

Cole, T.J., Bellizzi, M.C., Flegal, K.M., and Dietz, W.H. Establishing a standard definition for childhood overweight and obesity worldwide: international survey. *BMJ* 320:1240–1243, 2000.

Dowse, G.K., Gareeboo, H., Alberti, K.G., Zimmet, P., Tuomilehto, J., and Purran, A. Changes in population cholesterol concentrations and other cardiovascular risk factor levels after five years of the non-communicable disease intervention programme in Mauritius. *BMJ* 311:1255–1259, 1995.

Ferraro, K.F., Thorpe, R.J., and Wilkinson, J.A. The life course of severe obesity: does childhood overweight matter? *J Gerontol Psychol Sci Soc Sci* 58:S110–S119, 2003.

Flegal, K.M. and Trioano, R.P. Changes in the distribution of body mass index of adults and children in the U.S. population. *Int J Obes Relat Metab Disord* 24:807–818, 2000.

Friel, S., Chopra, M., and Satcher, D. Unequal weight: equity oriented policy responses to the global obesity epidemic. *BMJ* 335:1241–1243, 2007.

Goran, M.I. Metabolic precursors and effects of obesity in children: a decade of progress, 1990–1999. *Am J Clin Nutr* 73:158–171, 2000.

Hawkes, C., Eckhardt, C., Ruel, M., and Minot, N. Diet quality, poverty and food policy: a new research agenda for obesity prevention in developing countries. *UN Standing Committee Nutr News* 29:13–19, 2004.

Jing, J., Ed. *Feeding China's Little Emperors: Food, Children and Social Change.* Stanford, CA: Stanford University Press, 2000.

Martorell, R. Obesity in the developing world. In Caballero, B. and Popkin, B.M. *The Nutrition Transition: Diet and Disease in the Developing World.* London: Academic Press, 2002, pp. 147–164.

Mokdad, A.H., Marks, J.S., Stroup, D.F., and Geberding, J.L. Actual causes of death in the United States, 2000. *JAMA* 291:1238–1245, 2004.

Montiero, C.A., Conde, W.L., Lu, B., and Popkin, B.M. Obesity and inequities in health in the developing world. *Int J Obesity* 28:1181–1186, 2004a.

Montiero, C.A., Conde, W.L., and Popkin, B.M. The burden of disease from undernutrition and overnutrition in countries undergoing the nutrition transition. *Am J Public Health* 94:433–444, 2004b.

Norum, K.R. et al. Nutrition and food policy in Norway: effects on reduction of coronary heart disease. *Nutr. Rev.* 55: 532–539, 1997.

O'Dea, K. and Piers, L.S. Diabetes. In Caballero, B. and Popkin, B.M., Eds., *The Nutrition Transition: Dynamics of Diet and Disease in the Developing World.* San Diego: Academic Press, 2002, pp. 1656–1690.

Ogden, C.L., et al. Prevalence and trends in overweight among US children and adolescents, 1999–2000. *JAMA* 288:1728–1732, 2002.

Olshansky, S.J., Passaro, D.J., Hershow, R.C., Layden, J., Carnes, B.A., Brody, J., Hayflick, L., Butler, R.N., Allison, D.B., and Ludwig, D.S. A potential decline in life expectancy in the United States in the 21st century. *N Engl J Med* 352:1138–1145, 2005.

Popkin, B.M. The nutrition transition and its health implications in lower-income countries. *Public Health Nutr* 1(1):5–21, 1998.

Popkin, B.M. The dynamics of the dietary transition in the developing world. In Caballero, B. and Popkin, B.M., Eds., *The Nutrition Transition: Dynamics of Diet and Disease in the Developing World.* San Diego: Academic Press, 2002, pp. 111–128.

Popkin, B.M. The nutrition transition: an overview of world patterns of change. *Nutr Rev* 62:S140–S143, 2004.

Popkin, B.M. Global nutrition dynamics: the world is shifting rapidly toward a diet linked with noncommunicable diseases. *Am J Clin Nutr* 84:289–298, 2006.

Reardon, T. and Berdegué, J.A. The rapid rise of supermarkets in Latin America: challenges and opportunities for development. *Dev Policy Rev* 20:371–388, 2002.

Reddy, K.S. Cardiovascular diseases. In Caballero, B. and Popkin, B.M., Eds., *The Nutrition Transition: Dynamics of Diet and Disease in the Developing World.* San Diego: Academic Press, 2002a, pp. 191–204.

Reddy, K.S. Cardiovascular diseases in the developing countries: dimensions, determinants, dynamics and directions for public health action. *Public Health Nutr* 5:2317, 2002b.

Schmidhuber, J. and Shetty, P. Nutrition transition, obesity and noncommunicable disease: drivers, outlook and concerns. *UN Standing Committee Nutr News* 29:13–19, 2004.

Schwarz, M.B. and Brownell, K.D. The need for courageous action to prevent obesity. In Crawford, D. and Jefferey, R.W., *Obesity Prevention and Public Health.* Oxford, U.K.: Oxford University Press, 2005, pp. 307–330.

Seidell, J.C. and Visscher, T.L.S. Public health aspects of overnutrition. In Gibney, M.J., Margetts, B.M., Kearney, J.M., and Arab, L., *Public Health Nutrition.* Oxford, U.K.: Blackwell Science, 2004, pp. 167–177.

World Health Organization. *Obesity: Preventing and Managing the Global Epidemic.* WHO Technical Report Series, 894. Geneva: World Health Organization, 2000.

World Health Organization. Overweight and obesity fact sheet. 2003. Retrieved February 10, 2008, from http://www.who.int/dietphysicalactivity/publications/facts/obesity/en/.

World Health Organization. The WHO child growth standards. 2006. Available at http://www.who.int/childgrowth/en/. Accessed March 10, 20008.

Zlotnik, H. Demographic trends. In Caballero, B. and Popkin, B.M., *The Nutrition Transition: Diet and Disease in the Developing World.* San Diego: Academic Press, 2002, pp. 71–108.

10 Diseases of Nutrient Deficiencies

Barbara Elaine Golden

CONTENTS

ABSTRACT

Essential nutrients vary enormously in their body content and requirements. They are each crucial for normal metabolism, they interact with one another, and they act differently when metabolism is altered by either nutrient deficiencies or excesses or by disease. Their deficiencies can also affect the metabolism of invading microorganisms. In poverty, numerous nutrient deficiencies and numerous infections coexist and are synergistic, forming a vicious spiral that leads to malnutrition. The most common form of this, affecting about half the children living in poverty in the developing world, is chronic stunting. It is preventable. This chapter focuses on nutrient deficiencies and their effects. Specific micronutrients, important to health, are included: iron, zinc and vitamin A, which have deficiencies that together appear to be responsible for nearly 20% or 2 million deaths a year in young children; folic acid and vitamin B_{12}, which also contribute to the ubiquitous problem of anemia; and selenium, thiamin, and vitamin D as examples for which we have learned much from history and geography. For each nutrient, advice is given on estimation of nutrient status. For several nutrients, however, the tests available are not yet sufficiently specific. Zinc is an example; we still have no reliable measure of its status for either clinical use or population studies. Finally, the major problem of childhood malnutrition is outlined. Our knowledge and understanding of nutrient metabolism is sparse, but still progressing.

OVERVIEW

Among other challenges, poverty means inadequate intakes of monotonous diets of poor quality (imbalance of essential nutrients, excessive antinutrients or toxins). One meal may be the only meal possible. If it is breast milk for a young infant, then dietary quality is excellent for the first 6 months. However, in poverty in much of the developing world, the meal may only be maize meal porridge for all age groups. This represents poor quality, being deficient in several essential nutrients (e.g., lysine and zinc) and rich in antinutrients (e.g., phytate). Poverty also dictates that meals are often contaminated, especially when fed from a bottle to an infant.

The almost inevitable consequence is frequent infections and malnutrition. These become a vicious spiral: Malnutrition reduces barriers and resistance to infections, and the consequent increased severity and duration of infections exacerbates malnutrition.

What has this to do with micronutrient deficiencies? It is obvious to all that malnourished children receive too little dietary energy and protein. However, not so obvious are the needs for and roles of individual essential nutrients, including micronutrients, the vitamins and essential trace elements (Geissler and Powers 2006).

When a diet is of good quality and adequate quantity and provided hygienically to an individual in good health, then the risks of deficiencies or toxicities of micronutrients are minimized. In the face of poverty, the risks of deficiency of many micronutrients are enormous. However, our knowledge of the world population's micronutrient status is poor, as is our diagnosis and management of individual deficiencies. This is, among other reasons, because of:

- ignorance of risk factors for specific micronutrient deficiencies
- ignorance and lack of reliable tests for status of specific micronutrients
- inability to interpret such tests
- lack of understanding of factors affecting supply and distribution of micro-nutrients within the body
- lack of understanding of needs and roles of micronutrients and how these change with nutritional status and disease
- reductive adaptation, by which nutrient deficiency limits growth and there-fore nutrient requirements (World Health Organization 1999a)

The chapter is devoted mainly to a few micronutrient deficiencies for which we have some knowledge and less understanding. Our knowledge comes from historical writings, geographic differences, scientific basic studies, and human intervention studies.

IRON

Iron, in humans, is the most abundant of the essential trace elements, comprising around 5 g of an adult. It is mainly in hemoglobin in red blood cells. Iron deficiency is purported to be the most prevalent of all deficiencies. This is because anemia, as we define it, based on blood concentration of hemoglobin, is very common, especially in late infancy and pregnancy. Are our definitions of anemia correct? Are the low hemoglobins all because of iron deficiency? We know the answer to the second question is "No." In large-scale studies in infants and young children, much of the anemia was not associated with low iron stores (Duque et al. 2007).

Malaria is common and causes hemolytic anemia. Hemoglobin is not lost; its iron is retrieved and reused. Thus, iron available for hemoglobin synthesis may be adequate even though the anemia may be severe. Other infections are also associated with a decrease in blood hemoglobin concentration without loss of iron.

Folic acid, vitamin B_{12}, and vitamin A deficiencies (discussed separately) are relatively common causes of anemia in young children.

In severely malnourished children, the red blood cell picture tends to be mixed, showing signs of multiple deficiencies often compounded by malaria and other infections. Less-malnourished children comprise the vast majority of children living in poverty. Their anemias tend also to be less severe but probably no less complex, affected by multiple nutrient deficiencies and infections.

In pregnancy, hemodilution and major changes in nutrient transport and metabolism occur; anemia is common even using a lower cutoff for hemoglobin concentration. Lactation is more clearly a drain on iron reserves.

Anemia in the elderly, as in malnourished children, tends to show a mixed picture, often with less evidence of iron than vitamin deficiencies.

Thus, all anemia is not due to iron deficiency but there is little doubt that it is common. Why? There is no shortage of iron in the Earth's crust. However, inorganic iron and ferric salts are very poorly absorbed by humans. The form of iron that is most bioavailable to us is that found in animal tissues, within heme. Most of the world's children consume very little in the way of animal tissues. Most today have not been exclusively breastfed—breast milk iron is also absorbed well. Iron in plants is mostly as ferric salts. Thus, it is likely that iron deficiency is very common in young children and detrimental to their health and development (see Chapters 11 and 12).

Iron toxicity is usually considered only from iron overdose, as in toddlers consuming their mother's iron supplements or patients on excess oral or parenteral iron or having numerous blood transfusions. Iron absorption is normally highly controlled at the intestinal level. However, hemochromatosis is an inherited disorder in which iron absorption is increased, and the resultant excess stored iron causes organ damage, particularly in the liver, pancreas, testis and heart. Surprisingly, it was found that plasma ferritin, a measure of iron stores, was very high in severely malnourished children, especially those with edematous malnutrition and those who died (Srikantia 1958). This could be due to an acute-phase response in which iron is normally sequestered in the liver. However, there is little evidence of such a host response to infection in these ill children. The high plasma ferritin is more likely to reflect genuine iron overload. This implies the risk of "free" iron available for both the growth of pathogens and to act as a prooxidant, possibly responsible for generalized cell damage, via lipid peroxidation of membranes (Golden and Ramdath 1987). Therefore, iron therapy is contraindicated in the early management of severely malnourished children (World Health Organization [WHO] 1999a).

IRON STATUS

Anemia prevalence is not a good measure of the extent of the problem. The same is true for plasma iron concentration. This is affected, more often than not, by the presence of an acute-phase response, usually to invasion by bacteria, viruses, and the like as described. As part of this, circulating iron is bound to proteins within the liver so plasma iron decreases. Iron stores, mainly in the liver, can be estimated indirectly by measuring plasma ferritin concentration. However, this increases during an acute-phase response. Thus, a "normal" value may have been a low value before the child developed pneumonia, for example. Finally, another indirect measure of iron status is the plasma concentration of transferrin receptor. An increase in this protein implies that the iron supply for hemoglobin synthesis in the bone marrow is deficient. It is supposed to be independent of an acute-phase response, but this is unlikely.

In conclusion, iron status is probably best assessed today by measuring iron stores (plasma ferritin) when there is no evidence of an acute-phase response.

ZINC

Zinc is half as abundant as iron in the body. Whereas iron is required for a few proteins, zinc is ubiquitous: It is required for the structure or function of at least 200 metalloenzymes, within cytoplasm, nucleus, and membranes (Hallberg et al. 2000). Without zinc, neither cell hyperplasia, requiring DNA synthesis, nor hypertrophy, requiring protein synthesis, can occur.

When a zinc-deficient diet is fed to rats, they experience fluctuating anorexia, with spells of reasonable intake separated by spells of very poor intake; their growth is severely restricted. When such a diet is force-fed to rats, they die within only a few days.

In human diets, zinc accompanies protein; it is almost impossible to provide a protein-deficient diet that is not zinc deficient (Golden and Golden 1981). Zinc is highly available from breast milk and animal protein. Before the cause of acrodermatitis enteropathica, an inherited defect of intestinal zinc absorption, was discovered, it was treated effectively with breast milk.

In poverty, most of the monotonous plant-based diets contain little zinc but much phytate, which further reduces intestinal absorption of zinc. Many millions of infants and children grow less quickly than their genetic potential for many reasons. One reason is zinc deficiency; the extent of its contribution is unclear but may be considerable in, for example, rural India. When it limits growth, the children are smaller and require less zinc. This is an example of reductive adaptation. Providing zinc as a supplement allows growth to resume, but this can only continue while there is sufficient energy and all other essential nutrients to sustain it. As the child grows, more of everything is required.

The effects of zinc deficiency are, like zinc itself, ubiquitous. They appear first in tissues with normally high turnover rates. Thus, intestinal mucosal function suffers: Malabsorption is general and permeability is increased. Diarrhea, or *enteropathica*, is the clinical outcome. Skin epithelium, the epidermis, thins, and it becomes less effective as a barrier. Wound healing is also impaired. Minor trauma, which tends to occur around orifices and on extremities, leads to chronic skin lesions, or *acrodermatitis*. These features manifest themselves late; before this, the child will have "adapted" to a low-zinc diet by not growing and hence not requiring so much. There are other costs of this adaptation. Not only are the barriers to infection reduced, but also the ability to exhibit either nonspecific immunity, in the form of inflammation or an acute-phase response, or specific immunity, especially cell-mediated immunity, is reduced. Infections pass unnoticed because of the lack of clinical signs (of the child's response); they are not treated, and the child suffers severe, prolonged morbidity and much increased mortality from infections that were initially relatively innocuous. Zinc deficiency is rarely diagnosed in such circumstances.

ZINC STATUS

One of the most important and fascinating aspects of zinc in human health is that even today, it is almost impossible to assess zinc status reliably "in the field." It cannot be done on clinical grounds because the effects of its deficiency are nonspecific, growth failure, until very advanced. It cannot be done easily in the laboratory. Kinetic studies using stable isotopes are probably the gold standard but are confined to the research laboratory. Hair zinc analysis is fraught with methodological difficulties, and neither plasma nor cell zinc concentrations inform zinc status reliably. A retrospective diagnosis of zinc deficiency can be made following a positive effect of a period of zinc supplementation. This has led to large-scale zinc supplementation studies of infants and children with various infections. In most, a positive effect has been demonstrated (Umeta et al. 2000, Brooks et al. 2005). However, it is clear that, as with studies of several other micronutrients, effects are seen if deficiencies are made good, but not if the subjects are not deficient in the first place. Indeed, although zinc toxicity effects are rare, in one study these may have accounted for increased mortality in the supplemented group (Doherty et al. 1998). More is not necessarily better.

SELENIUM

Selenium was a little-known trace element in human health until recently, when two questions came under scrutiny: the role of oxidative stress in disease and the cause of Keshan disease.

Selenium is essential for the pivotal antioxidant glutathione peroxidase. This enzyme helps protect cells from the damaging effects of oxidation, for example, by ultraviolet radiation of skin, by exogenous toxins in the liver, or from any excess of "normal" oxidative processes that are used, for example, to damage and kill invading bacteria. Glutathione peroxidase acts on lipid peroxides and hydrogen peroxide to produce harmless hydroxy acids and water, respectively (Geissler and Powers 2006). In the 1980s, the "free-radical" theory of the etiology of edematous malnutrition was supported by evidence of deficiencies in several antioxidant systems, including glutathione peroxidase (Golden and Ramdath 1987). Since then, evidence has accumulated for a role of oxidative stress in several other common, chronic diseases, such as atherosclerosis and arthritis.

Keshan disease was described to the Western world also in the 1980s. It is characterized by a cardiomyopathy present in thousands of youngish women and children living in a specific region in southwest China in which soil selenium is particularly low. In large studies, selenium supplements appeared to prevent, although not treat, the disease. Later, it was observed that there were temporal fluctuations in Keshan disease (Geissler and Powers 2006). Enteroviruses were implicated, especially Coxsackie virus B4. This led to mouse experiments in which it was shown that increased oxidative stress, induced by deficiency of either selenium or vitamin E in mice infected with nonvirulent Coxsackie virus (B3/0), could induce mutations in the virus, causing it to become virulent (Beck et al. 2003). Indeed, high intakes of polyunsaturated fatty acids (PUFAs) or iron had similar effects, also explained by increased oxidative stress. Thus, it appears that Keshan disease is not a direct effect of selenium deficiency; rather, it is due to the effect of oxidative stress on infecting viruses. This shows how complex interactions are, not only between micronutrients themselves, but also between micronutrients and human metabolism and even the metabolism of invading microorganisms.

Another pair of interesting selenoenzymes are thyrodoxin and iodothyronine deiodinase. Both are involved in thyroid metabolism, which of course is also highly dependent on iodine supply.[1] Selenium deficiency exacerbates the effects of iodine deficiency (Geissler and Powers 2006). In the face of both deficiencies, giving iodine without selenium does not adequately treat the features of iodine deficiency disorders (IDD). This was also observed in south China, where the iodine-deficient belt overlaps the selenium-deficient belt.

Another important role of selenium, also related to its antioxidant function, is in prevention of a variety of malignancies, including the increasingly common prostate cancer.

Thus, this micronutrient is indeed essential for optimal health. However, its physiological range of intake is small. Intakes only three to four times higher are associated with physical signs of toxicity. Excess selenium is excreted in breath as dimethyl selenide, and hair, skin, and nail lesions occur (Geissler and Powers 2006). This suggests that at lower intakes, closer to physiological, metabolism may also be disturbed. Thus, yet again, it is important that the public are not led to believe that more is necessarily better in terms of selenium supplements.

SELENIUM STATUS

Selenium status is best estimated from red blood cell selenium concentration or glutathione peroxidase activity or content. They are generally closely related to one another and reflect status over the previous few months, the average life of red blood cells. Plasma selenium tends to fluctuate widely in relation to meals.

THIAMIN

Clinical thiamin deficiency, or beriberi, was probably first described in 2600 B.C. However, it came to the fore relatively recently, at the end of the nineteenth century, mainly in Eastern and South Eastern Asia, when rice was first effectively milled, yielding polished white rice. This became the sole diet of poor, laboring populations in China and Japan, and epidemics of beriberi followed. It was not a problem in Africa while traditional root crops or maize remained the staple (WHO 1999b).

Studies were performed in several isolated groups, including sailors and those in mental institutes and labor camps. They were fed largely rice and developed beriberi. It was shown that this could be prevented by providing either other foods as well as rice, or unpolished rice, or just by adding the otherwise discarded polishings themselves. Thus, it became clear that an essential nutrient was in the discarded polishings. Thereafter, in 1926, the water-soluble, heat-labile vitamin was isolated and 10 years later was given its chemical formula and name, thiamin (Williams 1961).

Since then, there have continued to be many outbreaks of thiamin deficiency from which we have learned much. Usually, they have been in poor communities living mainly on polished rice. However, the same disease has occurred in other isolated groups living mainly on white bread made from highly milled wheat. Wernicke-Korsakov syndrome, which was initially described in the 1880s, occurs in a small proportion of thiamin-deficient alcoholics, those who have an inherited abnormality of the enzyme transketolase, which prevents its binding to thiamin diphosphate (WHO 1999b). An alcohol diet means a high carbohydrate diet, and this requires a higher-than-usual thiamin intake; alcohol also inhibits the intestinal absorption of thiamin. Recently, beriberi was diagnosed in northeastern Thailand, but thiamin intake appeared to be adequate; however, antithiamin factors, in the raw, fermented fish that they ate and the betel nuts that they chewed, tipped the balance toward deficiency (Vimokesant et al. 1975). There have also been several recent outbreaks of beriberi in refugee camps. Again, the underlying problem has been a monotonous diet of poor quality, with inadequate thiamin relative to other nutrients. Finally, although breastfeeding provides the best-quality diet for an infant, when that infant's mother is thiamin deficient, her breast milk does not supply sufficient thiamin for her infant, and the consequence is beriberi.

Beriberi varies considerably in presentation. Chronic low-grade beriberi is easily overlooked. It is probably common in malnourished communities in which many essential nutrients are deficient. "Wet" and "dry" beriberi are more obvious, acute presentations but often occur together. Wet beriberi is heart failure with edema, while dry beriberi is chronic peripheral neuropathies or other neurological problems. Wernicke-Korsakov syndrome is an example of the latter. It is as yet unclear why thiamin deficiency presents so variably. It is probably related to the wide variety of other factors pertaining, such as alcohol, other nutrient deficiencies, antinutrients, and perhaps infections.

THIAMIN STATUS

The diagnosis of thiamin deficiency can be made crudely from the dietary history. Plasma thiamin does not reflect tissue thiamin content. However, erythrocyte transketolase activity (ETKA) does reflect tissue content. The assay is usually performed before and after addition of thiamin pyrophosphate. A positive percent change in activity, the thiamin pyrophosphate effect (TPPE), reflects the transketolase enzyme that lacks its coenzyme, thiamin, and hence, thiamin status (Brin and Ziporin 1965).

FOLIC ACID

Folic acid is another water-soluble vitamin whose recent claim to fame has been that its deficiency is associated with increased risks of neural tube defects and, more recently, increased risks of colorectal cancer and hyperhomocystinemia, which in turn is associated with atherosclerosis. It has been estimated that almost 10% of the populations of industrialized countries have low folate stores (Geissler and Powers 2006).

What appears to be in contrast, however, is the use of antifolate drugs to treat malignancies, epilepsy, and some bacterial infections, including malaria. Such treatment can produce what might be termed iatrogenic folate deficiency.

Folic acid's discovery started in 1931 when Wills (1931) found that yeast cured the megaloblastic anemia of pregnancy. She suggested that yeast contained a new hemopoietic factor. Numerous active factors were discovered in the next few decades: Snell (Snell and Peterson 1940) called one of these "folic acid" because it came from leaves. Stokstad isolated and chemically identified this same folic acid in 1943, and it was synthesized in 1945 (Shane and Carpenter 1997). Research has continued since.

A wholly natural diet contains very little folic acid. Now, synthesized folic acid and 5-formyl tetrahydrofolate (folinic acid), are used to fortify foods and as supplements and to treat side effects of cytotoxic antifolates like methotrexate. Surprisingly, in colorectal cancer, a combination of calcium levofolinate and fluorouracil, another cytotoxic antifolate, is more effective than fluorouracil alone (Murakami et al. 1998). Folic and folinic acids are stable compounds comprising a nonreduced pteridine ring joined to para-aminobenzoic acid joined to glutamate. They are very bioavailable. In contrast, food folates, which are a mixture of reduced folate polyglutamates, are fairly unstable and less bioavailable. They are hydrolyzed by two enzymes in the intestinal mucosa, one of which is zinc dependent, to monoglutamyl, or "free" folate. Thereafter, absorption seems to be an active process as it can be inhibited by anti-folates (e.g., sulfasalazine, a drug commonly used in inflammatory bowel disease). After absorption, it circulates as 5-methyl tetrahydrofolate (THF) monoglutamate attached to albumin. This is reconjugated in tissues to 5-methyl THF polyglutamate derivatives. These have one main role in all cells, that of picking up and depositing one-carbon units. They act as such in two interrelated processes, the methylation cycle and the DNA cycle.

The DNA cycle is relatively simple. By providing one-carbon units (viz. methyl groups), both purines and pyrimidines are synthesized, which are then incorporated in DNA (and RNA). The major folate product is 5,10-methylene THF, which is reduced by 5,10-methylene THF reductase (MTHFR) to 5-methyl THF, and the cycle restarts (WHO 2004). Thus, the vitamin is essential for cell replication; hence, its deficiency causes megaloblastic anemia, in which too few, large, immature red cells circulate accompanied by too few white cells and too few platelets. The intestinal mucosa's normally high turnover rate is reduced; malabsorption is another outcome of folate deficiency. Although there is an association between low folate status and many cancers, it is particularly strong for colorectal cancer. Recent research has focused on the possible mechanisms. There are clear possibilities. In particular, a common polymorphism of MTHFR seems to protect against colorectal cancer. Men who were homozygous for the polymorphism and had adequate folate status had a threefold lower risk compared with men whose genotype was heterozygous or wild type. The polymorphism was not protective if they had low folate status. It appears that the forms of folate within the cells affect the risk of disruption of DNA integrity and hence cancer (Choi and Mason 2000).

However, in population studies of folic acid supplementation, although the results have been described as "promising," they are not clearly positive.

In the methylation cycle, 5-methyl THF is used by methionine synthase to remethylate homocysteine to methionine. This requires vitamin B_{12} (see vitamin B_{12} discussion).

Excess methionine is degraded back to homocysteine. A raised plasma concentration of homocysteine is now considered to be a major independent risk factor for atherosclerosis and, indeed, an (imperfect) index of folate status. Homocysteine is normally safely cleared by cystathionine synthase, with help from vitamin B_6 (pyridoxine). Absence of the enzyme may be inherited, a consequence of an inborn error of metabolism, homocystinuria. The characteristic features include early atherosclerosis.

Thus, these three vitamins (folate, B_{12}, and B_6) are essential for two major linked metabolic cycles that involve numerous methylation reactions, including DNA and RNA syntheses.

Neural tube defects (NTDs), in which the brain or spinal cord does not form properly before day 27 of pregnancy, are six times more likely when folate status is poor. In fact, they are also more likely when vitamin B_{12} status is poor; this risk has been exposed in populations receiving folate-fortified flour (Ray et al. 2007). Like the causes of NTDs, the mechanisms underlying these findings are still unclear. It may be presumed that they involve the DNA and methylation cycles, however. So far, it appears that there are some genetic mutations associated with NTDs, and there is no doubt that having one affected child increases the risk of having a second affected child. Thus, folic acid supplementation is advised for all women "at risk" of becoming pregnant. Fortification has been considered for this purpose, but this means that the whole population receives considerably more folic acid than necessary. Because folic acid and vitamin B_{12} are so intimately related in terms of their metabolic roles, giving excess folic acid can mask vitamin B_{12} deficiency (discussed in the next section). With our present knowledge, it would seem that fortification with both vitamins is desirable. The argument about folic acid fortification still continues in the United Kingdom. However, the United States and several other countries do fortify flour, and the question now is whether the level of fortification is sufficient (Brent and Oakley 2007).

Liver is an excellent source of folic acid, although the usual sources are dark green leafy vegetables, beans, nuts, and milk. As folates are very susceptible to storage, processing, and the like, fortification of cereals has become the norm for preventing folate deficiency, particularly for women of childbearing age, as discussed. Folate deficiency occurs in those on restricted diets (e.g., those living in poverty, refugees) or in patients with malabsorption or who are unable to eat normally. It is exacerbated by increased demand for new tissue synthesis, as in pregnancy and infancy. Patients on antifolate drugs are also at risk, particularly those on long-term treatment, for example, for epilepsy. The risk is even greater if pregnancy occurs. Although anemia is the classical sign, hyperhomocystinemia occurs earlier and should alert one to the need for supplementation. "Toxic" effects of folic acid excess are limited to the indirect masking of vitamin B_{12} deficiency. It may antagonize some of the effects of antifolate drugs, but folate deficiency is more likely than toxic effects from supplementation.

FOLIC ACID STATUS

Currently, erythrocyte folate concentration can be measured efficiently using a radioligand-binding test. It reflects folate status over a few months, the average life of a red blood cell. The formiminoglutamate (FIGLU) test, in which FIGLU is metabolized by a folate-dependent enzyme, is a sensitive test of functional folate status but is rarely indicated.

VITAMIN B_{12}

The core of vitamin B_{12} is a cobalt atom within a corrin ring attached to various side groups. It is absent from uncontaminated plant food but elaborated by bacteria and present in animal tissues, especially liver (Geissler and Powers 2006). It is heat stable so is not destroyed in cooking, unlike folates. Outright deficiency rarely occurs other than in strict vegans. Unlike most essential nutrients, deficiency in the mother leads to profound deficiency in the fetus; the fetus is not spared. Secondary deficiency, however, is relatively common in patients with pernicious anemia or any other cause of atrophic gastritis in which lack of intrinsic factor production by the gastric mucosa prevents absorption of the vitamin in the terminal small intestine.

The early features of deficiency are due to loss of methionine synthase activity as vitamin B_{12} is its cofactor. Thus, the DNA cycle is impaired, and megaloblastic anemia, indistinguishable from that due to folic acid deficiency, ensues. This is part of the methyl trap hypothesis in which folate is trapped as its 5-methyl derivative in the DNA cycle. A further consequence of B_{12} deficiency is reduced synthesis of folylpolyglutamates, the active form of folate within cells. Thus, folate deficiency becomes part and parcel of vitamin B_{12} deficiency (Shane and Stokstad 1985). Later, neurological signs, subacute combined degeneration (SCD) of the spinal cord, occur, due to interference with the methylation cycle by which methionine converts to S-adenosyl methionine (SAM), which supplies methyl groups to myelin basic protein, which insulates nerve cells. Without this, demyelination occurs, leading to ataxia, paralysis, and eventually death. This also eventually occurs in folic acid deficiency, but much later.

VITAMIN B$_{12}$ STATUS

Vitamin B_{12} status is currently best estimated as serum vitamin B_{12}, which can be measured using a radioligand-binding assay.

VITAMIN A

The history of our knowledge about the fat-soluble vitamin A was beautifully described by George Wolf (1996). It goes back to Egyptian papyri written 1 to 2 millennia B.C. Even then, liver was recognized as an effective cure for eye diseases, used both topically and orally. There have been numerous repeats of those findings right up to the mid-nineteenth century, when night blindness beset those on board a ship of the Austrian Navy on a 3-year voyage of discovery. This was again cured by boiled ox liver. Soon after, Bitot gave his name to the white spots at the conjunctiva-cornea junction that were associated with night blindness. Mori, in 1904, described xerophthalmia and keratomalacia in Japanese children. Yet again, liver, and particularly cod liver oil, was the cure. Eventually, around 1910, research in rats by two separate groups identified the active fat-soluble factor—"A" (Wolf 1996). This was isolated and characterized in the 1930s after realizing that the bioactive yellow pigments found in plants, butter fat, and egg yolk could be converted into a colorless bioactive form in animal liver. The former are now termed *provitamin A carotenoids,* while the latter is *retinol.*

Several decades before, research into the physiology of vision was active, and by 1881, rod and cone vision were distinguished; it was considered that in night blindness, the visual purple, which was limited to the peripheral rods used for night vision, was altered. Most of the advances in knowledge on this subject during the twentieth century were made by Wald (1968). He showed that the visual purple, or rhodopsin, was a protein (opsin) combined with retinaldehyde, an active form of vitamin A. He went on to demonstrate that 11-*cis*-retinaldehyde is isomerized to all-*trans*-retinaldehyde by light (photons), and this reaction triggers a nerve impulse sent to the brain, which registers it as light.

Thus, the classical signs of vitamin A deficiency are night blindness due to abnormal rod physiology in the retina and xerophthalmia due to epithelial damage, particularly that of the cornea. In developing countries, these comprise the single most common preventable cause of blindness in children (Geissler and Powers 2006).

Numerous other functions of vitamin A have also become apparent. Vitamin A deficiency increases susceptibility and reduces immunity to infections. The former can be related to the need for vitamin A for epithelial integrity. Deficiency results in mucosal atrophy with loss of goblet cells and mucus (Wolbach and Howe 1925).

Although several studies of vitamin A supplementation in children with infections have not met with the expected results, such as reduction in incidence, morbidity, or mortality, it seems that this has been at least in part due to selecting children who were not vitamin A deficient. Indeed, one study showed a slight increase in respiratory infections in supplemented children (Grotto et al. 2003). It is clear yet again that more is not necessarily better.

Infection itself increases vitamin A loss in urine and may precipitate deficiency signs. Thus, in poverty in developing countries, measles is well known to precipitate vitamin A deficiency and, of interest, edematous malnutrition. In malnutrition, however, retinol's transport protein, retinol-binding protein, is low, in part due to zinc deficiency. This traps retinol in the liver, thereby reducing its tissue bioavailability. Thus, malnutrition, zinc deficiency, and infection each exacerbate vitamin A deficiency. In malnutrition, especially when complicated by infection, both zinc and vitamin A supplementation are indicated (WHO 1999a).

Finally, an association between vitamin A deficiency and anemia has been demonstrated repeatedly. Vitamin A supplementation alone has often resulted in a rise in hemoglobin. The cause is still unclear. There is evidence, however, that vitamin A deficiency modulates iron metabolism in a similar way to the effect of infection, which of course often coexists. Thus, when vitamin A deficiency is present, iron, like zinc, is trapped in the liver and spleen, rendering it unavailable for erythropoiesis (Semba and Bloem 2002).

Although retinol has not been shown to have any influence on the risk of malignancy, carotenoids appear to have separate effects as antioxidants and, perhaps, as protection against some cancers (Geissler and Powers 2006).

Vitamin A deficiency is common; WHO estimated that nearly 3 million children under 4 years were clinically affected in the 1990s, and around 250 million were subclinically affected (WHO 1995). The highest prevalence is in Southeast Asia in association with poverty and malnutrition. Breastfeeding protects, provided the mother is replete, but weaning and future diets, especially in rural areas, often contain little in the way of plant carotenoids, which still require conversion to retinol, and even less in the way of animal sources of preformed retinol. Vitamin A deficiency rates tend to fluctuate with season because the plants that are good sources of carotenoids are vulnerable to drought and heat (WHO 1995).

Vitamin A toxicity is not uncommon in industrialized countries, due mainly to easy access to supplements, which are advertised widely. Acute intoxication can be fatal. Chronic overdosing damages tissues, among them the central nervous system, bones, liver, and skin. Liver, the richest natural source of vitamin A, is contraindicated in pregnancy; however, in poverty in the developing world, vitamin A deficiency is a greater risk. Synthetic retinoids are useful, both orally and topically, for certain severe skin conditions; they are teratogenic.

More common than retinol excess is chronic carotenoid excess. Unlike retinoids, these do not appear to carry significant risks other than hypercarotinemia and yellowed skin.

Vitamin A Status

Vitamin A status is not easy to assess reliably. Clinical signs present late. A dietary history is, as usual, important. In relatively small population studies, conjunctival impression cytology estimates the conjunctival epithelial damage, and this has proved useful. Serum retinol is more commonly used, but this falls with infection, with malnutrition, and with zinc deficiency, so there are likely to be many false positives. Where available, the best estimate of individual status is the relative dose response test (RDRT), which measures the increase in plasma retinol after a dose of vitamin A.

VITAMIN D

Vitamin D is a hormone precursor and an unusual vitamin as it is only conditionally essential in the diet. The condition is insufficient conversion, by the sun's ultraviolet B (UV-B) rays, of 7-dehydrocholesterol to cholecalciferol in the skin. Unfortunately, this often pertains. Therefore, a knowledge of sources of dietary vitamin D is necessary; indeed, reference nutrient intakes (RNIs) and recommended daily allowances (RDAs) have been laid down for those at most risk, including infants, the elderly, and those unable to enjoy exposure to the sun (Geissler and Powers 2006). There is little doubt that rickets was present in the first century A.D., but the first good descriptions were published in the midseventeenth century. In the nineteenth and early twentieth centuries, when air pollution due to industry was severe in cities like Glasgow in northern Britain, rickets was rampant. In 1919, Mellanby showed that cod liver oil cured rickets in dogs, and this was quickly followed by human studies with similar results. In 1922, the dietary component "essential" for bone health was named vitamin D (Rajakumar 2003).

Since then, much has been learned about vitamin D's numerous roles in metabolism. Cholecalciferol itself is inactive. It is converted in the liver to 25-hydoxycholecalciferol, or calcidiol and then by 1-hydroxylase in the kidney to 1,25-dihydroxycholecalciferol, or calcitriol, the most active form of the vitamin. Calcitriol's primary activity is to raise plasma calcium by increasing its intestinal absorption, its renal reabsorption, and its release from bone. It has numerous other effects, mainly in nuclei via genes but also via cell surface receptors. These include effects on hormones such as parathormone, calcitonin, and insulin and on cell proliferation and differentiation, particularly in lymphocytes and monocytes, cells involved in immune processes. Although not confirmed, it has been implicated in the metabolic syndrome, in cancer, and in atherosclerosis. Excessive vitamin A has been shown to inhibit some nuclear actions of vitamin D by interfering with the dimerization of the vitamin D, which is necessary for vitamin D's activity.

Today, in temperate and subarctic regions, there is usually insufficient exposure to UV-B during the winter. In certain cultures, women are rarely exposed to UV-B because their skin is covered almost entirely, or they are indoors. Breast milk normally contains little vitamin D. If a combination of any of these three occurs, the risk of vitamin D deficiency is high. Thus, a rise in the incidence of rickets has occurred recently in the United Kingdom. It has been confined largely to the young children of immigrant (African and Asian) women in cities. All three risk factors have operated: northern latitude, maternal deficiency from lack of sun exposure, and breastfeeding with particularly low milk vitamin D content.

In infants, rickets is characterized by overgrowth of cartilage. This results in a soft cranium (craniotabes) and swellings at the ends of bones. Growth faltering occurs. Later, when the child stands up, the long bones bend, resulting in bow legs or knock-knees.

In adults, especially women who have had numerous pregnancies combined with lactation, and the elderly, who have reduced 7-dehydrocholesterol in their skin as well as limited sun exposure, vitamin D deficiency takes the form of osteomalacia. This is characterized by bone pain and muscle weakness, features similar to those in chronic fatigue syndrome. It is not the same as osteoporosis, although they may coexist, and both involve, inter alia, loss of bone mineral.

Prevention of these debilitating diseases is the aim. Vitamin D drops should be prescribed for vulnerable infants as described. Foods rich in vitamin D are few but include oily fish, eggs, liver, and full-fat milk. In today's industrialized countries, vitamin D intake is also from fortified foods and, often, supplements. For these, synthetic vitamin D is used, manufactured from cholesterol, which is converted to 7-dehydrocholesterol and irradiated. Supplements carry risks, however. As for most other micronutrients, more is not better. Toxicity occurs with excess and takes the form of raised blood calcium concentrations, tissue calcification, and calcium phosphate stones in the kidney.

Vitamin D Status

For women, whose skin sees little of the sun, who live in northern latitudes, and who are breastfeeding their infants, alarm bells should be rung for both mothers and all their children.

The most specific biochemical test of status is the plasma concentration of calcidiol, 25-hydroxycholecalciferol. At the same time, plasma alkaline phosphatase and parathormone concentrations will be high in the face of a true deficiency.

OTHER NUTRIENTS

Having focused on a selection of micronutrients, the impression should not be that they are more important or exciting than other essential nutrients. Indeed, our knowledge of what we mean by essential nutrients and what they are remains fairly limited. We have learned that vitamin D and niacin (which can be synthesized in the body from tryptophan) are not strict essential nutrients. We also realize that human milk, the "best" food for us during our most vulnerable period, contains compounds with functions that remain uncertain. Recent research suggests that, for preterm infants at least, some long-chain polyunsaturated fatty acids, polyamines, and nucleotides present in their mother's milk are essential.

Besides micronutrients, specific fatty acids, amino acids, and minerals are also essential nutrients (see Chapters 6–8). The role of individual essential amino acids tends to be overlooked as they are supplied in protein. Protein deficiency is recognized as important, but protein quality, its content, and balance of amino acids tend to be ignored. For growth, sufficient of each of the essential amino acids is necessary. The protein of many plant cereals, for example, maize meal, is lysine deficient; by itself, it is inadequate to support optimum growth. Consuming vast quantities may get over the lysine problem, but the excess other amino acids, which cannot be incorporated into body protein, have to be catabolized and excreted, a major task for young or malnourished infants' kidneys.

The various roles and requirements of minerals also tend to be overlooked. Thus, it is essential, especially during growth, that there is adequate dietary calcium and phosphate for the skeleton. Not surprisingly, calcium deficiency can present like vitamin D deficiency. Potassium is of fundamental importance and is concentrated intracellularly. Experimental deficiency of potassium results in edema, among other features. Correct management of potassium depletion in severe malnutrition is crucial to recovery (Golden 1988). Potassium metabolism is interrelated with that of magnesium, whose metabolism is also related to that of calcium. Sulfates are needed for connective tissue structure, its growth, and metabolism. Sodium and chloride are also essential for homeostasis. Without a single essential nutrient involved in synthesis of new tissue, growth is limited to the extent necessary to allow the available supply to meet the demand. Thus, growth achievement is a useful index of nutritional status, not only of energy and protein supply but also of supply of each and every essential nutrient.

NUTRIENT INTERACTIONS

It is clear that for much of the developing world living in poverty, nutrient deficiencies are common and rarely occur singly or in isolation. For example, poor growth is not simply due to energy, protein, lysine, iodine, or zinc deficiency; anemia is not simply due to iron, folate, or vitamin A deficiency. Duque et al. (2007) concluded from studies in Mexico that although iron and zinc were the most prevalent deficiencies in young children, anemia associated with low iron stores accounted for less than 50% of cases. Similarly, the effects of deficiencies are not independent of one another, although this is seen more clearly with the effects of supplementation. Not only do nutrients interact with other nutrients, but also other dietary components (including antinutrients and toxins) affect their bioavailability. Furthermore, diseases, particularly infections, cause nutrient losses and alter demand for particular nutrients as part of the host immune responses.

There have been numerous studies of the effects of iron and zinc fortification on the absorption and nutrient status of each. The results have not been clear-cut. In general, if given in meals there appears to have been little effect, while if given in water, there were significant inhibitory effects, each on the other (Whittaker 1998). Which food to fortify, what form of the fortificant is used, and the level of fortification are all important.

More recently, large randomized, placebo-controlled, double-blind trials of iron and zinc supplementation in infants were conducted in Southeast Asia (Wieringa et al. 2007). These studies showed that iron supplementation alone had no effect on zinc status (plasma zinc), either iron or the combination of iron and zinc supplementation reduced anemia prevalence, but zinc supplementation alone reduced iron status (plasma ferritin). Although vitamin A was given to most of the infants, other erythropoietic nutrients were not considered, including folic acid, which may have been relevant, as it was in Mexico (Duque et al. 2007).

Thus, consideration must also be paid to whether nutrients should be provided as single or multiple supplements.

MALNUTRITION

Malnutrition, defined as "bad nutrition," includes obesity, individual nutrient deficiencies, and the various "underweight" syndromes associated with poverty in the developing world.

Underweight children under 5 years old comprise about a quarter of the world's children, about 150 million at present. These numbers are falling slowly except in Africa (de Onis et al. 2004). Underweight syndromes used to be differentiated on the basis of their major clinical features. The Wellcome classification (Anonymous 1970) redefined marasmus as "under 60% weight-for-age, without oedema;" kwashiorkor as "60–80% weight for age with oedema;" and marasmic-kwashiokor as "under 60% weight for age with oedema." However, weight for age (child's weight as a percentage of reference weight of children of the same age) changes due to both change in linear growth (mainly skeletal) and change in soft tissue growth (lean and adipose tissues).

REDUCTIVE ADAPTATION

Reductive adaptation has been mentioned. It is the body's response to a supply of energy and nutrients insufficient to meet their demand; the demand falls. Thus, physical activity and growth rates fall. Weight may be lost as adipose tissue gives up its stored energy to more vital functions and muscle gives up its nutrients to replace inevitable losses. Two cell functions, the Na-K-ATPase (adenosine triphosphatase) pump and protein synthesis, normally consume about two thirds of maintenance energy. These functions decline, thereby reducing the demand for energy further.

The costs of reductive adaptation are large. The young child stops exploring the environment, and milestones regress or are not reached. The child loses reserve capacity, so cannot deal with stress (e.g., a large protein meal). The child's barriers to infection atrophy. The inflammatory response is reduced, as are the complement output and cell-mediated immune responses. Homeostasis is compromised. The child cannot keep warm in a cold room or vice versa. The cell potassium falls while sodium rises. This cannot continue, and eventually death is the only outcome. It is usually precipitated by overwhelming infection, even in the hospital.

STUNTING

Stunting, or linear growth faltering, is a clear expression of reductive adaptation. It tends to occur most in infancy when normal growth in length is greatest. Its extent can be described by comparing the child's height (or length, under 2 years) with WHO standard or reference values for children of the same age. Below 90% or −2 standard deviation score (SDS or Z score) implies significant stunting, while below 85% or −3 SDS implies severe stunting (WHO 1999a). In many developing countries, its prevalence is around 40% in those under 5 years old. Although puberty is late, so that growth can continue for longer in the girls, adult height falls far short of that in adults who were well nourished as children, which led in the past to the largely false belief that the variation in adult heights in different countries was genetic rather than environmental in origin.

Stunting, at the population level, is associated with impaired mental development. Studies in Jamaica showed that mental development could be accelerated in stunted toddlers with either extra milk or psychosocial stimulation. Sixteen years later, a positive effect of stimulation remained (Walker et al. 2006). Physical development is also impaired in children who are stunted; as short adults, their physical capacity is reduced. Short females tend to have more complicated pregnancies and births and children with low birth weights.

Reversal of stunting usually requires a major change of environment. However, this did occur in boarding school children in New Guinea when they were supplemented with milk (Malcolm 1970). Doubt still remains regarding which nutrient deficiencies are most responsible for stunting. Zinc deficiency is a good candidate in theory. However, cows' milk is a poor source of zinc, so it is probably not the most important deficiency. Other candidate nutrients include calcium, phosphate, sulfate, iodine, and specific essential amino acids required for cartilage and bone growth and structure (Golden 1991, Anonymous 1994).

WASTING

Wasting occurs when lean and adipose tissue grow too slowly or if the supply of energy or nutrients is so low relative to demand that the body has to use its own dispensable tissues to supply the demands of less-dispensable tissues like brain and heart. Acute loss of fluid or nutrients often adds greatly to the demand side. Infections (e.g., gastroenteritis or pneumonia) are the usual cause of increased demands. Such situations occur frequently in young children living in poverty. Wasting can be assessed as weight for height/length (discussed in the Stunting section) or as body mass index (BMI) for age. When compared with WHO standard or reference values, below 80% or −2 SDS implies significant wasting, while below 70% or −3 SDS implies severe wasting.

Wasting can be reversed rapidly, over a few weeks. However, if a child has undergone reductive adaptation and the wasting is severe, then mortality tends to be high, up to 50% in many African hospitals (Jackson et al. 2006). By managing severe malnutrition according to the principles outlined in the most recent WHO manual (WHO

1999a), mortality has fallen dramatically. The logistics of the original protocol have been adapted to meet different circumstances, but the principles have not changed.

EDEMATOUS MALNUTRITION

Edematous malnutrition tends to arise over a few days in children or adults who are already malnourished. It often presents during or just after an infection when signs of vitamin A deficiency also tend to present. It also occurs in particular regions and seasons, usually damp regions and damp seasons, just as infections do. The features are variable apart from the edema. They include misery, apathy or irritability, anorexia, skin and hair lesions, and fatty hepatomegaly. Edematous malnutrition is a form of severe malnutrition, and mortality in hospital tends to be higher than that from severe wasting.

The cause remains unknown. Research was devoted to the "protein deficiency theory" over several decades (Waterlow 1984). Plasma albumin tends to be very low, and this was thought to reduce plasma oncotic pressure sufficiently to induce edema. However, there is no evidence that children who develop edematous malnutrition have been consuming less protein than those who develop nonedematous malnutrition (Gopalan 1968). Also, plasma albumin does not change significantly with loss of edema (Golden et al. 1980). Finally, giving protein to children or adults with edematous malnutrition does not help. In a study in young children, increased energy intake was associated with loss of edema but not increased protein intake (Golden 1982). In a study in adults, increased protein intake was associated with increased anorexia and death (Collins et al. 1998).

There are several other hypotheses to explain edematous malnutrition; some encompass others. The most likely is the oxidative stress theory: Antioxidant status is insufficient to counteract the damaging effects of free radicals on cell structures, particularly membranes, and hence cell functions. This can explain several features, in particular the association with infections and numerous indices of poor antioxidant status and increased oxidative stress (Golden and Ramdath 1987). However, as yet, there is no good evidence of improvement with antioxidant supplements or evidence that individuals with particularly low antioxidant status are at high risk of developing edematous malnutrition.

The management of edematous malnutrition differs little from that for severe wasting. To the best of our knowledge, both are managed with most success by adhering to the principles outlined in the WHO manual (WHO 1999a).

GLOBAL BURDEN OF DISEASE

WHO, in its 2002 World Health Report, concluded that micronutrient deficiencies were of the utmost importance. Compared with all others, iron deficiency was estimated to be 9th, zinc deficiency 11th, and vitamin A deficiency 13th from the top of the list of risk factors for the world's disease burden (DALYs, disability-adjusted life years) in 2000. Together, these three deficiencies were estimated to be responsible for over 2 million (19%) of the total 10.8 million deaths per year in young children; in contrast, malaria accounted for less than a million deaths. Iodine deficiency is diminishing rapidly due to the effective iodized salt programs for populations living in iodine-deficient regions.

Thus, in theory, providing iron, zinc and vitamin A as supplements or in fortified foods to vulnerable populations could be a cost-effective remedy. In practice, it is very important to choose appropriate contents of each in a supplement to prevent damaging interactions. Iron is a particularly reactive element, so the choice of compound used in a supplement or, more important, in fortification, is vital to its effectiveness and its effects on other nutrients.

Meanwhile, it is clear that poverty underlies most deficiencies by removing choice and variety.

Breastfeeding is a good start, but thereafter it is vital to remove the vicious spiral of malnutrition and infection affecting so many millions of children in poverty in developing countries. One spark of hope is that the mortality of the world's young children appears overall to be falling, to under 10 million in 2006 (UNICEF 2007).

CONCLUSIONS

Most of the world's children live in poverty in the developing world. They consume monotonous, low-quality diets deficient in many essential nutrients as well as energy. They are subject to numerous infections, and their susceptibility and resistance to these infections are reduced by their nutrient deficiencies. The vicious spiral created leads to malnutrition, which in turn leads to either death or stunting, impaired physical and mental capacities, poor employment prospects, and chronic ill health. This does not bode well for the next generation.

For individual nutrient deficiencies, although many associations have been made between indices of their status and specific diseases, mechanisms still remain unclear. Our limited understanding, particularly of nutrient interactions, means that we must proceed cautiously when considering supplementation and fortification. More is not necessarily better.

STUDY TOPICS

Poverty
Malnutrition
Causes of anemia
Deficiencies
    Iron
    Zinc
    Selenium
    Thiamin
    Folic acid
    Vitamin A
    Vitamin D
Nutrient deficiencies in general
Deficiency diseases

Anemia
Stunting
Acrodermatitis enteropathica
Keshan disease
Hypothyroidism
Beriberi
Neural tube defects
Cancers
Xerophthalmia, night blindness
Rickets, osteomalacia

NOTE

1. Iodine deficiency disorders (IDDs) were, until very recently, a major public health problem responsible for a huge burden of brain damage. However, where it is available, salt
fortification with iodine has almost eliminated the problem in fetal and infant development (ICCIDD/UNICEF/WHO 2007).

REFERENCES

Anonymous. 1970. Classification of infantile malnutrition. *Lancet* 2 (7667) (Aug
8):302–303.
Anonymous. 1994. Causes and mechanisms of linear growth retardation. Proceedings of
an IDECG workshop. London, January 15–18, 1993. *European Journal of Clinical
Nutrition* 48 (Suppl 1) (Feb):S1–S216.
Beck, M.A., O.A. Levander, and J. Handy. 2003. Selenium deficiency and viral infection.
Journal of Nutrition 133 (5 Suppl 1) (May):1463S-1467S.
Brent, R.L. and G.P. Oakley Jr. 2007. Further efforts to reduce the incidence of neural tube
defects. *Pediatrics* 119 (1) (Jan):225–226.
Brin, M. and Z.Z. Ziporin. 1965. Evaluation of thiamin adequacy in adult humans. *Journal of
Nutrition* 86:319–324.
Brooks, W.A., M. Santosham, A. Naheed, D. Goswami, M.A. Wahed, M. Diener-West, A.S.
Faruque, and R.E. Black. 2005. Effect of weekly zinc supplements on incidence of pneu-
monia and diarrhoea in children younger than 2 years in an urban, low-income population
in Bangladesh: randomised controlled trial. *Lancet* 366 (9490) (Sep 17–23):999–1004.
Choi, S.W. and J.B. Mason. 2000. Folate and carcinogenesis: An integrated scheme. *Journal
of Nutrition* 130 (2) (Feb):129–132.
Collins, S., M. Myatt, and B. Golden. 1998. Dietary treatment of severe malnutrition in adults.
American Journal of Clinical Nutrition 68 (1) (Jul):193–199.
de Onis, M., M. Blossner, E. Borghi, E.A. Frongillo, and R. Morris. 2004. Estimates of
global prevalence of childhood underweight in 1990 and 2015. *Journal of the American
Medical Association* 291 (21) (Jun 2):2600–2606.
Doherty, C.P., M.A. Sarkar, M.S. Shakur, S.C. Ling, R.A. Elton, and W.A. Cutting. 1998. Zinc
and rehabilitation from severe malnutrition: Higher-dose regimens are associated with
increased mortality. *American Journal of Clinical Nutrition* 68 (3): 742–748.

Duque, X., S. Flores-Hernandez, S. Flores-Huerta, I. Mendez-Ramirez, S. Munoz, B. Turnbull, G. Martinez-Andrade, et al. 2007. Prevalence of anemia and deficiency of iron, folic acid, and zinc in children younger than 2 years of age who use the health services provided by the Mexican social security institute. *BMC Public Health* 7:345–358.

Geissler, C. and H.J. Powers. 2006. *Human Nutrition*, 11th ed. Edinburgh: Elsevier Churchill Livingstone.

Golden, M.H. 1982. Protein deficiency, energy deficiency, and the oedema of malnutrition. *Lancet* 1 (8284) (Jun 5):1261–1265.

Golden, M. 1988. The effects of malnutrition in the metabolism of children. *Transactions of the Royal Society of Tropical Medicine and Hygiene* 82 (1):3–6.

Golden, M.H. 1991. The nature of nutritional deficiency in relation to growth failure and poverty. *Acta Paediatrica Scandinavica—Supplement* 374:95–110.

Golden, M.H., B.E. Golden, and A.A. Jackson. 1980. Albumin and nutritional oedema. *Lancet* 1 (8160) (Jan 19):114–116.

Golden, M.H.N. and B.E. Golden. 1981. Trace elements. potential importance in human nutrition with particular reference to zinc and vanadium. *British Medical Bulletin* 37:31–36.

Golden, M.H. and D. Ramdath. 1987. Free radicals in the pathogenesis of kwashiorkor. *Proceedings of the Nutrition Society* 46 (1) (Feb):53–68.

Gopalan, C. 1968. Kwashiorkor and marasmus: evolution and distinguishing features. In *Calorie Deficiencies and Protein Deficiencies*, R. A. McCance and E.M. Widdowson, Eds. London: Churchill, pp. 48–58.

Grotto, I., M. Mimouni, M. Gdalevich, and D. Mimouni. 2003. Vitamin A supplementation and childhood morbidity from diarrhea and respiratory infections: a meta-analysis. *Journal of Pediatrics* 142, (3) (Mar):297–304.

Hallberg, L., B. Sandstrom, A. Ralph, and J. Arthur. 2000. Iron, zinc and other trace elements. In *Human Nutrition and Dietetics*, 10th ed., J.S. Garrow, W.P.T. James, and A. Ralph, Eds. London: Churchill Livingstone, 177–192.

ICCIDD/UNICEF/WHO. 2007. *Assessment of Iodine Deficiency Disorders and Monitoring Their Elimination. A Guide for Programme Managers*. Geneva: World Health Organization.

Jackson, A.A., A. Ashworth, and S. Khanum. 2006. Improving child survival: malnutrition task force and the paediatrician's responsibility. *Archives of Disease in Childhood* 91 (8) (Aug):706–710.

Malcolm, L.A. 1970. Growth retardation in a New Guinea boarding school and its response to supplementary feeding. *British Journal of Nutrition* 24 (1) (Mar):297–305.

Murakami, Y., H. Fujii, A. Ichimura, A. Murata, N. Yamashita, H. Takagi, and K. Tauchi. 1998. Effects of levofolinate calcium on subacute intravenous toxicity of 5-fluorouracil in rats. *Journal of Toxicological Sciences* 23 (Suppl 1) (May):11–29.

Rajakumar, K. 2003. Vitamin D, cod-liver oil, sunlight, and rickets: A historical perspective. *Pediatrics* 112 (2) (August 1):e132–e135.

Ray, J.G., P.R. Wyatt, M.D. Thompson, M.J. Vermeulen, C. Meier, P.Y. Wong, S.A. Farrell, and D.E. Cole. 2007. Vitamin B_{12} and the risk of neural tube defects in a folic-acid-fortified population. *Epidemiology* 18 (3) (May):362–366.

Semba, R.D. and M.W. Bloem. 2002. The anemia of vitamin A deficiency: epidemiology and pathogenesis. *European Journal of Clinical Nutrition* 56 (4) (Apr):271–281.

Shane, B. and K.J. Carpenter. 1997. E. L. Robert Stokstad (1913–1995). *Journal of Nutrition* 127 (2) (Feb):199–201.

Shane, B. and E.L. Stokstad. 1985. Vitamin B_{12}-folate interrelationships. *Annual Review of Nutrition* 5:115–141.

Snell, E.E. and W.H. Peterson. 1940. Growth factors for bacteria: additional factors required by certain lactic acid bacteria. *Journal of Bacteriology* 39:273–285.

Srikantia, S.G. 1958. Ferritin in nutritional oedema. *Lancet* 1 (7022) (Mar 29):667–668.

Umeta, M., C.E. West, J. Haidar, P. Deurenberg, and G.A.J. Hautvast. 2000. Zinc supplementation and stunted infants in Ethiopia: a randomised controlled trial. *Lancet* 355:2021–2026.

UNICEF. 2007. *The State of the World's Children 2008.* New York: United Nations Children's Fund (UNICEF).

Vimokesant, S.L., D.M. Hilker, S. Nakornchai, K. Rungruangsak, and S. Dhanamitta. 1975. Effects of betel nut and fermented fish on the thiamin status of north eastern Thailand. *American Journal of Clinical Nutrition* 28 (12) (Dec):1458–1463.

Wald, G. 1968. Molecular basis of visual excitation. *Science* 162 (850) (Oct 11):230–239.

Walker, S.P., S.M. Chang, C.A. Powell, E. Simonoff, and S.M. Grantham-McGregor. 2006. Effects of psychosocial stimulation and dietary supplementation in early childhood on psychosocial functioning in late adolescence: follow-up of randomised controlled trial. *British Medical Journal* 333 (7566) (Sep 2):472.

Waterlow, J.C. 1984. Kwashiorkor revisited: the pathogenesis of oedema in kwashiorkor and its significance. *Transactions of the Royal Society of Tropical Medicine and Hygiene* 78:436–441.

Whittaker, P. 1998. Iron and zinc interactions in humans. *American Journal of Clinical Nutrition* 68 (2 Suppl) (Aug):442S–446S.

Wieringa, F.T., J. Berger, M.A. Dijkhuizen, A. Hidayat, N.X. Ninh, B. Utomo, E. Wasantwisut, and P. Winichagoon. 2007. Combined iron and zinc supplementation in infants improved iron and zinc status, but interactions reduced efficacy in a multicountry trial in Southeast Asia. *Journal of Nutrition* 137 (2) (Feb):466–471.

Williams, R.R. 1961. *Toward the Conquest of Beriberi.* Cambridge, MA: Harvard University Press.

Wills, L. 1931. Treatment of "pernicious anemia of pregnancy" and "tropical anemia" with special reference to yeast extract as a curative agent. *British Medical Journal* 1:1059–1064.

Wolbach, S.B. and P.R. Howe. 1925. Tissue change following deprivation of fat-soluble vitamin A. *Journal of Experimental Medicine* 42:753–777.

Wolf, G. 1996. A history of vitamin A and retinoids. *FASEB Journal* 10 (9) (Jul):1102–1107.

World Health Organization. 1995. *Global Prevalence of Vitamin A Deficiency. Micronutrient Deficiency Information System,* No 2 (WHO/NUT/95.3). Geneva: World Health Organization.

World Health Organization. 1999a. *Management of Severe Malnutrition: A Manual for Physicians and Other Senior Health Workers.* Geneva: World Health Organization.

World Health Organization. 1999b. *Thiamin Deficiency and Its Prevention and Control in Major Emergencies.* Geneva: World Health Organization.

World Health Organization. 2002. *The World Health Report 2002.* Geneva: World Health Organization.

World Health Organization. 2004. *Folate and Folic Acid.* Geneva: World Health Organization.

11 Effects of Growth Retardation and Nutrient Deficiencies on Cognitive Function and Behavior in Infants and Children

Susan P. Walker and Julie M. Meeks Gardner

CONTENTS

ABSTRACT

This chapter presents the evidence linking inadequate nutrition in children with poor developmental outcomes. Children are most likely to experience undernutrition in the early years, coinciding with the period of greatest brain development. Chronic undernutrition can result in linear growth retardation or "stunting," which affects about one third of children under 5 years in developing countries. There are other markers of poor nutrition, including deficiencies of micronutrients and insufficient breastfeeding, that may also affect children's development. There is very strong evidence that iodine deficiency causes cognitive deficits, leading to cretinism as well as less-devastating deficits in development. There is also strong evidence that chronic undernutrition leading to linear growth retardation and iron deficiency anemia have long-lasting effects on children's cognitive development and behavior. There is an urgent need for strategies to reduce the numbers of children who become undernourished and the negative impact on children's development as well as to strengthen efforts to prevent iodine deficiency in mothers and children. It also appears important that efforts are made to reduce iron deficiency anemia among women and children. Most studies suggest that breastfeeding is associated with a small benefit in IQ, but problems with the study designs and interpretation have not been fully addressed. Several other nutrients, such as zinc, vitamin A, and vitamin B$_{12}$, may affect children's development, but the evidence is inconsistent, and further research is needed.

INTRODUCTION

The growth potential of young children is similar across different populations (WHO Multicentre Growth Reference Study Group 2006). Faltering in linear growth results from chronic undernutrition and high levels of infections (Waterlow 1992). Reduction in growth rates occurs early, sometimes beginning in utero, and is most common in the first 2 years of life. The first years are also the most critical period for child development; thus, children are most likely to experience undernutrition during the period when the brain is developing rapidly and can be affected by the quality of the environment, including nutrition.

Linear growth retardation or stunting is defined as height-for-age less than or equal to 2 standard deviations (SD) of reference values. One third of children under 5 in developing countries are stunted (UNICEF 2004), with as many as 50% of children stunted in some countries. Thus, the implications for the development of children in many countries are enormous. Children's nutritional intake may be inadequate in micronutrients as well as energy and protein. In this chapter, we describe the evidence linking undernutrition, defined in terms of linear growth retardation, with child development, and then the role of insufficient breastfeeding and specific micronutrients.

LINEAR GROWTH RETARDATION (STUNTING)

STUNTING AND COGNITIVE DEVELOPMENT

Several studies have shown cross-sectional associations (height-for-age and development measured at the same time) between linear growth retardation and poor child development. These have been reviewed in detail (Grantham-McGregor et al. 2007) and are addressed briefly here. In young children, height-for-age is associated with poor development, for example, as shown in studies in Guatemala (Lasky et al. 1981), Jamaica (Powell and Grantham-McGregor 1985), and Kenya (Sigman et al. 1989). Many cross-sectional studies in school-aged children have also shown that being stunted is associated with poorer cognitive ability and lower school achievement levels (Grantham-McGregor et al. 2007).

Longitudinal Studies of Stunting and Cognitive Development

Several prospective cohort studies of the association of stunting in early childhood (by age 2 or 3 years) with later cognitive functioning have been conducted (in longitudinal studies, stunting or height-for-age was assessed in early childhood and children remeasured after varying intervals). Details of the studies are given in Table 11.1. After controlling for social background covariates, stunting was associated with deficits in IQ (Berkman et al. 2002, Walker et al. 2005), nonverbal intelligence (Martorell et al. 1992, Mendez & Adair 1999), and other cognitive domains (Walker et al. 2005). Follow-up in some cases continued to late adolescence (Walker et al. 2005) and adulthood (Martorell et al. 1992). Stunting is also associated with poor educational outcomes, such as fewer grades completed (Martorell et al. 1992) and increased dropout (Daniels and Adair 2004, Walker et al. 2005). Deficits in school achievement have also been found (Martorell et al. 1992, Chang et al. 2002, Walker et al. 2005).

Two cohorts from South Africa and Brazil were included in a recent analysis (Grantham-McGregor et al. 2007), and stunting was associated with reasoning ability at age 7 years (South Africa) and school grades attained at age 18 years (Brazil). Deficits were thus consistently found in all longitudinal studies investigating the effect of early childhood stunting. The magnitude of the difference between stunted and nonstunted children varied, but effects were generally moderate to large.

STUNTING AND SOCIAL–EMOTIONAL OUTCOMES

In cross-sectional studies, underweight and stunted children were less happy and more apathetic and fussy (Meeks Gardner et al. 1999), and they showed lower levels of play and exploratory behavior (Graves 1976, 1978, Meeks Gardner et al. 1999) and more anxious attachment (Graves 1976, 1978, Valenzuela 1990). At school age, children malnourished during the first 2 years of life had attention deficits, more

TABLE 11.1

Linear Growth Retardation and Cognitive Outcomes: Prospective Cohort Studies

Reference	Sample	Outcomes	Covariates	Results
Berkmann et al. 2002	Peru: random sample from periurban shanty town; 239 children followed from birth to 2 years old: 143 assessed at 9 years	IQ WISC-R at age 9 years	Paternal education, school type, grade level, tester	Severely stunted in second year of life lower IQ ($p = .011$) than combined group (never stunted and not severely stunted)
Martorell et al. 1992	Guatemala: 4 villages; 243 subjects with height measurement at 3 years assessed at 18–26 years	Reasoning ability, numeracy, literacy, general knowledge years at school	Village, maternal education, household wealth	Height at 3 years significantly correlated with schooling, literacy, knowledge, numeracy; correlated with reasoning ability in males only
Mendez and Adair, 1999	Philippines: birth cohort from randomly selected administrative units; 2131 children with cognition at 8 years and height at 2 years (69% of birth cohort)	Philippines nonverbal intelligence test at age 8 and 11 years	Grade level, several social background variables, birth weight, maternal height, sex, dietary fat	Children stunted at age 2 years had significantly lower scores at 8 years; at 11 years, severely stunted had significantly lower scores but not those moderately stunted
Daniels and Adair, 2004	Same cohort as above; 1997–2191 children with schooling data	Age at enrollment, grade repetition, highest grade attained	Parity, maternal height, several social background variables	1 SD greater height for age at 2 years associated with reduced risks of late enrollment (boys 32%, girls 40%), grade repetition (boys 14%, girls 22%), and dropout in primary (boys 26%, girls 34%) or secondary school (boys 33%, girls 9%, n.s.)
Walker et al, 2005	Stunted and nonstunted children aged 9–24 months from poor neighborhoods in Kingston, Jamaica; 103 of 129 stunted followed up at 17 years, 64 of 84 nonstunted	IQ, reasoning ability, vocabulary, verbal analogies, memory, reading, math, school dropout	Maternal verbal IQ, education, occupation, housing, and hunger at 17 years; home environment on enrollment	Stunted participants had significant deficits compared to nonstunted in all tests except one memory test and more likely to drop out of school

n.s., not significant; WISC-R, Wechsler Intelligence Scale for Children—Revised.

aggressive behavior, and poorer social relationships than classmates or neighborhood controls (Richardson et al. 1972, Galler and Ramsey 1989).

The Jamaican follow-up study of stunted children provided the most information on possible long-term social-emotional effects of childhood growth retardation. Children stunted in early childhood had more conduct disorder at age 11 years (Chang et al. 2002) and were more inhibited and less attentive in a test session than nonstunted children (Fernald and Grantham-McGregor 1998). At 17 years, stunted participants reported more symptoms of anxiety and depression and lower self-esteem than participants who were never stunted. Parents of the stunted group reported more problems with hyperactive behavior than parents of nonstunted participants (Walker et al. 2007).

IMPACT OF SUPPLEMENTATION ON DEVELOPMENT

If undernutrition contributes to poor child development, then interventions to improve nutrition in undernourished children or reduce undernutrition in children at high risk of growth retardation would be one approach to reducing the cognitive and behavioral consequences. A number of randomized trials have been reported in which food supplements were given in an attempt to improve children's nutritional status and development.

Supplementation Trials with Undernourished Children

Indonesia

In Indonesia, 20 day care centers on tea plantations were randomized to treatment or no treatment (75 supplemented, 38 control children aged 6–20 months). Snacks were provided 6 days a week for 90 days. Supplementation benefited motor but not mental development when measured at the end of intervention (Husaini et al. 1991). Sixty-six supplemented children and 36 controls were remeasured at age 9 years. Benefits were found for one of four cognitive tests only in children who were less than 18 months at the beginning of the intervention. No differences were found in arithmetic or verbal comprehension (Pollitt et al. 1997).

Indonesia

Two cohorts of nutritionally at-risk children were enrolled from 24 day care centers in Indonesia. One cohort comprised 53 children aged 12 months and the other 83 children aged 18 months. Children were randomized to energy plus micronutrients, skimmed milk plus micronutrients, or skimmed milk only. The supplement was provided 6 days a week for 1 year. In the 12-month cohort, children who received the energy intervention had benefits to motor and mental development, increased vocalizations, longer duration of play, and decreased fussing but no differences in sociability. There were no significant benefits to the 18-month cohort (Pollitt and Schurch 2000).

Jamaica

There were 129 stunted children aged 9 to 24 months who were identified by house-to-house survey of poor neighborhoods in Kingston, Jamaica. Children were randomized to (1) supplementation, (2) psychosocial stimulation with weekly home visits, (3) supplement and stimulation, or (4) control. Interventions were provided for 2 years. Stimulation and supplementation benefited the children's developmental levels (Grantham-McGregor et al. 1991) and had additive but not interactive effects. There were no benefits of supplementation to child exploration or affect assessed after 6 months of intervention (Meeks Gardner et al. 1999). Children were reassessed at 7 to 8 years, and small benefits of supplementation on a range of educational and cognitive tests were seen but no significant differences on any one test (Grantham-McGregor et al. 1997). No benefits from supplementation were detected at 11 or 17 years to cognitive functioning or behavior (Chang et al. 2002, Walker et al. 2005, 2006).

Supplementation Trials to Prevent Undernutrition in High-Risk Populations
Guatemala
Four villages in Guatemala were randomized to high-energy and protein supplement (atole) or low-energy supplement. Supplements were provided ad libitum twice daily at centers for pregnant and lactating women and children up to age 7 years. Atole benefited motor but not mental development at 24 months and perceptual organization and verbal skills at age 4 and 5 years but not 3 and 6 years (Pollitt et al. 1993). Children of lower socioeconomic status benefited the most. Children of participants supplemented with atole in pregnancy who were supplemented through at least age 2 years performed significantly better on 4 of 6 psychoeducational tests and on 2 of 7 information processing tests at age 13–19 years (Pollitt et al. 1993). In a subsample of children at age 6 to 8 years, those who received higher levels of supplementation from birth to 2 years were more socially involved, happier and more angry, and less anxious (Barrett et al. 1982).

Colombia
High-risk families in Bogota, Columbia, were randomized to receive supplementation for varying periods: from pregnancy to infants aged 6 months (n = 57), infants aged 6 to 36 months (n = 60), from pregnancy to infants aged 36 months (n = 57), or control (n = 54). At 4 and 8 months, supplemented infants showed less apathy. At 36 months, supplement benefited developmental levels in those supplemented from pregnancy through 36 months (Waber et al. 1981). Benefits for the two groups supplemented for shorter periods were inconsistent. At follow-up at age 5 to 8 years, supplementation benefited scores on reading readiness but not arithmetic or basic knowledge; however, these results have only been reported in a conference abstract.

These studies provide consistent evidence that food-based supplements that provide additional energy and protein (and varying amounts of micronutrients) have concurrent benefits for child development. Limited or no benefits were found after supplementation ended in two studies in which supplementation was given to undernourished children. The most substantial benefits to later development have been found in the study conducted in Guatemala among participants whose mothers were supplemented in pregnancy and who were supplemented at least until age 2 years.

In summary, early childhood linear growth retardation is associated with poor development in childhood and with later cognitive deficits and poorer educational outcomes. Undernutrition is also associated with altered behavior in early childhood. Information on later behavior suggests problems with attention, social relationships, and psychological functioning. The benefits demonstrated in food supplementation trials provide further evidence that childhood undernutrition contributes to underachievement among children in developing countries. There is an urgent need for strategies to reduce the numbers of children who become undernourished and the impact on children's development.

INSUFFICIENT BREASTFEEDING

Breastfeeding provides several important benefits to both mothers and their infants, especially in terms of reduced infectious diseases and mortality (World Health Organization [WHO], 2003). However, the benefits of breastfeeding to children's development are much less clear. Such benefits could result from direct effects of the nutrients in breast milk, especially the fatty acid composition, or there may be indirect benefits from improved growth or immune response or from the closer mother-child interactions associated with breastfeeding (Grantham-McGregor et al. 1999).

The question has been widely investigated in over 50 studies, a number of reviews, and a meta-analysis. Many studies reported short or long-term benefits to psychomotor development with breastfeeding. The meta-analysis included 11 studies (Anderson et al. 1999), and the results indicated that breastfeeding was associated with significantly higher cognitive scores than formula feeding (3.16 points adjusted for covariates). This benefit was greater among preterm infants (5.18 points) compared with term infants (2.26 points). The IQ advantage increased with duration of breastfeeding, reaching a maximum at 4–6 months. However, a later review (Rey 2003) pointed out a number of flaws within the 11 included studies, suggesting that the conclusions were probably not valid. Other studies have not found a relationship between breastfeeding and development after adjustment for confounders (Clark et al. 2006).

In one systematic review (Jain et al. 2002), although 27 (68%) of the 40 studies concluded that breastfeeding promotes intelligence, methodological flaws were found in most of the studies. Only 2 were high-quality studies of full-term infants. One concluded that breastfeeding had a significant beneficial effect (Johnson et al. 1996), while the other did not (Wigg et al. 1998); thus, the reviewers cautioned readers attempting to interpret the results.

One of the major problems with the studies is that many have not adjusted for important covariates. In general, mothers in developed countries who choose to breastfeed and are successful at it have a number of other advantages that are likely to lead to better developmental outcomes among their children. These include higher socioeconomic status, better education, greater intelligence, less depression, and enhanced home environments. Most studies of breastfeeding and development have been carried out in developed countries where breastfeeding is highly confounded with socioeconomic status (Grantham-McGregor et al. 1999). Two reports from developing countries suggest a complicated relationship between development and breastfeeding. In Chile, exclusive breastfeeding for under 2 months or more than 8 months was associated with poorer mental and motor outcomes at 5½ years (Clark et al. 2006). Another report from Honduras indicated that infants who were exclusively breastfed for 6 months crawled sooner and were more likely to be walking at 12 months compared with infants exclusively breastfed for 4 months when additional high-quality, complementary foods were also offered (Dewey et al. 2001). These studies require replication before firm conclusions are drawn.

Other problems with the studies included variation in the definitions of *breastfeeding* (e.g., definitions depending on the duration and exclusivity of breastfeeding), small sample sizes in some studies, and observers were not always blind to the participants' status (Drane and Loggeman 2000).

Despite the problems with the studies, some reviewers have argued that the consistency of the findings suggest that a small benefit to cognition from breastfeeding is likely (e.g., Grantham-McGregor et al. 1999, American Academy of Pediatrics 2005). Uauy and Peirano (1999) concluded that while high-quality scientific research should be demanded before accepting findings, the suggestive positive results should be taken into account for policy recommendations. The additional 2 to 5 IQ points would be a small difference for individuals but may be of great importance at a population level.

IODINE

Iodine deficiency can lead to cretinism and irreversible mental retardation (Black 2003b) and is the most common preventable cause of mental retardation. A global program to reduce iodine deficiency primarily through salt iodization has had substantial success; nonetheless, iodine deficiency continues to threaten the development of children in many parts of the world.

Iodine is a constituent of the thyroid hormones thyroxine and triiodothyronine, which influence the development of the central nervous system (Pharoah and Connolly 1995). Maternal iodine deficiency during pregnancy affects brain development in utero (Hetzel and Mano 1989), and deficiency in infancy and early childhood may also affect development.

In addition to the most serious effects of cretinism, iodine deficiency in the subclinical range can also affect mental development. Studies comparing children living in iodine-deficient areas with those living in iodine-sufficient areas have generally shown children in iodine-deficient regions to have lower cognitive development and school achievement. In a 1994 meta-analysis of 18 studies, IQ scores averaged 13.5 points lower in iodine-deficient groups (Bleichrodt and Born 1994). A second meta-analysis of 37 studies reported a very similar deficit of 12.5 IQ points (Qian et al. 2005). However, communities with iodine-deficient soil are often more isolated and poorer. In a study adjusting for possible confounding factors, subclinical prenatal iodine deficiency remained associated with poor infant development (Choudhury and Gorman 2003).

IODINE SUPPLEMENTATION STUDIES

Peru

Children of mothers from three iodine-deficient villages in Peru supplemented in pregnancy had higher developmental levels compared with unsupplemented controls from the same villages (Pretell et al. 1972). However, the difference was not significant.

Ecuador

Pregnant women in one iodine-deficient village in Ecuador were treated with iodized oil either during months 4 to 7 of pregnancy (group 1) or before conception (group 2). Compared with control children from an untreated village, children in group 2 had higher IQ scores, but there were no differences between group 1 and controls (Ramirez et al. 1969). In a second study, a greater percentage of children of mothers treated before the sixth month of pregnancy had normal developmental levels than was the case in children of untreated mothers (Ramirez et al. 1972).

Zaire

Pregnant women in an area of severe iodine deficiency in Zaire were randomized to either iodized oil or a vitamin placebo on average at the 28th week of pregnancy (Thilly et al. 1980). Infants of treated mothers had significantly higher developmental levels than infants of placebo-treated mothers.

Papua New Guinea

In a double-blind randomized controlled trial in 16 villages in Papua New Guinea, iodine treatment in pregnancy led to a significant reduction in cretinism (Pharoah et al. 1971). Children in 5 of the villages were followed up in later studies. Children whose mothers received iodized oil had better cognitive and fine motor skills than control children (Connolly et al. 1979).

China

Treatment in China was given to pregnant women, infants, and children aged 1 to 3 years. Scores of the 1- to 3-year-old children prior to treatment were used as control values. At age 2 years, children whose mothers received iodine during the first or second trimester of pregnancy had higher developmental scores than the controls (Cao et al. 1994). In another trial, children whose mothers received iodine in early pregnancy (first and second trimester) had higher psychomotor scores compared with a group who received iodine later in pregnancy or with children who first received iodine at age 2 years. Later in childhood, there was also a trend of better cognitive performance in those who received iodine before the third trimester (O'Donnell et al. 2002).

These studies provided strong evidence of the impact of iodine deficiency, particularly during pregnancy, on children's development. Improvements in iodine status in school-aged children may also benefit cognitive ability (van den Briel et al. 2000). There is a clear need to strengthen efforts to prevent iodine deficiency in mothers and children.

IRON DEFICIENCY ANEMIA

Numerous studies have shown that iron deficiency anemia (IDA) in infants and children is associated with poorer concurrent mental and motor function and behavioral changes. These have been discussed in detail in recent reviews (Grantham-McGregor and Ani 2001, Lozoff et al. 2006). IDA is associated with many socioeconomic factors that also affect development, so it is difficult to draw conclusions about causality from these cross-sectional studies. Here, we focus on long-term follow-up studies of previously anemic children and supplementation trials that provided stronger evidence of the effect of IDA on child development.

FOLLOW-UP STUDIES OF IDA

Follow-up studies have shown poorer developmental outcomes in infants and children who experienced IDA despite treatment for IDA and adequate iron status at follow-up. A meta-analysis estimated that for each 10 g/L decrease in hemoglobin, subsequent IQ later in childhood was lower by 1.73 points (Stoltzfus et al. 2005). In a population study in one county in the state of Florida, children with low hemoglobin in infancy were more likely to have been placed in special education by age 10 years (Hurtado et al. 1999).

The longest follow-up of IDA children was conducted in Costa Rica. In this study, children with moderate IDA or chronic iron deficiency at age 12 to 23 months were assessed at age 11 to 14 years and compared with children who were never iron deficient. The previously iron-deficient children had poorer educational achievement and motor ability and had several behavioral problems, including more anxiety and depression, difficulties with attention, and social problems (Lozoff et al. 2000). The children were tested again at age 19 years, and the cognitive difference between the iron deficient and those never iron deficient had increased (Lozoff et al. 2006b).

STUDIES OF THE IMPACT OF TREATMENT ON DEVELOPMENT

There have been several reports of the effects of treatment of IDA on the development of infants and children. However, few of these have been randomized trials, and many were of short duration (up to 2 weeks) and therefore may not have allowed sufficient time for benefits to development to occur. These have been reviewed in detail elsewhere (Grantham-McGregor and Ani 2001). Studies in which treatment was given for at least 2 months have had varied results, with some showing that children with IDA continued to have poorer development. In other studies, substantial benefits have been reported. For example, in an Indonesian study children aged 12 to 18 months with mild-to-moderate IDA showed substantial gains in mental and motor development after 4 months of treatment compared with untreated children (Idjradinata and Pollitt 1993). Studies with older children have generally shown some benefits to cognitive ability and school achievement with treatment (e.g., Soemantri et al. 1985, Seshadri and Gopaldas 1989).

STUDIES OF THE BENEFITS OF PREVENTIVE SUPPLEMENTATION

A number of trials have been conducted of the effects of giving iron-fortified formula to healthy infants compared with usual formula or cow's milk. These have shown no benefit (Morley et al. 1999), short-term benefits to motor development (Moffatt et al. 1994), or significant benefits to overall developmental scores (Williams et al. 1999). In a trial of iron supplementation of healthy breast-fed infants, those who received iron had better visual acuity and motor development at 1 year, but there were no benefits to mental development (Friel et al. 2003). In Chile, well-nourished infants who received iron-fortified formula for 6 months from age 6 months had benefits to recognition memory and social-emotional outcomes compared with infants on non-fortified formula but had no benefits on the mental and motor scales of the Bayley Scales of Infant Development (Lozoff et al. 2003).

Four trials of preventive iron supplementation have been conducted in developing countries among populations at high risk of both stunting and anemia. All studies showed benefits to children's motor development, although in one this was only when iron and zinc were both given (Stoltzfus et al. 2001, Black et al.. 2002, 2004a, Lind et al. 2004). None of the studies showed benefits to mental development, although one benefited language, and two showed benefits to social emotional outcomes.

IDA is associated with concurrent cognitive delays and behavioral changes, and follow-up studies indicated that there are long-lasting effects. Some studies of treatment lasting at least 2 months have shown benefits to development, with the findings from older children being more consistent. Preventive supplementation has also led to developmental benefits, primarily to motor and socioemotional outcomes. Although there are some conflicting findings, the body of research on IDA indicates the need for strategies to reduce the prevalence of IDA in young children.

ZINC

Zinc deficiency is thought to be highly prevalent in undernourished children. It has been linked to reduced activity and play (Sazawal et al. 1996, Bentley et al. 1997) and may exacerbate undernourished children's poor development. Zinc deficiency is associated with growth retardation (Brown et al. 1998) and increased rates of diarrhea and pneumonia in children (Zinc Investigators Collaborative Group 1999), which could indirectly link zinc deficiency with poor development. However, evidence of associations between zinc and development is inconsistent in young children (Hamadani et al. 2002, Black 2003a, Lozoff and Black 2004) and among school-aged children (Lozoff and Black 2004).

More than a dozen studies have reported the effects of zinc supplementation on young children's activity, development, or behavior. Unfortunately, the findings vary considerably, with some studies showing benefits, others no change, and others negative effects. It is difficult to draw conclusions from such a disparate set of findings. The only two studies reporting activity levels indicated a benefit with zinc supplementation (Friel et al. 1993, Sazawal et al. 1996). Some aspects of behavior improved in two of the studies (Ashworth et al. 1998, Black et al. 2004a), but there was no effect or a detrimental effect on development in three others (Hamadani et al. 2001, 2002, Black et al. 2004b).

It is likely that other nutrient deficiencies confounded these trials since children with zinc deficiency because of a poor diet could be expected to have deficiencies of other micronutrients. When zinc was given with or without additional iron, there were benefits to motor development and behavior in one study (Black et al. 2004a) but not in the other (Lind et al. 2004). The only study with a benefit to global developmental scores showed this effect only when zinc supplementation was combined with a psychosocial stimulation intervention (Meeks Gardner et al. 2005). It is possible that even if children become more exploring and active with zinc supplements, improvements to development may only occur if their environments provide adequate stimulation.

The role of zinc deficiency in children's development remains to be clarified. Zinc status is difficult to measure, and response to supplementation is often used as an indicator of deficiency. Possible explanations for the lack of response in development in some studies could be that the children were not initially zinc deficient, that additional deficiencies affected development, or even that zinc given as supplements produced imbalances in other micronutrients.

VITAMIN A

Vitamin A deficiency is associated with increased frequency and severity of infections, increased mortality, and vision impairment. Effects on vision would almost certainly have detrimental effects on children's development, especially where facilities for the visually impaired are lacking.

In one study, Indonesian mothers were supplemented with iron and folate only or iron, folate, and vitamin A during pregnancy (Schmidt et al. 2004). Vitamin A produced no benefits to the infants' mental or motor scores at 6 or 12 months of age. However, there was a suggestion of an association as maternal vitamin A status in pregnancy was related to infant mental and motor development outcomes. Two studies looked at vitamin A supplementation of young children and their developmental outcomes. In one, Indonesian newborns given vitamin A had slightly higher scores on Bayley's Scales of Infant Development at age 3 years compared with control children (Humphrey et al. 1998). It remains unknown if the mechanism was through a reduction in severe morbidity or by some direct effect of vitamin A. In the second study, Bangladeshi 3 year olds given vitamin A supplements at birth, did not show developmental benefits compared with control children (van Dillen et al. 1996). These findings suggest that vitamin A deficiency probably has only small, direct negative effects on young children's development.

VITAMIN B$_{12}$

Several studies have shown links between vitamin B$_{12}$ deficiency and poor cognitive functioning among the elderly, including dementia and neurocognitive deficits. Among children, there have been only a few cross-sectional reports. The mechanism linking B$_{12}$ deficiency may be through changes in neuroanatomy or neurotransmission (Georgieff and Rao 2001). In an early case study, infants of mothers who had pernicious anemia (caused by vitamin B_{12} deficiency) or were vegan (that is, did not consume foods of animal origin, the source of vitamin B_{12}) were at risk for delayed developmental milestones (Lampkin and Saunders 1969). Observational studies among children with B_{12} deficiency in the Netherlands showed that infants whose mothers were on a macrobiotic diet (restricted in animal foods) and who were weaned onto macrobiotic diets had delayed motor and language development compared with children who were omnivores (Dagnelie and van Staveren 1994). Although they started to consume omnivorous (foods of plant and animal origin) or lactovegetarian (plant-based with milk products) diets by age 6 years, their cognitive functioning was impaired compared with the consistently omnivorous at age 10 to 18 years (Louwman et al. 2000). These data are consistent with cross-sectional findings among Guatemalan schoolchildren (Allen at al. 1999, Penland et al. 2000.) The findings suggest that B_{12} deficiency is likely to be a risk factor for developmental delays, but further evidence is needed, preferably from supplementation trials.

CONCLUSIONS

There is very strong evidence that iodine deficiency causes cognitive deficits. There is also strong evidence that chronic undernutrition leading to linear growth retardation and IDA have long-lasting affects on children's cognitive development and behavior. Most studies suggest that breastfeeding is associated with a small benefit in IQ, but problems with confounding have not been fully addressed. Several other nutrients, such as zinc, may affect children's development, but the evidence is inconsistent, and further research is needed.

REFERENCES

Allen, L.H., Penland, J.G., Boy, E., DeBaessa, Y., and Rogers L.M. 1999. Cognitive and neuromotor performance of Guatemalan schoolers with deficient, marginal and normal plasma vitamin B-12 [abstract]. *FASEB Journal* 13:A544.

American Academy of Pediatrics. 2005. Policy statement: breastfeeding and the use of human milk. *Pediatrics* 115:496–506.

Anderson, J.W., Johnstone, B.M., and Remley, D.T. 1999. Breast-feeding and cognitive development: a meta-analysis. *American Journal of Clinical Nutrition* 70:525–535.

Ashworth, A., Morris, S.S., Lira, P.I.C., and Grantham-McGregor, S.M. 1998. Zinc supplementation and behavior in low birth weight term infants in NE Brazil. *European Journal of Clinical Nutrition* 52:223–227.

Barrett, D.E., Radke-Yarrow, M., and Klein, R.E. 1982. Chronic malnutrition and child behavior: effects of early caloric supplementation on social and emotional functioning at school age. *Developmental Psychology* 18:541–556.

Bentley, M.E., Caulfield, L.E., Ram, M., Santizo, M.C., Hurtado, E., Rivera, J.A., Ruel, M.T., and Brown, K.H. 1997. Zinc supplementation affects the activity patterns of rural Guatemalan infants. *Journal of Nutrition* 127:1333–1338.

Berkman, D.S., Lescano, A.G., Gilman, R.H., Lopez, S.L., and Black, M.M. 2002. Effects of stunting, diarrhoeal disease, and parasitic infection during infancy on cognition in late childhood: a follow-up study. *Lancet* 359, 564–571.

Black, M.M. 2003a. The evidence linking zinc deficiency with children's cognitive and motor functioning. *Journal of Nutrition* 133:1473S-1476S.

Black, M.M. 2003b. Micronutrient deficiencies and cognitive functioning. *Journal of Nutrition* 133(11 Suppl 2):3927S-3931S.

Black, M.M., Sazawal, S., Black, R.E., Khosla, S., Sood, M., Juyal, R.C., et al. 2002. Micronutrient supplementation leads to improved development and behaviour among infants born small-for-gestational-age. *Pediatric Research* 51:2565.

Black, M.M., Baqui, A.H., Zaman, K., Persson, L.A., El Arifeen, S., Le, K., McNary, S.W., Parveen, M., Hamadani, J.D., and Black, R.E. 2004a. Iron and zinc supplementation promote motor development and exploratory behaviour among Bangladesh infants. *American Journal of Clinical Nutrition* 80:903–910.

Black, M.M., Sazawal, S., Black, R.E., Khosla, S., Kumar, J., and Menon, V. 2004b. Cognitive and motor development among small-for-gestational-age infants: impact of zinc supplementation, birth weight, and caregiving practices. *Pediatrics* 113:1297–1305.

Bleichrodt, N. and Born, M.P. 1994. A meta-analysis of research on iodine and its relationship to cognitive development. In: *The Damaged Brain of Iodine Deficiency,* J.B. Stanbury, Ed. New York: Cognizant Communication.

Brown, K.H., Peerson, J.M., and Allen, L.H. 1998. Effect of zinc supplementation on children's growth: a meta-analysis of intervention trials. *Bibliotheca nutritio et dieta* 54, 76–83.

Cao, X.Y., Jiang, X.M., Dou, Z.H., Rakeman, M.A., Zhang, M.L., O'Donnell, K., et al. 1994. Timing of vulnerability of the brain to iodine deficiency in endemic cretinism. *New England Journal of Medicine* 331:1739–1744.

Chang, S.M., Walker, S.P., Grantham-McGregor, S., and Powell, C.A. 2002. Early childhood stunting and later behaviour and school achievement. *Journal of Child Psychology and Psychiatry* 43:775–783.

Choudhury, N. and Gorman, K.S. 2003. Subclinical prenatal iodine deficiency negatively affects infant development in northern China. *Journal of Nutrition* 133:3162–3165.

Clark, K.M., Castillo, M., Calatroni, A., Walter, T., Cayzzo, M., Pino, P., and Lozoff, B. 2006. Breast-feeding and mental and motor development at 5½ years. *Ambulatory Pediatrics* 6: 65–71.

Connolly, K., Pharoah, P.O., and Hetzel, B.S. 1979. Fetal iodine deficiency and motor performance during childhood. *Lancet* 2;1149–1151.

Dagnelie, P.C. and van Staveren, W.A. 1994. Macrobiotic nutrition and child health: results of a populaton-based, mixed-longitudinal cohort study in the Netherlands. *American Journal of Clinical Nutrition* 59(Suppl):1187S–96S.

Daniels, M.C. and Adair, L.S. 2004. Growth in young Filipino children predicts schooling trajectories through high school. *Journal of Nutrition* 134:1439–1446.

Dewey, K.G., Cohen, R.J., Brown, K.H., and Rivera, L.L. 2001. Effects of exclusive breastfeeding for four versus six months on maternal nutritional status and infant motor development: results of two randomized trials in Honduras. *Journal of Nutrition* 131:262–267.

Drane, D.L. and Loggeman, J.A. 2000. A critical evaluation of the evidence on the association between type of infant feeding and cognitive development. *Paediatric and Perinatal Epidemiology* 14:349–356.

Fernald, L.C. and Grantham-McGregor, S.M. 1998. Stress response in school-age children who have been growth retarded since early childhood. *American Journal of Clinical Nutrition* 68:691–698.

Friel, J.K., Andrews, W.L., Matthew, J.D., Long, D.R., Cornel, A.M., Cox, M., McKim, E., and Zerbe, G.O. 1993. Zinc supplementation in very-low-birth-weight infants. *Journal of Pediatric Gastroenterology and Nutrition* 17:97–104.

Friel, J.K., Aziz, K., Andrews, W.L., Harding, S.V., Courage, M.L., and Adams, R. 2003. A double-masked, randomized controlled trial of iron supplementation in early infancy in healthy full-term infants. *Journal of Pediatrics* 143:582–586.

Galler, J.R. and Ramsey, F. 1989. A follow-up study of the influence of early malnutrition on development: behavior at home and at school. *Journal of the American Academy of Child and Adolescent Psychiatry* 28: 254–261.

Georgieff, M.K. and Rao, R. 2001. The role of nutrition in cognitive development. In *Handbook of Developmental Cognitive Neuroscience*, C.A. Nelson and M. Luciana, Eds. Cambridge, MA: MIT Press, pp. 491–504.

Grantham-McGregor, S.M., Powell, C.A., Walker, S.P., and Himes, J.H. 1991. Nutritional supplementation, psychosocial stimulation, and mental development of stunted children: The Jamaican Study. *Lancet* 338: 1–5.

Grantham-McGregor, S., Walker, S., Chang, S., and Powell, C. 1997. Effects of early childhood supplementation with and without stimulation on later development in stunted Jamaican children. *American Journal of Clinical Nutrition* 66:247–253.

Grantham-McGregor, S.M., Fernald, L.C., and Sethuraman, K. 1999. Effects of health and nutrition on cognitive and behavioural development in children in the first three years of life. Part 1: Low birth weight, breastfeeding and protein-energy malnutrition. *Food and Nutrition Bulletin* 20:53–75.

Grantham-McGregor, S. and Ani, C. 2001. A review of studies on the effect of iron deficiency on cognitive development in children. *Journal of Nutrition* 131:649S–68S.

Grantham-McGregor, S., Cheung, Y.B., Cueto, S., Glewwe, P., Richter, L., and Strupp, B. 2007. Developmental potential in the first 5 years for children in developing countries. *Lancet* 369:60–70.

Graves, P.L. 1976. Nutrition, infant behavior, and maternal characteristics: a pilot study in West Bengal, India. *American Journal of Clinical Nutrition* 29:305–319.

Graves, P.L. 1978. Nutrition and infant behavior: a replication study in the Katmandu Valley, Nepal. *American Journal of Clinical Nutrition* 31:541–551.

Hamadani, J., Fuchs, G.J., Osendarp, S.J.M., Khatun, F., Huda, S.N., and Grantham-McGregor, S.M. 2001. Randomized controlled trial of the effect of zinc supplementation on the mental development of Bangladeshi infants. *American Journal of Clinical Nutrition* 74:381–386.

Hamadani, J.D., Fuchs, G.J., Osendarp, S.J., Huda, S.N., and Grantham-McGregor, S.M. 2002. Zinc supplementation during pregnancy and effects on mental development and behaviour of infants: a follow-up study. *Lancet* 360:290–294.

Hetzel, B.S. and Mano, M. 1989. A review of experimental studies of iodine deficiency during fetal development. *Journal of Nutrition* 119:145–181.

Humphrey, J.H., Agoestina, T., Juliana, A., Septiana, S., Widjaja, H., Cerreto, M.C., Wu, L.S.F., Ichord, R.N., Katz, J., and West, K.P. 1998. Neonatal vitamin A supplementation: effect on development and growth at 3 years of age. *American Journal of Clinical Nutrition* 68:109–117.

Hurtado, E.K., Claussen, A.H., and Scott, K.G. 1999. Early childhood anemia and mild or moderate mental retardation. *American Journal of Clinical Nutrition* 69:115–119.

Husaini, M.A., Karyadi, L., Husaini, Y.K., Karyadi, D., and Pollitt, E. 1991. Developmental effects of short-term supplementary feeding in nutritionally-at-risk Indonesian infants. *American Journal of Clinical Nutrition* 54, 799–804.

Idjradinata, P. and Pollitt, E. 1993. Reversal of developmental delays in iron-deficient anaemic infants treated with iron. *Lancet* 341:1–4.

Jain, A., Concato, J., and Leventhal, J.M. 2002. How good is the evidence linking breastfeeding and intelligence? *Pediatrics* 109: 1044–1053.

Johnson, D.L., Swank, P.R., Howie, V.M., Baldwin, C.D., and Owen, M. 1996. Breast feeding and children's intelligence. *Psychology Report* 79:1179–1185.

Lampkin, B.C. and Saunders, E.F. 1969. Nutritional vitamin B_{12} deficiency in an infant. *Journal of Pediatrics* 75:1053–1055.

Lasky, R.E., Klein, R.E., Yarbrough, C., Engle, P., Lechtig, A., and Martorell, R. 1981. The relationship between physical growth and infant behavioral development in rural Guatemala. *Child Development* 52:219–226.

Lind, T., Lonnerdal, B., Stenlund, H., Gamayani, I.L., Ismail, D., Seswandhana, R., and Persson, L.A. 2004 A community-based randomized controlled trial of iron and zinc supplementation in Indonesian infants: effects on growth and development. *American Journal of Clinical Nutrition* 80:729–736.

Louwman, M.W.J., van Dusseldorp, M., van der Vijver, F.R.J., Thomas, M.G., Schneede, J., Ueland, P.M., Refsum, H., and van Staveren, W.A. 2000. Signs of impaired cognitive function in adolescents with marginal cobalamin status. *American Journal of Clinical Nutrition* 72:762–769.

Lozoff, B. and Black, M. 2004. Impact of micronutrient deficiencies on behavior and development. In *Micronutrient Deficiencies during the Weaning Period and the First Years of Life*, J. Pettifor and S. Zlotkin, Eds. Basel: Karger, pp. 119–135.

Lozoff, B., Jimenez, E., Hagen, J., Mollen, E., and Wolf, A.W. 2000. Poorer behavioral and developmental outcome more than 10 years after treatment for iron deficiency in infancy. *Pediatrics* 105:E51.

Lozoff, B., De Andraca, I., Castillo, M., Smith, J., Walter, T., and Pino, P. 2003. Behavioral and developmental effects of preventing iron-deficiency anemia in healthy full-term infants. *Pediatrics* 112:846–854.

Lozoff, B., Beard, J., Connor, J., Felt, B., Georgieff, M., and Schallert, T. 2006. Long-lasting neural and behavioral effects of iron deficiency in infancy. *Nutrition Reviews* 64:S34–S43.

Lozoff, B., Jimenez, E., and Smith, J.B. 2006b. Double burden of iron deficiency and low socio-economic status: a longitudinal analysis of cognitive test scores to 19 years. *Archives of Pediatric and Adolescent Medicine* 160:1108–1113.

Martorell, R., Rivera, J., Kaplowitz, H., and Pollitt, E. 1992. Long-term consequences of growth retardation during early childhood. In: *Human Growth: Basic and Clinical Aspects*, M. Hernandez and J. Argente, Eds. Amsterdam: Elsevier Science, pp. 143–149.

Meeks Gardner, J.M., Grantham-McGregor, S.M., Himes, J., and Chang, S. 1999. Behaviour and development of stunted and nonstunted Jamaican children. *Journal of Child Psychology and Psychiatry* 40:819–827.

Meeks Gardner, J., Powell, C.A., Baker-Henningham, H., Walker, S., Cole, T.J., and Grantham-McGregor, S. 2005. Zinc supplementation and psychosocial stimulation: effects on the development of undernourished Jamaican children. *American Journal of Clinical Nutrition* 82:399–405.

Mendez, M.A. and Adair, L.S. 1999. Severity and timing of stunting in the first two years of life affect performance on cognitive tests in late childhood. *Journal of Nutrition* 129:1555–1562.

Moffatt, M.E.K., Longstaffe, S., Besant, J., and Dureski, C. 1994. Prevention of iron deficiency and psychomotor decline in high risk infants through iron fortified infant formula: a randomized clinical trial. *Journal of Pediatrics* 125:527–534.

Morley, R., Abbott, R., Fairweather-Tait, S., MacFayden, U., and Sterman, M.B. 1999. Iron fortified follow on formula from 9 to 18 months improves iron status but not development or growth: a randomised trial. *Archives of Diseases in Childhood* 81:247–252.

O'Donnell, K.J., Rakeman, M.A., Zhi-Hong, D., Xue-Yi, C., Mei, Z.Y., DeLong, N., et al. 2002. Effects of iodine supplementation during pregnancy on child growth and development at school age. *Developmental Medicine and Child Neurology* 44:76–81.

Penland, J., Allen, L.H., Boy, E., DeBaessa, Y., and Rogers, L.M. 2000. Adaptive functioning, behaviour problems and school performance of Guatemalan children with deficient, marginal and normal plasma B12 [abstract]. *FASEB Journal* 14:A561.

Pharoah, P.O., Buttfield, I.H., and Hetzel, B.S. 1971. Neurological damage to the fetus resulting from severe iodine deficiency during pregnancy. *Lancet* 1:308–310.

Pharoah, P.O. and Connolly, K.J. 1995. Iodine and brain development. *Developmental Medicine and Child Neurology* 38:464–469.

Pollitt, E., Gorman, K.S., Engle, P.L., Martorell, R., and Rivera, J. 1993. Early supplementary feeding and cognition: effects over two decades. *Monogaphs of the Society for Research in Child Development* 58:1–99.

Pollitt, E. and Schurch, B. 2000. Developmental pathways of the malnourished child. Results of a supplementation trial in Indonesia. *European Journal of Clinical Nutrition* 54:2–113.

Pollitt, E., Watkins, W.E., and Husaini, M.A. 1997. Three-month nutritional supplementation in Indonesian infants and toddlers benefits memory function 8 y later. *American Journal of Clinical Nutrition* 66:1357–1363.

Powell, C.A. and Grantham-McGregor, S.M. 1985. The ecology of nutritional status and development in young children in Kingson, Jamaica. *American Journal of Clinical Nutrition* 41:1322–1331.

Pretell, E.A, Torres, T., Zenteno, V., and Cornejo, M. 1972. Prophylaxis of endemic goiter with iodized oil in rural Peru. *Advances in Experimental Medicine and Biology* 30:246–265.

Qian, M., Wang, D., and Watkins, W.E. 2005. The effects of iodine on intelligence in children: a metanalysis of studies conducted in China. *Asia Pacific Journal of Clinical Nutrition* 14:32–42.

Ramirez, I., Fierro-Benitez, R., Estrella, E., Jaramillo, C., Diaz, C., and Urresta, J. 1969. Iodized oil in the prevention of endemic goiter and associated defects in the Andean region of Ecuador. Effects on neuromotor development and somatic growth before 2 years. In: *Endemic Goiter,* J. Stanbury, Ed. Washington, DC: Pan American Health Organization, pp. 341–359.

Ramirez, I., Fierro-Benitez, R., Estrella, E., Jaramillo, C., Diaz, C., and Urresta, J. 1972. The results of prophylaxis of endemic cretinism with iodized oil in rural Andean Ecuador. In: *Human Development and the Thyroid Gland,* J.B. Stanbury and R.L. Kroc, Eds. New York: Plenum Press, pp. 223–237.

Rey, J. 2003. Breastfeeding and cognitive development. *Acta Paediatrica* Suppl 442:11–8.

Richardson, S.A., Birch, H.G., Grabie, E., and Yoder, K. 1972. The behavior of children in school who were severely malnourished in the first 2 years of life. *Journal of Health Social Behavior* 13:276–284.

Sazawal, S., Bentley, M., Black, R.E., Dhingra, P., George, S., and Bhan, M.K. 1996. Effect of zinc supplementation on observed activity in low socioeconomic Indian preschool children. *Pediatrics* 98:1132–1137.

Schmidt, M.K., Muslimatun, S., West, C.E., Schultink, W., and Hautvast, J.G.A. 2004. Mental and psychomotor development in Indonesian infants of mothers supplemented with vitamin A in addition to iron during pregnancy. *British Journal of Nutrition* 91:279–285.

Seshadri, S. and Gopaldas, T. 1989. Impact of iron supplementation on cognitive functions in preschool and school-aged children: the Indian experience. *American Journal of Clinical Nutrition* 50:675–686.

Sigman, M., Neumann. C., Baksh, M., Bwibo, N., and McDonald, M.A. 1989. Relationship between nutrition and development in Kenyan toddlers. *Journal of Pediatrics* 115:357–364.

Soemantri, A.G., Pollitt, E., and Kim, I. 1985. Iron deficiency anemia and educational achievement. *American Journal of Clinical Nutrition* 42:1221–1228.

Stoltzfus, R.J., Kvalsvig, J.D., Chwaya, H.M., Montresor, A., Albonico, M., Tielsch, J.M., et al. 2001. Effects of iron supplementation and anthelmintic treatment on motor and language development of preschool children in Zanzibar: double blind, placebo controlled study. *British Medical Journal* 323:1389–1393.

Stoltzfus, R.J., Mullany, L., and Black, R.E. 2005. *Iron Deficiency Anaemia. Comparative Quantification of Health Risks: Global and Regional Burden of Disease Attributable to Selected Major Risk Factors,* Vol. 1. Geneva: World Health Organization, pp. 163–209.

Thilly, C.H., Lagasse, R., Roger, G., Bourdoux, P., and Ermans, A.M. 1980. Impaired fetal and postnatal development and high perinatal death-rate in a severe iodine deficient area. In: *Thyroid Research VIII,* J.R. Stockigt and S. Nagataki, Eds. Oxford, U.K.: Pergamon, pp. 20–23.

Uauy, R. and Peirano, P. 1999. Breast is best: human milk is the optimal food for brain development. *American Journal of Clinical Nutrition* 70:433–434.

UNICEF. 2004. *The State of the World's Children 2005: Childhood Under Threat.* New York: UNICEF.

Valenzuela, M. 1990. Attachment in chronically underweight young children. *Child Development* 61:1984–1996.

van den Briel, T., West, C.E., Bleichrodt, N., van de Vijver, F.J.R., Ategbo, E.A., and Hautvast, J.G.A.T. 2000 Improved iodine status is associated with improved mental performance of schoolchildren in Benin. *American Journal of Clinical Nutrition* 72:1179–1185.

van Dillen, J., de Francisco, A., and Overweg-Plandsoen, W.C. 1996. Long-term effect of vitamin A with vaccines [letter]. *Lancet* 347:1705.

Waber, D.P., Vuori-Christiansen, L., Ortiz, N., Clement, J.R., Christiansen, N.E., Mora, J.O., Reed, R.B., and Herrera, M.G. 1981. Nutritional supplementation, maternal education, and cognitive development of infants at risk of malnutrition. *American Journal of Clinical Nutrition* 34:807–813.

Walker, S.P., Chang, S.M., Powell, C.A., and Grantham-McGregor, S.M. 2005. Effects of early childhood psychosocial stimulation and nutritional supplementation on cognition and education in growth-stunted Jamaican children: prospective cohort study. *Lancet* 366:1804–1807.

Walker, S.P., Chang, S.M., Powell, C.A., Simonoff, E., and Grantham-McGregor, S.M. 2006. Effects of psychosocial stimulation and dietary supplementation in early childhood on psychosocial functioning in late adolescence: follow-up of randomised controlled trial. *British Medical Journal* 333:472–474.

Walker, S.P., Chang, S.M., Powell, C.A., Simonoff, E., and Grantham-McGregor, S.M. 2007. Early childhood stunting is associated with poor psychological functioning in late adolescence and effects are reduced by psychosocial stimulation. *Journal of Nutrition* 137:2464–2469.

Waterlow, J.C. 1992. *Protein Energy Malnutrition.* London: Edward Arnold, pp. 187–203.

Wigg, N.R., Tong, S., McMichael, A.J., Baghurst, P.A., Vimpani, G., and Roberts, R. 1998. Does breastfeeding at 6 months predict cognitive development? *Australian and New Zealand Journal of Public Health* 22:232–236.

Williams, J., Wolff, A., Daly, A., MacDonald, A., Aukett, A., and Booth, I.W. 1999. Iron supplemented formula milk related to reduction in psychomotor decline in infants for inner city areas: randomised study. *British Medical Journal* 318:693–698.

World Health Organization. 2003. *Global Strategy for Infant and Young Child Feeding.* Geneva: WHO.

WHO Multicentre Growth Reference Study Group. 2006. Assessment of differences in linear growth among populations in the WHO Multicentre Growth Reference Study. *Acta Paediatrica* 450 suppl:56–65.

Zinc Investigators Collaborative Group. 1999. Prevention of diarrhea and pneumonia by zinc supplementation in children in developing countries: Pooled analysis of randomized controlled trials. *Journal of Pediatrics* 135:689–697.

12 Animal Source Foods: Effects on Nutrition and Function in Children in Developing Countries

Monika Grillenberger

CONTENTS

ABSTRACT

The prevalence of micronutrient deficiencies is high in children in developing countries. The majority of them live on a diet high in cereals, tubers, and legumes but low in animal source foods. Meat and milk are good sources of protein and are likely to be the only unfortified foods that can provide enough vitamin B_{12}, calcium, iron, and zinc that are essential for optimal growth, health, and cognitive and psychomotor development in children. A number of observational studies in developed and developing countries have shown associations between a low intake of animal source foods and low intakes of important nutrients, impaired growth, and health. Positive associations of the consumption of animal source foods with improved growth, cognitive function, activity, and school performance in young children have been found in several cross-sectional studies in developing countries. The supplementation of rural Kenyan schoolchildren with a snack containing meat almost eliminated their previously high vitamin B_{12} deficiency and increased their lean body mass. The children receiving the supplement with meat were also more cognitively able, physically active, and social initiating than children provided with snacks that did not contain animal source foods. The provision of milk improved height gain, particularly in those children who were stunted. In addition to providing considerable amounts of easily available heme iron, meat tissue is known to have an enhancing effect on nonheme iron absorption from other food components in the same meal. Several studies have shown that the inclusion of meat alone or in addition to other enhancers of iron bioavailability can improve nonheme iron absorption and as a consequence helps reduce iron deficiency. An increase in the consumption of animal source foods by children in developing countries offers a viable food-based approach to increase the intake of highly bioavailable nutrients and improve their functional outcome.

INTRODUCTION

Micronutrient malnutrition is widely prevalent among children in developing countries and is probably the main nutritional problem in the world posing important public health problems with long-term effects on human capital and national economic growth. Besides iron deficiency anemia, vitamin A deficiency and iodine deficiency disorders, deficiencies of zinc, vitamin B_{12}, folate, and others are increasingly recognized, and many population groups do not suffer from single, but from simultaneous deficiencies of multiple nutrients (ACC Sub-Committee on Nutrition 2000), which can impair growth and immunity with an increased risk of morbidity, mortality, and poor cognitive development, school performance, and physical activity. Maternal micronutrient malnutrition during pregnancy and lactation impairs not only maternal health but also neonatal growth and health.

Micronutrient deficiencies that often coexist with protein energy malnutrition are caused not only by high morbidity rates and insufficient food quantity, but also by the low quality of the diet common in many areas in developing countries. Diets predominantly consist of cereals, tubers, and legumes, with varying amounts of legumes, vegetables, and fruits grown on small farm holdings. Animal source foods, such as meat, fish, fowl, or eggs, are compact and efficient sources of micronutrients, yet in developing countries, their availability is limited, and they are eaten infrequently, particularly by women and young children. The intake of protein from meat is usually extremely low, providing only 15% of dietary protein compared with around 60% in developed countries (Higgs and Pratt 1998). Animal milks are also consumed in low quantities, often in tea or in fermented form. This type of diet is associated with low intakes of several vitamins and minerals and poor mineral bioavailability.

Some but not all studies supplementing children with nutrients, particularly iron, zinc, and vitamin A, were successful in improving growth and health. However, to address multiple micronutrient deficiencies, approaches are needed that are able to improve the intake of all limiting micronutrients at the same time. An increase in the consumption of animal source foods by children in developing countries offers a viable food-based approach to increase the intake of highly bioavailable nutrients and improve growth, health, and function.

ANIMAL SOURCE FOODS IN THE NUTRITION OF CHILDREN

NUTRITIONAL VALUE OF ANIMAL SOURCE FOODS

Animal source foods, such as meat and milk, are nutrient-dense foods that provide protein of high biological value, energy, and fat and are likely to be the only unforti-fied foods that can provide enough calcium, iron, and zinc for infants and children. They are more energy dense than plant foods, as well as a good source of fat-soluble vitamins and essential fatty acids. Vitamin B_{12} requirements must be met by animal source foods because there is none in plants. Meat is a good source of high-quality protein, iron, zinc, vitamin B_{12}, niacin, vitamin B_6, and heme iron. Furthermore meat, poultry, and fish contain additional other nutrients important for optimal health, such as riboflavin, taurine, selenium, and the long-chain polyunsaturated fatty acids, pen-taenoic and hexaenoic acids (Neumann et al. 2002). Milk is high in calcium, phos-phorus, and also contains appreciable amounts of preformed vitamin A, vitamin B_{12}, riboflavin, folate, vitamin D, zinc, and small amounts of iron. The bioavailability of micronutrients is generally higher from animal source foods than from plant foods.

IMPORTANCE OF ANIMAL SOURCE FOODS IN CHILDREN'S DIETS

The utility of including animal source foods in a child's diet has been described by Neumann et al. (2002). A sample diet including maize and beans was compared to a diet including meat. To meet the average daily requirements for energy, iron, or zinc, a child would need to consume 1.7–2.0 kg of maize and beans in 1 day. To consume these amounts is far more than a child can tolerate, while the requirements could be met by 60 g (2 oz) of meat per day. Likewise, the inclusion of milk and milk products is necessary to meet calcium requirements, which can hardly be met by a cereal-based diet without animal source foods.

Several studies have been conducted among affluent populations to examine the adequacy of the diet of certain subgroups who avoid the consumption of animal source foods. Generally, a vegetarian diet is considered a healthy alternative to an omnivorous diet that is high in saturated fat and cholesterol and low in fiber (Phillips and Segasothy 1999, Hoffmann et al. 2001). Despite the low content of animal source foods in the diet and the apparent lower bioavailability of some minerals, the mineral status of most adult vegetarians appears to be adequate (Gibson 1994). A number of comparative studies of vegetarian and omnivorous children also showed no differ-ence in nutrient intake, nutrient status, and growth (Herbert 1985, Jacobs and Dwyer 1988, Tayter and Stanek 1989, Sabate et al. 1991, Nathan et al. 1997, American Academy of Pediatrics Committee on Nutrition 1998, Hebbelinck et al. 1999, Thane and Bates 2000, Leung et al. 2001). However, to meet nutritional requirements, con-siderable care must be taken for true vegan diets, which include no animal prod-ucts, especially for children who have higher energy and nutrient needs than adults. Children are also at a greater risk of nutrient deficiency, especially during periods of physiological stress and accelerated growth (Mangels and Messina 2001). Nutrients of concern for children raised on diets that do not contain any animal source foods are iron, zinc, iodine, vitamin B_{12}, vitamin D, and calcium (Institute of Medicine 2001).

Negative associations between the vegetarian or vegan diets and the health of children in developed countries have been found in some studies. Children in New Zealand who avoid drinking cows' milk had low dietary calcium intakes and poor bone health (Black et al. 2002). British vegan children were smaller and had lower body weights compared with nonvegan children (Sanders and Purves 1981). Lower rates of growth have also been reported in children reared on vegan (Sanders and Purves 1981, O'Connell et al. 1989) and macrobiotic diets (Sanders and Purves 1981, Dagnelie et al. 1994). A macrobiotic diet is in some ways similar to a diet characteristic of children in developing countries. It consists primarily of cereals (mainly rice), vegetables, legumes, marine algae, small amounts of cooked fruit, and occasionally fish. No meat or dairy products are used. Rickets were also observed in children reared on vegetarian (Curtis et al. 1983) and macrobiotic diets (Dagnelie et al. 1990). The Dutch infants consuming macrobiotic diets also had poorer nutritional status and were more likely to have deficiencies of riboflavin, vitamin B_{12}, and iron, with consequent anemia (Dagnelie et al. 1990). After 2 years of increased consumption of fish or dairy products, their linear growth velocity improved (Dagnelie et al. 1994). The same children followed up in adolescence showed impaired cobalamin status and low bone mineral density (Rutten et al. 2005). Adequate milk consumption in children has been associated with better bone density in adulthood (Hirota et al. 1992, Murphy et al. 1994, Teegarden et al. 1999). In Nepal, xerophthalmia in young children was less likely to occur if they had relatively high meat or fish intakes when they were 13 to 24 months of age (Gittelsohn et al. 1997).

ANIMAL SOURCE FOODS: FUNCTIONAL OUTCOMES IN CHILDREN IN DEVELOPING COUNTRIES

Diets of children from poorer regions in developing countries usually do not contain animal source foods, and parents do not have the choice to include a variety of healthy foods, fortified foods, or supplements into their children's diets as have families who live on vegan or vegetarian diets in developed countries. A large number of studies have been carried out investigating the relation of micronutrient deficiencies, particularly of iron, zinc, and vitamins A and B_{12} to functional outcomes in children. These are described in Chapters 10 and 11 of this book. The inclusion of animal source foods in diets gives the chance to increase the intake of iron, zinc, and vitamins A and B_{12} and prevents deficiencies and consequent health impairment. Several cross-sectional studies have shown that consumption of animal source foods is associated with improved nutritional status and growth among children in developing countries. Findings from the Human Nutrition Collaborative Support Program, a longitudinal observational study in Egypt, Kenya, and Mexico, suggest that low intake of animal protein is associated with low intakes of available

TABLE 12.1

Selected Micronutrient Contents of Meat and Milk (Relative Amount/kcal) and Functions Affected by Deficiency

	Iron	Zinc	Vitamin B_{12}	Vitamin A
Animal source foods				
Meat	+++	+++	+++	+
Milk	+	+	+++	++
Functional areas affected				
Anemia	+++	0	+++	+
Immunodeficiency	++	+++	+	+
Intrauterine malnutrition	+	++	−	0
Cognition	+++	0	++	0
Activity	+++	++	0	0
Work capacity	+++	0	0	0

Source: Modified from Neumann, C., D. Harris, and L. Rogers, 2002, *Nutrition Research* 22:193–220.

zinc, iron, and vitamin B_{12}, and that the intake of animal source foods is strongly associated with improved growth, cognitive function, activity, school performance, pregnancy outcome, and morbidity in young children (Neumann and Bwibo 1987, Allen et al. 1992b, Calloway et al. 1992). However, intervention studies that supplement animal source foods and examine the relation of their intake to functional outcomes in children are rare. Findings of studies that examined the importance of animal source foods in the diets of children and the impact of animal source foods on neonatal growth and health, micronutrient status, growth, cognitive development, school performance, physical activity, emotional state, and social interactions during free play are presented in more detail in the following sections. An overview of the relationship between meat and milk, micronutrients, and human function is given in Table 12.1.

NEONATAL GROWTH AND HEALTH

An inadequate diet during pregnancy and lactation with consequent micronutrient deficiencies, particularly of iron, zinc, and vitamin B_{12}, can impair pregnancy and postnatal outcomes of children. This relationship and the observations of some studies investigating the relationship of maternal intake of animal source foods during pregnancy and lactation have been reviewed by Neumann et al. (2002). It has been observed that maternal intake of animal source foods during pregnancy predicted gestational age, pregnancy weight gain, birth weight, and birth length, and a high-quality diet including animal source foods during pregnancy and lactation predicted postnatal growth.

Micronutrient Status

Micronutrient malnutrition, which is widely prevalent among children in developing countries, has detrimental effects on their health and functional outcome (see Chapters 10 and 11). In Kenyan schoolchildren, the high prevalence of low plasma vitamin B_{12} concentrations at the beginning of a food supplementation study was predicted by a low intake of animal source foods (McLean et al. 2007). Children received three isoenergetic food supplements that were based on maize and beans and contained meat or no animal source foods or were served with milk during school break. The snacks including meat or milk significantly increased plasma vitamin B_{12} concentrations after 1 and 2 years, but no significant effects on other indicators of micronutrient status were detected, possibly because of variation introduced by the presence of malaria and other infections (Siekmann et al. 2003, McLean et al. 2007). After 1 year, the prevalence of severe plus moderate vitamin B_{12} deficiency fell from 80.7 to 64.1% in the children receiving meat and from 71.6 to 45.1% in the children receiving milk. After 2 years, the supplementation with meat or milk almost completely eliminated low plasma vitamin B_{12} concentrations (McLean et al. 2007). The improved vitamin B_{12} status of children fed meat or milk may reduce the risk of consequences of vitamin B_{12} deficiency, such as megaloblastic anemia, poor motor and cognitive function, and subsequent poor school performance (Siekmann et al. 2003).

The children in the Kenyan study were severely depleted of vitamin A, assessed by plasma retinol concentrations or on the basis of low liver retinol stores. Plasma retinol improved during the first year of intervention but fell to even lower values than baseline by the end of the second year. The majority of children had depleted liver retinol after 2 years, which was unaffected by the food supplements (Zubieta et al. 2005).

Growth

There is a strong probability that growth is limited by multiple simultaneous deficiencies in children in many areas in developing countries. Micronutrient deficiencies may also contribute indirectly to growth retardation through anorexia or increased morbidity (Rivera et al. 2003). Stunting has detrimental consequences not only during childhood, but also lifelong, including low physical activity, impaired motor and mental development, lowered immunocompetence, reduced work productivity, greater severity of infections, and increased mortality (Norgan 2000). More details on the effects of nutrient deficiencies on stunting can be found in Chapter 11. Animal source foods provide high amounts of bioavailable micronutrients that are needed for optimal growth, such as iron, zinc, calcium, and vitamin A.

Stunted Jamaican children were found to consume significantly fewer servings of dairy products and fruits than the nonstunted children (Walker et al. 1990). The percentage of protein intake from animal sources was positively associated with growth in Peruvian children (Graham et al. 1981), and because the protein intakes were considered adequate, it was suggested that other nutrients contained in the animal products, such as heme iron, zinc, and vitamin B_{12}, may explain the observed effect. Linear growth was also positively associated with intake of animal source foods in another cohort of Peruvian children with low intakes of complementary foods (Marquis et al. 1997). Animal protein intake was correlated with height-for-age in Korean children, whereas fat intake was a more important factor for weight-for-age and weight-for-height (Paik et al. 1992). In a study in Latin American children, availability of dairy products, oils, and meats were negatively related to underweight, and protein, total fat, total energy, and animal fat were negatively related to stunting; it was concluded that animal source foods are important to support the normal growth of children (Uauy et al. 2000).

A number of controlled studies showed positive effects of supplementation with milk or milk products on children's weight (Malcolm 1970, Golden and Golden 1981, Lampl et al. 1987, Heikens et al. 1989, Walker et al. 1991, Schroeder et al. 1995), height (Malcolm 1970, Baker et al. 1980, Zumrawi et al. 1981, Lampl et al. 1987, Heikens et al. 1989, Walker et al. 1991, Schroeder et al. 1995), and bone health (Köhler et al. 1984, Lampl et al. 1987, Bonjour et al. 1997, Cadogan et al. 1997, Merrilees et al. 2000), whereas others did not show any effect on growth (Elwood et al. 1981, Köhler et al. 1984, Chan et al. 1995). However, interpretation of findings from these studies is complicated by the inability to accurately determine if increases in energy, protein, or micronutrients were responsible for the outcome observed. Controlled supplementation studies with other animal source foods than milk are scarce. A trial in which dry fish powder was added to fermented maize porridge did not improve growth or micronutrient status of Ghanaian children (Lartey et al. 1999).

A controlled food supplementation study in Kenyan schoolchildren showed that weight gain was significantly higher in children receiving any of three isoenergetic food supplements (with or without animal source foods). The supplementation with the snack plus milk resulted in improved height gain in children who were more stunted, probably because there is more scope for improvement in their growth. Children receiving the meat supplement gained more mid-upper-arm muscle area (MMA) (i.e., lean body mass) than those in the other groups. These observed positive effects of the food supplementation with meat could have been due to the higher intakes of complete protein or bioavailable zinc and iron. Zinc supplementation has been shown in some studies to increase lean body mass in children, possibly through its effects on protein metabolism or stimulation of appetite (Grillenberger et al. 2003, Neumann et al. 2007). Iron and protein are essential to myoglobin synthesis in striated muscle (Neumann et al. 2007). Children who received the snack with meat were more active than the other children, and since a positive association was found between MMA and percentage of time spent in high levels of physical activity in children who received the meat supplement (Neumann et al. 2007), this could be another explanation why the children in the meat group gained more muscle mass than the other children. No effect of the food supplements was found on measures of body fat, probably because no extra energy was available to be stored (Grillenberger et al. 2003).

The study in Kenyan children further showed that growth was positively predicted by energy and nutrients that are provided in high amounts and a bioavailable form in meat and milk, such as heme iron, preformed vitamin A, calcium, and vitamin B_{12}. In contrast, nutrients predominantly found in plant foods and dietary components that inhibit micronutrient absorption, such as fiber and phytate, negatively predicted the children's growth (Grillenberger et al. 2006).

COGNITIVE FUNCTION AND SCHOOL PERFORMANCE

Malnutrition can impair cognitive development, school performance, and physical activity, which has a negative impact on productivity, not only of individuals, but also collectively, of societies and whole nations, particularly in developing countries and among disadvantaged communities in affluent nations (Neumann et al. 2002). More details on the effects of nutrient deficiencies on cognitive function can be found in Chapter 11. Animal source foods are a good source of bioavailable nutrients, such as iron, zinc, and vitamin B_{12}, needed for optimal cognitive function, activity, attendance, and school performance. Findings from the Human Nutrition Collaborative Support Program, a longitudinal observational study in Egypt, Kenya, and Mexico on the relation of intake of animal source foods and child development, indicated that the mother's intake of animal source foods had positive effects on infant alertness, newborn's orientation, and habituation behavior. Animal source foods positively predicted children's developmental outcomes, behavior, verbal ability, and involvement in classroom activities (Neumann et al. 2002).

Findings from another study in Kenya showed that supplementation with food, including animal source foods, had positive effects on the children's cognitive performance (Whaley et al. 2003). However, these effects differed across domains of cognitive functioning, and the different types of supplements did not show the same beneficial effects. Children receiving a snack with meat significantly performed better on the Raven's Progressive Matrices than any other group. The Raven's is widely used as a culturally reduced test of fluid intelligence, thus tapping into on-the-spot reasoning and problem-solving ability as opposed to accumulated factual knowledge. This suggests that increasing energy intake alone is not sufficient for improving cognitive performance, and that the quality of the diet is important. In addition, meat and milk supplementation do not appear to be interchangeable—the snack containing meat had a greater impact than the supplement with milk in problem-solving ability. In contrast, supplementation of any kind had no impact on verbal performance, a skill that is generally believed to illustrate the accumulation of factual knowledge and is in the category of crystallized intelligence. The results of the intervention study indicate that both diet quality and diet quantity are important predictors of arithmetic performance as children supplemented with meat or energy performed better than the children who did not receive supplementation. In addition, children supplemented with energy performed better than the children supplemented with milk. These findings confirm results of past studies that suggested that animal source foods were an important correlate of cognitive performance, but this was the first experimental demonstration of the efficacy of meat supplementation for child cognitive performance (Whaley et al. 2003). The improved cognitive performance in the children receiving meat may be linked to greater intake of vitamin B_{12} and more available iron and zinc as a result of the presence of meat, which in addition to providing iron and zinc increases their absorption from fiber and phytate-rich plant staples. Meat, through its intrinsic micronutrient content and other constituents and high-quality protein, may facilitate specific mechanisms, such as speed of information processing, that are involved in learning tasks such as problem-solving capacity. Children receiving the snack with milk performed the poorest in the test of problem-solving ability. A possible explanation is that milk, with its high casein and calcium contents, impedes iron absorption—iron is intimately involved with cognitive function (Neumann et al. 2007).

For school performance, as measured by end-of-term test scores, the children who received the snack with meat had the greatest percentage increase in zonal end-term total test scores. The greatest percentage increase in arithmetic subtest scores was also seen in these children (Neumann et al. 2007).

PHYSICAL ACTIVITY, EMOTIONAL STATE, AND SOCIAL INTERACTIONS DURING FREE PLAY

The Human Nutrition Collaborative Support Program, a longitudinal observational study in Egypt, Kenya, and Mexico, found that children who were better nourished (in terms of both energy and animal source foods) were more active and happy and showed more leadership behavior during free play, whereas poorly nourished children appeared more anxious on the school playground (Espinosa et al. 1992). Mexican boys who consumed poor-quality diets were apathetic in the classroom (Allen et al. 1992a). Contrary to cognitive abilities, which were more highly associated with the level of intake of animal source food than with overall energy intake, a diet providing sufficient energy seemed more important than diet quality for activity level, positive emotion, and leadership on the playground (Sigman et al. 2005).

The results of a food supplementation study in Kenyan children show that both diet quantity and quality are important for children's activities on the school playground. Children who received any type of snack during school break (with or without animal source foods) were more active and showed more leadership behavior and social initiation behaviors during playing than did children who did not receive extra food. However, children who received a snack containing meat showed the greatest increase in percentage time in high levels of physical activity and in initiative and leadership behaviors compared with children who received a vegetarian snack with equal levels of energy but less protein, iron, and zinc. Children who received a snack with milk performed the most poorly of the three intervention groups, probably because it provided less of the important nutrients for child behavior than did the meat supplement (Sigman et al. 2005).

MEAT CONSUMPTION AND DIETARY IRON BIOAVAILABILITY

Iron deficiency anemia is highly prevalent in developing countries (ACC Sub-Committee on Nutrition 2000) and can be partly attributed to the low iron bioavailability in the customary cereal and legume-based diet. *Bioavailability* is defined as the amount of a nutrient that is potentially available for absorption from a meal and, once absorbed, utilizable for metabolic processes and storage in the body. Iron status, the content of heme and nonheme iron, and the bioavailability of the two kinds of iron determine the amount of iron absorbed from a meal (Hallberg 1981). About 40% of the iron in meat is heme iron. Its absorption is not greatly influenced by other dietary components present in the meal, so that 15 to 35% is absorbed. In contrast, the absorption of nonheme iron found in plant sources can vary between 2 and 35%, depending on the presence of enhancers and inhibitors in a meal as well as an individual's iron status (Monsen et al. 1978, Dwyer 1991, Hallberg and Hulthén 2000). The most important inhibitors of iron absorption are phytates, polyphenols, and calcium. Ascorbic acid is probably the most efficient enhancer of nonheme iron absorption (Lynch 1997, Hallberg and Hulthén 2000). The absolute amount of ascorbic acid in the meal and the ratio between the concentration of ascorbic acid and iron absorption inhibitors may be more important than the molar ratio of ascorbic acid to iron (Lynch and Stoltzfus 2003), and sufficient amounts of ascorbic acid in a meal can counteract the inhibition of iron absorption by phytates (Siegenberg et al. 1991).

In addition to providing considerable amounts of highly bioavailable heme iron, meat tissue is known to have an enhancing effect on nonheme iron absorption from other food components in the same meal, even in the presence of dietary inhibitors (Hallberg and Hulthén 2000). Since this effect was first noted by Layrisse et al. (1968), there have been numerous studies on the effect of meat, fish, and poultry on iron absorption; however, the magnitude of the effect and the mechanisms involved have not yet been conclusively resolved (Hallberg and Hulthén 2000). It had been suggested that the so-called meat factor could be a protein per se, certain peptides or amino acids, especially those containing cysteine, or their metabolites or unidentified components in proteinaceous foods (Taylor et al. 1986). However, more recent findings indicate that protein and sulfhydryl groups from cysteine residues are not contributing to iron absorption (Bæch et al. 2003). It has also been suggested that meat factors stimulate gastric acid secretion and may chelate solubilized iron in the acid environment of the stomach, thereby maintaining iron solubility during intestinal digestion and absorption (Kim et al. 1993). Meat very effectively counteracts the inhibition of nonheme iron absorption by phytate and polyphenols (Lynch 1997, Hallberg and Hulthén 2000), and it seems that the inhibitory effects of phytate on mineral absorption are not seen in varied diets containing sufficient amounts of animal protein (Zheng et al. 1993). The bioavailability of nutrients in meat and milk are not equivalent. If meat and milk are consumed together in one meal, the calcium and casein in milk can form insoluble complexes with iron and zinc in meat and decrease their bioavailability (Neumann et al. 2002).

An increase in iron absorption through the addition of meat or ascorbic acid to meals has been shown in several studies. The addition of 50 and 75 g pork meat to a phytate-rich meal was found to increase nonheme iron absorption by 44% and 57%, respectively (Bæch et al. 2003). Iron absorption was increased by 85% if 20 g meat were added to a weaning meal of whole wheat gruel (Hallberg et al. 2003). An enhancing effect of about 140 and 165% on nonheme iron absorption was observed by adding 50 mg ascorbic acid or 75 g meat, respectively, to a simple Latin-American-type meal of maize, rice, and beans (Hallberg and Rossander 1984). In another study, in which 50 mg ascorbic acid were added to wheat rolls with no detectable phytate, an increase in iron absorption of 75% was found (Hallberg et al. 1989).

Besides increasing enhancers of iron absorption, such as meat or ascorbic acid, household dietary strategies to improve iron nutrition also focus on decreasing inhibitors, such as phytates. Food-processing methods that can be applied at the household level to reduce phytates include physical removal (extraction and dehulling) and enzymatic degradation by soaking, germination, malting, and fermentation. Some studies in developing countries testing those strategies have been able to demonstrate improvements in micronutrient intake (Gibson and Hotz 2001, Yeudall et al. 2002, Gibson et al. 2003). The consumption of polyphenol-containing drinks, particularly tea and coffee, with a meal is known to inhibit iron absorption (Disler et al. 1975, Bravo 1998, Prystai et al. 1999), and because this practice is common in many areas in developing countries, a behavior change might contribute to an improvement in iron nutrition. However, when these household dietary strategies were examined through simulations in a group of Kenyan schoolchildren, the application of household-level food-processing methods or the avoidance of polyphenol-containing drinks at meals was not found to decrease the prevalence of inadequate iron intake. The combined addition of meat and ascorbic acid to a meal seems to be more efficacious in reducing the prevalence of inadequate iron intake in these children (Grillenberger 2006).

IMPLICATIONS

Multiple micronutrient deficiencies seem to be the norm rather than the exception in rural areas in developing countries. When choosing a suitable nutrition intervention strategy to address micronutrient malnutrition, it therefore has to be considered that functional deficits may not be alleviated by the provision of single micronutrients. Study findings indicate that micronutrients contained in high amounts and in a bioavailable form in animal source foods are beneficial for micronutrient status, growth, cognitive function, and behavior. Further, the addition of meat alone or combined with ascorbic acid seems to be an efficacious approach to improve iron bioavailability. The promotion of animal source food consumption by children therefore seems to be a viable food-based approach to provide highly bioavailable nutrients simultaneously to children and at the same time to improve iron bioavailability.

Obviously, food-based approaches, such as an increase in the consumption of animal source foods, take time and are complex as they require interdisciplinary approaches (Demment et al. 2003). However, in contrast to pharmanutrient supplementation, they are not "top-down" approaches, but they involve the target population and are likely to be the most effective means of addressing the problem at its source. If locally available and familiar foods and preparation methods are used, the food-based approach is likely to be accepted by the population and therefore sustainable. It may also be more economically feasible without the risk of antagonistic interactions than, for example, pharmanutrient supplementation (Gibson and Ferguson 1998, Allen and Gillespie 2001). Food-based approaches are rather long-term solutions aimed at the prevention of micronutrient deficiencies than a short-term correction of the problem and can lead to more diverse diets with an improvement of the overall diet quality. Another advantage is the potential for households to benefit economically from increased production of high-value foods (Allen and Gillespie 2001). The question of how feasible it is for low-income households in developing countries to increase production and consumption of animal source foods is addressed in Chapter 24.

Despite the advantages that a food-based approach including animal source foods in the prevention of micronutrient deficiencies might have, there are also some critical aspects associated with an increase in the consumption of animal source foods. An increase in nutrition-related noncommunicable diseases such as diabetes, cancer, and cardiovascular disease is now apparent in large segments of many developing countries that are undergoing the so-called nutrition transition with changing lifestyle and dietary patterns that include increased meat consumption (Popkin 1994). However, the view of some researchers is that, particularly in low-income countries, the contribution of meat to improved nutrient intakes more than offsets the uncertain association with these diseases (Biesalski 2002, Hill 2002). In places where there is a choice of a large variety of healthy foods, a vegetarian diet has been associated with several health benefits and is definitely a worthwhile alternative, particularly if such a diet includes eggs and dairy products. However, diets that exclude animal source foods are not recommended for small children due to the energetic demands of their rapidly expanding large brain and generally high metabolic and nutritional demands relative to adults (Biesalski 2002, Milton 2003). Given that the diets of children in developing countries are generally very low in fat, the dietary fat provided by animal source foods is even advantageous because it is a concentrated source of energy and enhances absorption of fat-soluble nutrients. Further, it is not likely that any additional fat provided by animal foods would result in fat intakes that exceed current recommendations (Haskell and Brown 1997). However, the potential adverse health effects linked with an increased intake of animal source foods should not be ignored by policy makers who focus on livestock promotion (Popkin and Du 2003).

CONCLUSIONS

- Deficiencies of multiple micronutrients are widely prevalent among children in developing countries and are partly caused by poor diet quality.
- Vegan diets are not suitable for infants and children and can impair their health.
- Animal source foods are nutrient-dense foods that provide protein of high biological value, energy, fat, and micronutrients, such as highly bioavailable iron and zinc, calcium, vitamin A, and vitamin B_{12}.
- Positive associations between the intake of animal source foods and neonatal growth and health, improved micronutrient status, growth, cognitive function, physical activity, emotional state, social interactions, and school performance in children in developing countries have been found in several studies.
- The promotion of animal source food consumption by children in developing countries seems to be a viable food-based approach to provide highly bioavailable nutrients simultaneously to children and to improve their health and development.

REFERENCES

ACC Sub-Committee on Nutrition. 2000. *Fourth Report on the World Nutrition Situation*. Geneva: United Nations Administrative Committee on Coordination, Sub-Committee on Nutrition, and International Food Policy Research Institute.

Allen, L.H., J.R. Backstrand, A. Chávez, and G.H. Pelto. 1992a. *People Cannot Live by Tortillas Alone. The Results of the Mexico Nutrition CRSP*. USAID, University of Connecticut and Instituto Nacionál de la Nutrición Salvador Zubirán, Washington, DC.

Allen, L.H., J.R. Backstrand, E.J.I. Stanek, G.H. Pelto, A. Chávez, E. Molina, J.B. Castillo, and A. Mata. 1992b. The interactive effects of dietary quality on the growth of young Mexican children. *American Journal of Clinical Nutrition* 56:353–364.

Allen, L.H. and S.R. Gillespie. 2001. *What works? A Review of the Efficacy and Effectiveness of Nutrition Interventions*. Manila: ACC/SCN, Geneva, in collaboration with the Asian Development Bank.

American Academy of Pediatrics Committee on Nutrition. 1998. Soy protein-based formulas: recommendations for use in infant feeding. *Pediatrics* 101:148–153.

Bæch, S.B., M. Hansen, K. Bukhave, M. Jensen, S.S. Sorensen, L. Kristensen, P.P. Purslow, L.H. Skibsted, and B. Sandstrom. 2003. Nonheme-iron absorption from a phytate-rich meal is increased by the addition of small amounts of pork meat. *American Journal of Clinical Nutrition* 77(1):173–179.

Baker, I., P. Elwood, J. Hughes, M. Jones, F. Moore, and P. Sweetnam. 1980. A randomized controlled trial of the effect of the provision of free school milk on the growth of children. *Journal of Epidemiology and Community Health* 34:31–34.

Biesalski, H. 2002. Meat and cancer: meat as a component of a healthy diet. *European Journal of Clinical Nutrition* 56:S2–S11.

Black, R., S. Williams, I. Jones, and A. Goulding. 2002. Children who avoid drinking cow milk have low dietary calcium intakes and poor bone health. *American Journal of Clinical Nutrition* 76(3):675–680.

Bonjour, J., A. Carrie, S. Ferrari, H. Clavien, D. Slosman, G. Theintz, and R. Rizzoli. 1997. Calcium-enriched foods and bone mass growth in prepubertal girls: a randomised, double-blind, placebo-controlled trial. *Journal of Clinical Investigation* 99:1287–1294.

Bravo, L. 1998. Polyphenols: chemistry, dietary sources, metabolism, and nutritional significance. *Nutrition Reviews* 56:317–33.

Cadogan, J., R. Eastell, N. Jones, and M. Barker. 1997. Milk intake and bone mineral acquisition in adolescent girls: randomised, controlled intervention trial. *British Medical Journal* 315:1255–1260.

Calloway, D.H., S. Murphy, O. Balderston, O. Receveur, D. Lein, and M. Hudes. 1992. *Final Report: Functional Implications of Malnutrition, Across NCRSP Projects*. USAID, Washington, DC.

Chan, G., K. Hoffman, and M. McMurry. 1995. Effects of dairy products on bone and body composition in pubertal girls. *Journal of Pediatrics* 126:551–556.

Curtis, J.A., S.W. Kooh, D. Fraser, and M.L. Greenberg. 1983. Nutritional rickets in vegetarian children. *Canadian Medical Association Journal* 128(2):150–152.

Dagnelie, P.C., M. van Dusseldorp, W.A. van Staveren, and J.G.A.J. Hautvast. 1994. Effects of macrobiotic diets on linear growth in infants and children until 10 years of age. *European Journal of Clinical Nutrition* 48:S103–S112.

Dagnelie, P., F. Vergote, W. Van Staveren, H. van den Berg, P. Dingjan, and J. Hautvast. 1990. High prevalence of rickets in infants on macrobiotic diets. *American Journal of Clinical Nutrition* 51:202–208.

Demment, M.W., M.M. Young, and R.L. Sensenig. 2003. Providing micronutrients through food-based solutions: a key to human and national development. *Journal of Nutrition* 133(11):3879S–3885S.

Disler, P.B., S.R. Lynch, J.D. Torrance, M.H. Sayers, T.H. Bothwell, and R.W. Charlton. 1975. The mechanism of the inhibition of iron absorption by tea. *The South African Journal of Medical Sciences* 40:109–116.

Dwyer, J. 1991. Nutritional consequences of vegetarianism. *Annual Review of Nutrition* 11:61–91.

Elwood, P., T. Haley, S. Hughes, P. Sweetnam, O. Gray, and D. Davies. 1981. Child growth (0–5 years), and the effect of entitlement to a milk supplement. *Archives of Disease in Childhood* 56:831–835.

Espinosa, M.P., M.D. Sigman, C.G. Neumann, N.O. Bwibo, and M.A. McDonald. 1992. Playground behaviors of school-age children in relation to nutrition, schooling, and family characteristics. *Developmental Psychology* 28(6):1188–1195.

Gibson, R. 1994. Zinc nutrition in developing countries. *Nutrition Research Reviews* 7:151–73.

Gibson, R. and E. Ferguson. 1998. Nutrition intervention strategies to combat zinc deficiency in developing countries. *Nutrition Research Reviews* 10:1–18.

Gibson, R. and C. Hotz. 2001. Dietary diversification/modification strategies to enhance micronutrient content and bioavailability of diets in developing countries. *British Journal of Nutrition* 85(Suppl 2):159S–66S.

Gibson, R.S., F. Yeudall, N. Drost, B.M. Mtitimuni, and T.R. Cullinan. 2003. Experiences of a community-based dietary intervention to enhance micronutrient adequacy of diets low in animal source foods and high in phytate: a case study in rural Malawian children. *Journal of Nutrition* 133(11):3992S–3999S.

Gittelsohn, J., A. Shankar, K. West, R. Ram, C. Dhungel, and B. Dahal. 1997. Infant feeding practices reflect antecedent risk of xerophthalmia in Nepali children. *European Journal of Clinical Nutrition* 51:484–490.

Golden, B.E. and M.H.N. Golden. 1981. Plasma zinc, rate of weight gain and energy cost of tissue deposition in children recovering from severe malnutrition on cow's milk or soya protein based diet. *American Journal of Clinical Nutrition* 34:892–899.

Graham, G., H. Creed, W. MacLean, C. Kallman, J. Rabold, and E. Mellits. 1981. Determinants of growth among poor children: nutrient intake-achieved growth relationships. *American Journal of Clinical Nutrition* 34:539–54.

Grillenberger, M. 2006. Impact of animal source foods on growth, morbidity and iron bioavailability in Kenyan school children. Thesis/dissertation, Wageningen University, Netherlands.

Grillenberger, M., C.G. Neumann, S.P. Murphy, N.O. Bwibo, P. van't Veer, J.G.A.J. Hautvast, and C.E. West. 2003. Food supplements have a positive impact on weight gain and the addition of animal source foods increases lean body mass of Kenyan schoolchildren. *Journal of Nutrition* 133(11):3957S–3964S.

Grillenberger, M., C.G. Neumann, S.P. Murphy, N.O. Bwibo, R.E. Weiss, J.G.A.J. Hautvast, and C.E. West. 2006. Intake of micronutrients high in animal source foods is associated with better growth in rural Kenyan school children. *British Journal of Nutrition* 95:379–390.

Hallberg, L. 1981. Bioavailability of dietary iron in man. *Annual Review of Nutrition* 1:123–147.

Hallberg, L., M. Brune, and L. Rossander. 1989. Iron absorption in man: ascorbic acid and dose-dependent inhibition by phytate. *American Journal of Clinical Nutrition* 49:140–144.

Hallberg, L., M. Hoppe, M. Andersson, and L. Hulthén. 2003. The role of meat to improve the critical iron balance during weaning. *Pediatrics* 111:864–870.

Hallberg, L. and L. Hulthén. 2000. Prediction of dietary iron absorption: an algorithm for calculating absorption and bioavailability of dietary iron. *American Journal of Clinical Nutrition* 71:1147–1160.

Hallberg, L. and L. Rossander. 1984. Improvement of iron nutrition in developing countries: comparison of adding meat, soy protein, ascorbic acid, citric acid, and ferrous sulphate on iron absorption from a simple Latin American-type of meal. *American Journal of Clinical Nutrition* 39(4):577–583.

Haskell, M.J. and K.H. Brown. 1997. The nutritional value of animal products and their role in the growth and development of young children in developing countries [unpublished review]. Davis: University of California.

Hebbelinck, M., P. Clarys, and A. De Malsche. 1999. Growth, development, and physical fitness of Flemish vegetarian children, adolescents, and young adults. *American Journal of Clinical Nutrition* 70(Suppl):579S–585S.

Heikens, G.T., W.N. Schofield, S. Dawson, and S. Grantham-McGregor. 1989. The Kingston Project. I. Growth of malnourished children during rehabilitation in the community, given a high energy supplement. *European Journal of Clinical Nutrition* 43:145–160.

Herbert, J. 1985. Relationship of vegetarianism to child growth in South India. *American Journal of Clinical Nutrition* 42:1246–1254.

Higgs, J. and J. Pratt. 1998. Meat, poultry and meat products. In *Encyclopedia of Human Nutrition*, M.J. Sadler, J.J. Strain, and B. Caballero, Eds. San Diego, CA: Academic Press, p. 1275.

Hill, M. 2002. Meat, cancer and dietary advice to the public. *European Journal of Clinical Nutrition* 56:36S–41S.

Hirota, T., M. Nara, M. Ohguri, E. Manago, and K. Hirota. 1992. Effect of diet and lifestyle on bone mass in Asian young women. *American Journal of Clinical Nutrition* 55:1168–1173.

Hoffmann, I., M. Groeneveld, H. Boeing, C. Koebnick, S. Golf, N. Katz, and C. Leitzmann. 2001. Giessen Wholesome Nutrition Study: relation between a health-conscious diet and blood lipids. *European Journal of Clinical Nutrition* 55(10):887–895.

Institute of Medicine. 2001. *Dietary Reference Intakes for Vitamin A, Vitamin K, Arsenic, Boron, Chromium, Copper, Iodine, Iron, Manganese, Molybdenum, Nickel, Silicon, Vanadium and Zinc*. Washington, DC: National Academy Press.

Jacobs, C. and J. Dwyer. 1988. Vegetarian children: appropriate and inappropriate diets. *American Journal of Clinical Nutrition* 48(3, Suppl):811–818.

Kim, Y., C.E. Carpenter, and A.W. Mahony. 1993. Gastric acid production, iron status, and dietary phytate alter enhancement by meat of iron absorption in rats. *Journal of Nutrition* 123:940–946.

Köhler, L., G. Meeuwisse, and W. Mortensson. 1984. Food intake and growth of infants between six and twenty-six weeks of age on breast milk, cow's milk formula, or soy formula. *Acta Paediatrica Scandinavica* 73:40–48.

Lampl, M., F. Johnston, and L. Malcolm. 1987. The effects of protein supplementation on the growth and skeletal maturation of New Guinean school children. *Annals of Human Biology* 5:219–227.

Lartey, A., A. Manu, K.H. Brown, J.M. Peerson, and K.G. Dewey. 1999. A randomized, community-based trial of the effects of improved, centrally processed complementary foods on growth and micronutrient status of Ghanaian infants from 6 to 12 months of age. *American Journal of Clinical Nutrition* 70(3):391–404.

Layrisse, M., C. Martinez-Torres, and M. Roch. 1968. Effect of interaction of various foods on iron absorption. *American Journal of Clinical Nutrition* 21:1175–1183.

Leung, S., R. Lee, R. Sung, H. Luo, C. Kam, M. Yuen, M. Hjelm, and S. Lee. 2001. Growth and nutrition of Chinese vegetarian children in Hong Kong. *Journal of Paediatrics and Child Health* 37(3):247–253.

Lynch, S.R. 1997. Interaction of iron with other nutrients. *Nutrition Reviews* 55:102–110.

Lynch, S.R. and R.J. Stoltzfus. 2003. Iron and ascorbic acid: proposed fortification levels and recommended iron compounds. *Journal of Nutrition* 133:2978S–2984S.

Malcolm, L. 1970. Growth retardation in a New Guinea boarding school and its response to supplementary feeding. *British Journal of Nutrition* 24:297–305.

Mangels, A. and V. Messina. 2001. Considerations in planning vegan diets: infants. *Journal of The American Dietetic Association* 101(6):607–607.

Marquis, G., J. Habicht, C. Lanata, R. Black, and K. Rasmussen. 1997. Breast milk or animal-product foods improve linear growth of Peruvian toddlers consuming marginal diets. *American Journal of Clinical Nutrition* 66(5):1102–1109.

McLean, F.C., L.H. Allen, C.G. Neumann, J.M. Peerson, J.H. Siekmann, S.P. Murphy, N.O. Bwibo, and M.W. Demment. 2007. Low plasma vitamin B-12 in Kenyan school children is highly prevalent and improved by supplemental animal source foods. *Journal of Nutrition* 137:676–682.

Merrilees, M.J., E.J. Smart, N.L. Gilchrist, C. Frampton, J.G. Turner, E. Hooke, R.L. March, and P. Maguire. 2000. Effects of dairy food supplements on bone mineral density in teenage girls. *European Journal of Clinical Nutrition* 39:256–262.

Milton, K. 2003. The critical role played by animal source foods in human (homo) evolution. *Journal of Nutrition* 133(11):3886S–3892S.

Monsen, E.R., L. Hallberg, M. Layrisse, M. Hegsted, J.D. Cook, W. Mertz, and C.A. Finch. 1978. Estimation of available dietary iron. *American Journal of Clinical Nutrition* 31:134–41.

Murphy, S., K.-T. Khaw, H. May, and J. Compston. 1994. Milk consumption and bone mineral density in middle aged and elderly women. *British Medical Journal* 308:939–941.

Nathan, I., A. Hackett, and S. Kirby. 1997. A longitudinal study of the growth of matched pairs of vegetarian and omnivorous children, aged 7–11 years, in the north-west of England. *European Journal of Clinical Nutrition* 51:20–25.

Neumann, C.G. and N.O. Bwibo. 1987. *Final Report: Food Intake and Human Function, Kenya Project*. Human Collaborative Support Program, USAID, Office of Nutrition. Washington, DC.

Neumann, C., D. Harris, and L. Rogers. 2002. Contribution of animal source foods in improving dietary quality and function in children in the developing world. *Nutrition Research* 22:193–220.

Neumann, C.G., S.P. Murphy, C. Gewa, M. Grillenberger, and N.O. Bwibo. 2007. Meat supplementation improves growth, cognitive, and behavioral outcomes in Kenyan children. *Journal of Nutrition* 137:1119–1123.

Norgan, N.G. 2000. Long-term physiological and economic consequences of growth retardation in children and adolescents. *Proceedings of the Nutrition Society* 59(2):245–256.

O'Connell, J., M. Dibley, J. Sierra, B. Wallace, J. Marks, and R. Yip. 1989. Growth of vegetarian children: the Farm Study. *Pediatrics* 84(3):475–481.

Paik, H., S. Hwang, and S. Lee. 1992. Comparative analysis of growth, diet, and urinary N excretion in elementary school children from urban and rural areas of Korea. *International Journal of Vitamin and Nutrition Research* 62:83–90.

Phillips, P. and M. Segasothy. 1999. Commentary. Vegetarian diet: panacea for modern lifestyle diseases? *QJM: Monthly Journal of the Association of Physicians* 92(9):531–544.

Popkin, B.M. 1994. The nutrition transition in low-income countries: an emerging crisis. *Nutrition Reviews* 52:285–298.

Popkin, B.M. and S. Du. 2003. Dynamics of the nutrition transition toward the animal foods sector in China and its implications: a worried perspective. *Journal of Nutrition* 133(11):3898S–3906S.

Prystai, E.A., C.V. Kies, and J.A. Driskell. 1999. Calcium, copper, iron, magnesium and zinc utilization of humans as affected by consumption of black, decaffeinated black and green teas. *Nutrition Research* 19(2):167–177.

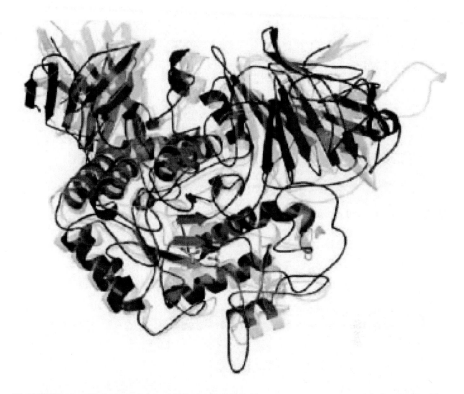

FIGURE 2.2 Phylogenetic conservation of GH 31 enzyme sequences and structures. The superposition of the human N-terminal of human maltase-glucoamylase (MGAM) (red), archael YicI (cyan), and bacterial MalA (yellow) structures is shown within a conserved $(\beta/\alpha)_8$ barrel structure. Glycerol molecules that indicate locations of glucose-binding areas in the active-site MGAM are represented as stick figures. Note the conservation of the tertiary protein structures, which diverged over more than 2.5 billion years. (From Sim, L., Quezada-Calvillo, R., Sterchi, E.E., Nichols, B.L., and Rose, D.R., *J Mol Biol*. 2008 Jan 18; 375(3):782–792. With permission.)

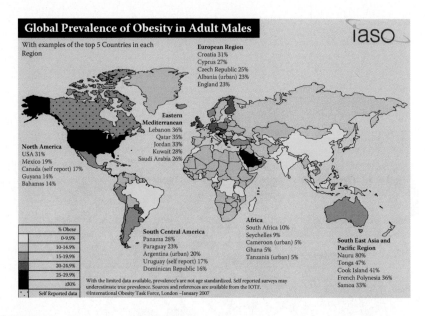

FIGURE 9.1 Global prevalence of obesity in adult males.

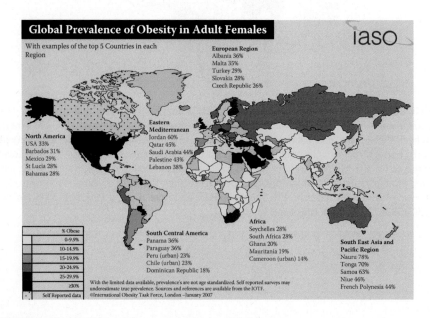

FIGURE 9.2 Global prevalence of obesity in adult females.

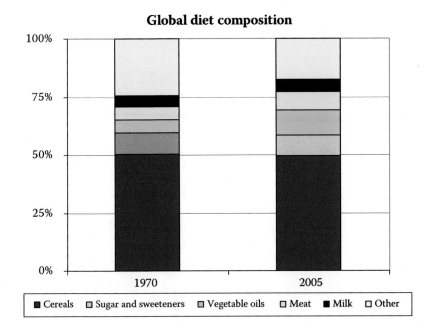

FIGURE 22.1 Global diet composition.

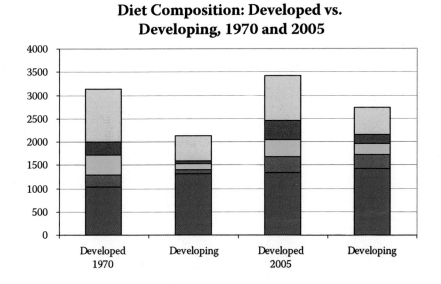

FIGURE 22.3 Diet composition: developed versus developing countries, 1970 and 2005.

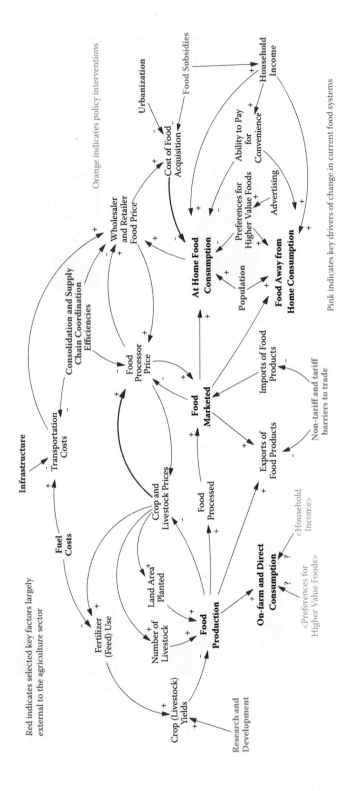

Red indicates selected key factors largely external to the agriculture sector

Orange indicates policy interventions

Pink indicates key drivers of change in current food systems

FIGURE 23.1 Conceptual diagram of a generic food system, indicating selected drivers, external factors, and interventions.

Rivera, J.A., C. Hotz, T. Gonzalez-Cossio, L. Neufeld, and A. Garcia-Guerra. 2003. The effect of micronutrient deficiencies on child growth: a review of results from community-based supplementation trials. *Journal of Nutrition* 133(11):4010S–4020S.

Rutten, R.A.M., M. van Dusseldorp, J. Schneede, L.C.P.G.M. de Groot, and W.A. van Staveren. 2005. Low bone mineral density and bone mineral content are associated with low cobalamin status in adolescents. *European Journal of Nutrition* 44(6):341–347.

Sabate, J., K. Lindsted, R. Harris, and A. Sanchez. 1991. Attained height of lacto-ovo vegetarian children and adolescents. *European Journal of Clinical Nutrition* 45:51–58.

Sanders, T. and R. Purves. 1981. An anthropometric and dietary assessment of the nutritional status of vegan preschool children. *Journal of Human Nutrition* 35:349–357.

Schroeder, D.G., R. Martorell, J.A. Rivera, M.T. Ruel, and J.-P. Habicht. 1995. Age differences in the impact of nutritional supplementation on growth. *Journal of Nutrition* 125:1051S–1059S.

Siegenberg, D., R.D. Baynes, T.H. Bothwell, B.J. Macfarlane, R.D. Lamparelli, N.G. Car, P. MacPhail, U. Schmidt, A. Tal, and F. Mayet. 1991. Ascorbic acid prevents the dose-dependent inhibitory effects of polyphenols and phytates on nonheme-iron absorption. *American Journal of Clinical Nutrition* 53:537–541.

Siekmann, J.H., L.H. Allen, N.O. Bwibo, M.W. Demment, S.P. Murphy, and C.G. Neumann. 2003. Kenyan school children have multiple micronutrient deficiencies, but increased plasma vitamin B-12 is the only detectable micronutrient response to meat or milk supplementation. *Journal of Nutrition* 133(11):3972S–3980S.

Sigman, M., S.E. Whaley, C.G. Neumann, N.O. Bwibo, D. Guthrie, R.E. Weiss, L.-J. Liang, and S.P. Murphy. 2005. Diet quality affects the playground activities of Kenyan children. *Food and Nutrition Bulletin* 26(2, Suppl 2):S202–S212.

Taylor, P.G., C. Martinez-Tores, E.L. Ramono, and M. Layrisse. 1986. The effect of cysteine-containing peptides released during meat digestion on iron absorption in humans. *American Journal of Clinical Nutrition* 43:68–71.

Tayter, M. and K. Stanek. 1989. Anthropometric and dietary assessment of omnivore and lacto-ovo-vegetarian children. *Journal of The American Dietetic Association* 89(11):1661–1663.

Teegarden, D., R. Lyle, W. Proulx, C. Johnston, and C. Weaver. 1999. Previous milk consumption is associated with greater bone density in young women. *American Journal of Clinical Nutrition* 69:1014–1017.

Thane, C. and C. Bates. 2000. Dietary intakes and nutrient status of vegetarian preschool children from a British national survey. *Journal of Human Nutrition and Dietetics* 13(3):149–162.

Uauy, R., C. Mize, and C. Castillo-Duran. 2000. Fat intake during childhood: metabolic responses and effects on growth. *American Journal of Clinical Nutrition* 72:1254S–1360S.

Walker, S.P., C.A. Powell, and S.M. Grantham-McGregor. 1990. Dietary intakes and activity levels of stunted and non-stunted children in Kingston, Jamaica, part 1, dietary intakes. *European Journal of Clinical Nutrition* 44:527–534.

Walker, S., C. Powell, S. Grantham-McGregor, J. Himes, and S. Chang. 1991. Nutritional supplementation, psychosocial stimulation, and growth of stunted children: the Jamaican study. *American Journal of Clinical Nutrition* 54(4):642–648.

Whaley, S.E., M. Sigman, C. Neumann, N. Bwibo, D. Guthrie, R.E. Weiss, S. Alber, and S.P. Murphy. 2003. The impact of dietary intervention on the cognitive development of Kenyan school children. *Journal of Nutrition* 133(11):3965S–3971S.

Yeudall, G., R. Gibson, C. Kayira, and E. Umar. 2002. Efficacy of a multi-micronutrient dietary intervention based on haemoglobin, hair zinc concentrations, and selected functional outcomes in rural Malawian children. *European Journal of Clinical Nutrition* 56(12):1176–1185.

Zheng, J., J.B. Mason, I.H. Rosenberg, and R.J. Wood. 1993. Measurement of zinc bioavailability from beef and ready-to-eat high-fiber breakfast cereal in humans: application of a
whole gut lavage technique. *American Journal of Clinical Nutrition* 58:902–907.

Zubieta, A.C., L.H. Allen, and N.O. Bwibo. 2005. *Intervention with Animal Source Foods to
Improve Vitamin A Status.* Davis, CA: University of California. Research Brief 05-02-
CNP. Global Livestock CRSP.

Zumrawi, F., J. Vaughan, J. Waterlow, and B. Kirkwood. 1981. Dried skimmed milk, breastfeeding and illness episodes - a controlled trial in young children in Karthoum Province,
Sudan. *International Journal of Epidemiology* 10:303–308.

13 Functional Foods

John W. Finley

CONTENTS

ABSTRACT

Functional food has become a term used to describe foods or food ingredients that provide health benefits beyond basic nutrition. Basic nutrition includes carbohydrates, proteins, fats, vitamins, and minerals normally associated with essential needs. Functional foods can range from additives such as soy sterols to formulated foods such as calcium-fortified orange juice, to pure products such as oatmeal. Each of these can deliver a benefit beyond basic nutrition. Soy protein and soy foods, which contain phytosterols, are believed to deliver multiple health benefits. These benefits include reduction of menopausal symptoms and reduction of coronary heart disease risk. Fats are often thought of as negative factors in nutrition; however, several functional foods are based on delivering healthier fats. It is clear that we need to include more omega-3 (N-3) fatty acids in our diets. There are currently many foods and supplements appearing in commerce that contain enhanced levels of N-3 fatty acids. Antioxidants have become an extremely broad and potentially very important segment of the functional food market. The natural antioxidants are primarily polyphenolic compounds, which include a very broad class of plant-based materials. These compounds have many in vivo functions, ranging from oxygen and free-radical scavenging to regulating the expression of various genes. The changing of gene expression or nutrigenomics is a critical function. One of the primary pathways that is impacted in nutrigenomics is the chronic inflammatory pathway. Chronic inflammation leads to many conditions, including coronary heart disease, Alzheimer's disease, and some cancers. Understanding how polyphenolics regulate the genes in the inflammatory pathways may eventually lead to improved screening for risk and personalized nutritional recommendations.

DEFINITION

WHAT IS A FUNCTIONAL FOOD?

Essentially all foods and ingredients in foods have functions. Food ingredients can provide texture and flavor, and others provide biological functions, which generally are related to health benefits. Foods function to provide us with necessary nutrients, health benefits, and the pleasure of eating. The Institute of Medicine's Food and Nutrition Board (IOM/NAS, 1994) defined *functional foods* as "any food or food ingredient that may provide a health benefit beyond the traditional nutrients it contains." In recent years, the term *functional food* has been used as a descriptor for foods that provide a biological or health-related benefit. This chapter focuses on the biological or health benefits of foods and ingredients. The term has evolved to mean a "food that provides value beyond basic nutrition." *Basic nutrition* is the delivery of carbohydrates, protein, fat, vitamins, and minerals that are essential for growth and maintenance of the body. Functional foods provide extra value, such as antioxidants that help protect the body from chronic diseases or components that help improve plasma cholesterol levels by increasing high-density lipoprotein (HDL) while decreasing low-density lipoprotein (LDL) cholesterol. Some functional foods act by providing bioactive ingredients that are absorbed and elicit specific reactions such as antioxidant activity, or they can induce specific gene expression. There are also functional foods that deliver fiber to the diet that is not absorbed. The fiber supplies substrate for production of short-chain fatty acids used for energy and as gut-lining growth factors and nutrient source.

Demonstration of bioactivity of foods or supplements is difficult since many of the activities delivered by bioactive ingredients have endpoint markers that may not appear until later. For example, antioxidants that prevent lipid oxidation in the short term can reduce lifetime risk of cardiovascular disease and some types of cancer. To demonstrate these benefits would require feeding the bioactive material to large numbers of subjects for many years, making any clinical trial prohibitively expensive. Model systems have therefore been developed to assess the benefits of a wide variety of bioactive ingredients.

ROLES OF BIOACTIVE INGREDIENTS

Bioactive ingredients are those present in a food or supplement that elicit a specific reaction in the body. For example, antioxidants are bioactive because they protect the body by controlling lipid oxidation. Another example of bioactivity is the reduction in plasma cholesterol when isoflavones are added to the diet. Demonstration of bioactivity of foods or supplements is difficult since many of the activities delivered by bioactive ingredients have endpoint markers that may not appear until later. For example, antioxidants that prevent lipid oxidation in the short term can reduce lifetime risk of cardiovascular disease and some types of cancer. Typically, potential bioactive ingredients are identified by their chemical characteristics. A large number of such ingredients have been identified based on chemical activity as antioxidants.

CLASSES OF BIOACTIVE INGREDIENTS

The broad spectrum of bioactive ingredients can be broken down by chemical classes. Figure 13.1 represents the major classes of bioactive ingredients found in functional foods. For example, the terpenoids include the carotenoids and tocopherols, which include β-carotene, vitamin A, and α-tocopherol (vitamin E). The phenolics represent an extremely diverse group of compounds, many of which exert significant bioactivity. The isoflavones from soy help reduce LDL cholesterol, and the polyphenolics, anthocyanins, and phenolic acids all deliver antioxidant activity along with altering the expression of various genes. For example, polyphenolics have been shown to reduce the expression of proinflammatory genes in in vitro systems. The carbohydrates generally deliver fiber, which enhances digestive health and includes the prebiotics (listed under microbial). The prebiotics are fermented by the probiotic bacteria, resulting in the production of short-chain fatty acids in the colon. These short-chain fatty acids are a preferred substrate for the colonic cells, resulting in

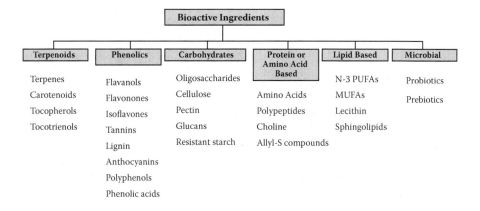

FIGURE 13.1 Categories of bioactive ingredients in functional foods. MUFAs, monounsaturated fatty acids; N-3 PUFAs, omega-3 polyunsaturated fatty acids.

more rapid turnover of colonic cells, thus helping maintain healthier tissue. Butyrate also enhances the health of other cells in the gastrointestinal (GI) tract. Proteins and peptides affect a number of activities, including many peptides that are angiotensin-converting enzyme (ACE) inhibitors, helping control plasma cholesterol levels. The influence of lipids is best exemplified with omega-3 (N-3) fatty acids. The N-3 fatty acids are reported to enhance neurofunction, lower cholesterol, and reduce chronic inflammation. The balance between N-3 and N-6 fatty acids is a major concern in Western diets. Higher levels of N-3 fatty acids in the Mediterranean diet are thought to be one of the important benefits of the diet.

SOURCES OF BIOACTIVE INGREDIENTS

Bioactive functional ingredients can come from a variety of sources, including plants, animals, and microbial sources. Some lipid-based materials, such as phosphatidyl-choline and sphingolipids, can be recovered from plants, animals, and microbial sources. The major dietary source of N-3 fatty acids is fish, which provide long-chain fatty acids. The long-chain fatty acids from fish are eicosapentaenoic acid (C20:5) and docosahexaenoic acid (C24:6). Plants such as flax and walnut provide α-linolenic acid (C18:3). Some of the C18:3 fatty acids are converted to the more important C20:5 and C24:6 fatty acids in vivo, but the conversion is limited. Generally, consumption of higher levels of the longer-chain fatty acids is preferred. Commercially available eggs are now marketed that provide the longer-chain fatty N-3 fatty acids. These are produced by feeding the chickens either fish oil or microalgae, which are rich in the long-chain N-3 fatty acids.

Plants provide the greatest variety of bioactive ingredients. The terpenes and the phenolics typically are from plant sources. The carbohydrates are also primarily found in plant-based products. Amino acids, proteins, and peptides can come from plants, animals, or microbial fermentation. Many interesting bioactive peptides have been recovered from milk.

Table 13.1 lists bioactive components in foods and provides examples of functional ingredients, their sources, and the purported benefits. Many of the functions include those that are anticancer, antioxidant, and anti-inflammatory. It is likely that these categories are all interrelated. Oxidative stress can lead to chronic inflammation, which can ultimately be seen as cancer or cardiovascular disease.

EXAMPLES OF FUNCTIONAL FOOD INGREDIENTS

There are many functional foods available, and more will emerge as the technology emerges for understanding how they work. Soy protein, the N-3 fatty acids, and the antioxidants constitute functional ingredients with significant documentation of the health benefits they deliver.

SOY, A REMARKABLY FUNCTIONAL FOOD

Soy is one of the most widely recognized and accepted functional foods. Soy protein is consumed in a number of forms, ranging from tofu, to soy milk, to inclusion

TABLE 13.1
Bioactive Components in Foods, Sources, and Purported Benefits

Bioactive Component	Sources	Benefits
Soy foods	Soybeans, soy protein	Reduce risk of CHD
Daidzein, genestein	Soy	Menopausal health, immunity, bone health
Carotenoids		
β-Carotene	Citrus, carrots, pumpkin, squash	Free radical scavenger, converted to vitamin A in vivo
Lutein, zeazanthin	Spinach, corn, eggs, citrus	Antioxidant, eye health
Lycopene	Tomatoes, watermelon	Antioxidant, prostate health
Dietary Fats (Fatty Acids)		
Monounsaturated fatty acids	Olive oil, canola oil	Reduced risk of CHD
Conjugated linoleic acid	Beef, lamb, dairy products	Immune function and improved body composition
Omega-3 fatty acids	Fish and fish oils	Reduced risk of CHD, improved mental acuity, vision
Dietary Fiber and Prebiotics		
Insoluble fiber	Cereal bran (wheat and corn)	Digestive health, cancer prevention
β-Glucans	Oatmeal, oat bran, barley	Reduces CHD risk
Soluble fiber	Psyllium, beans, apples	Reduces CHD risk
Whole grains	Wheat, brown rice, cereals	Reduces CHD risk, control serum glucose levels
Inulin and fructooligosaccharides	Whole grains, fruit, garlic, onions	Improved immunity, gastrointestinal health, mineral absorption
Probiotics	Yogurt, culture dairy	Immune function, gastrointestinal health
Sulfur Compounds		
Allyl sulfides	Onions, leeks, garlic	Immunity, heart health
Dithiothionones	Cruciferous vegetables	Immunity and detoxification
Sulforaphane	Broccoli, cauliflower, kale	Cellular antioxidant, detoxification
Phytosterols and satanols	Soy, wood, in fortified table spreads	Reduce LDL cholesterol
Flavonoids		
Anthocyanins: cyanidin, malvidin, delphinidin	Cane berries, cherries, grapes, red wine	Antioxidant defenses
Flavone-3-ols: catechins, epicatechins, procyanidins	Tea, chocolate, apples	Heart health, anti-inflammatory

(continued)

TABLE 13.1 (continued)
Bioactive Components in Foods, Sources, and Purported Benefits

Bioactive Component	Sources	Benefits
Flavonols: quercetin, kaempferol	Apples, broccoli, tea, onions, blueberries, red wine	Scavenge free radicals, antioxidant, anti-inflammatory
Pronathocyanidins	Cranberry, apple, strawberry, wine, cinnamon	Heart health, antioxidants, urinary tract health
Flavones: luteolin, apigenin	Celery, parsley, sweet peppers	Antioxidants, anticancer
Isoflavones: genestein, diadzein, glycitein	Soy and other beans	Menopausal health, bone health
Resveratrol	Wine, grapes, peanuts	Anticancer, longevity
Curcumin	Tumeric	Anti-inflammatory, anticancer

CHD, coronary heart disease; LDL, low-density lipoprotein.

in a wide range of bars. Soy delivers a range of health benefits, as reviewed by Messina et al. (2001), Hendrich and Murphy (2001), and Friedman and Brandon (2001). Raw soy protein contains several antinutritional components, including Bowman-Birk inhibitor, Kunitz inhibitor, and lectins. The lectins are more labile to heat, and the protease inhibitors are significantly reduced either by processing or refining of the protein.

The Bowman-Birk inhibitor is thought to be responsible for soy's protection against obesity, diabetes, and anticarcinogenic activity (Friedman and Brandon, 2001).

Isoflavones are major components found in soy foods that provide benefits in menopausal health and in bone health. The isoflavones in soy are present as aglycones, glucosides, malonylglucosides, and acetylglucosides. Hendrich and Murphy (2001) provided extensive data on the distribution of the isoflavones in soy. Table 13.2 summarizes the isoflavone contents in a variety of soy-based products.

The hypocholesterolemic influence of soy foods was first reported by Meeker and Kesten (1940). The mechanism has yet to be clearly defined. Carroll and Kurowska (1995) suggested that it was related to the amino acids and peptide structure, Setchell (1985) proposed isoflavones were factors, and Potter et al. (1993) suggested that the combination of saponins and soy protein may have been responsible. It is highly likely that the effects are a result of all of the components, not a single factor. Zhan and Ho (2005) did a meta-analysis that measured the benefits of soy on serum lipid profiles. The health benefits of soy are supported by several epidemiological studies showing benefits in reduced risk of breast, prostrate, and colon cancer (Messina et al., 1997; Hendrich and Murphy, 2001), improved bone health, and protection against cardiovascular disease (Hendrich and Murphy, 2001; McCue and Shetty, 2004).

The current status of the influence of soy isoflavones on breast cancer risk have recently been reviewed by Duffy et al. (2007), describing current proposed mechanisms for protection against breast cancer.

TABLE 13.2

Isoflavone Content of Representative Soy Foods (µg/g wet weight)

Source	Daidzein + Glycosides	Genistein + Glycosides	Glycitein + Glycosides	Total
Soybeans	986	1175	168	2329
Soy concentrate, ethanol washed	40	97	9	116
Soy isolate	216	521	59	796
Tofu	91	138	22	251
Soy milk, pasteurized	37	51	7	95
Soy milk, aseptic process	59	74	10	143
Soy burger, raw	37	66	12	115
Chicken analog	35	79	9	123

Source: Adapted from Hendrich and Murphy, 2001.

LIPIDS WITH HEALTH BENEFITS

A large body of evidence implicates the quality and quantity of dietary fats with risk of diseases, including cardiovascular disease, arthritis, some cancers, and Alzheimer's disease. Foods with beneficial fatty acid profiles represent a large segment of functional foods. Removing *trans* fats and saturated fats from foods represents a class of functional foods for which the enhancement is achieved by removing a hypercholesterolemic fat when they are replaced with a hypocholesterolemic fat that is rich in monounsaturated fatty acids and polyunsaturated fatty acids, particularly N-3 fatty acids.. The Mediterranean diet has long been considered beneficial because one of the primary sources of fat was olive oil, which is rich in monounsaturated oleic acid. Kris-Etherton et al. (1999) demonstrated that monounsaturated fatty acid diets lowered both cholesterol and triglyceride. It was concluded that the monounsaturated fatty acid diet was preferable to a low-fat diet for reducing cardiovascular risk. Chemically, it is lower melting than *trans* fats or saturated fatty acids such as palmitate and stearate. Thus, population of cell membranes with fatty acids that are more fluid (de la Lastra et al., 2001) results in healthier tissue. Increasing levels of unsaturation of fatty acids increases the likelihood of oxidation of the fatty acids. Oleic acid is a monounsaturated fatty acid and is therefore more stable to oxidation than polyunsaturated fatty acids (PUFAs), such as linoleic and linolenic. While the highly liquid PUFAs increase membrane fluidity, the risk of oxidation is much higher. Lipid oxidation is initiated by the removal of a hydrogen atom from the methylene side chain of an unsaturated hydrocarbon. The greater the degree of unsaturation, the more easily the hydrogen is removed. This is described more in the section on antioxidants.

Omega-3 fatty acids have clearly been associated with lower rates of chronic inflammation, reduced risk and treatment of heart disease, lower levels of arthritis, lower autoimmune disease, and reduced cancer risk (Simopoulos, 1999). Simopoulos discussed how the expression of multiple genes is influenced by higher ratios of N-3 fatty acids in the diet. The benefits of N-3 fatty acids are further supported by the observation that reduced cardiovascular risk is observed in populations who consume high levels of N-3 fatty acids. Lack of N-3 fatty acids in the diet can increase lipogenesis, leading to obesity, and is associated with increased risk of type 2 diabetes, high triglycerides, low HDL, and resultant higher risk of coronary heart disease. Clearly, increasing the levels of N-3 fatty acids in the diet has enormous potential for health improvement. One barrier is that because they are highly unsaturated, they are easily oxidized, resulting in off flavors and exposure to oxidation products that have negative health implications. Consuming more oily fish is one way to include these important fatty acids in the diet. Unfortunately, our food system and choices do not provide enough fish at a reasonable cost, and many consumers reject regular fish consumption. We are left with attempts to deliver N-3 fatty acids in formulated foods, which is a challenge for food technologists. One highly successful approach has been to add N-3-rich oil to poultry feed so the resulting eggs are rich in N-3 fatty acids. Technologists have developed several approaches to microencapsulation to deliver the N-3 oils in foods. Although expensive, the technology shows great promise for a new generation of healthier functional foods. Additional protection with antioxidants is also essential.

ANTIOXIDANTS

Antioxidants represent a broad group of compounds that stretch across several chemical classes of compounds found in foods and in traditional medicines. Oxidation of lipids in foods results in off flavors . If high levels of these oxidized materials are consumed over an extended period of time, they can be detrimental to health. Typically, oxidized foods have strong off flavors and are rejected. Examples of oxidation in foods in which off flavor occurs are in fish, which develops strong off flavors from the oxidation of the lipids. Another example is "warmed over" flavor in turkey. When cooked turkey is reheated, an off flavor is observed. In this case, many consumers either do not recognize the off flavor, and some actually prefer it. Crackers are another class of food that can develop rancid or oxidized off flavors. This flavor typically occurs when crackers sprayed with oils rich in unsaturated fats are stored too long, and the lipids become oxidized. Essentially, lipid oxidation proceeds through a chain reaction and continues until the chain is broken by an antioxidant. There is an increasing body of literature demonstrating that the breakdown products from lipid oxidation are detrimental and are correlated with conditions such as atherosclerosis (Kubow, 1993; Jessup et al., 2004).

Addition of antioxidants to foods prevents lipid oxidation or breaks the oxidative chain reaction before the food becomes inedible. Typically, vitamin E or C is used as a common antioxidant in foods. Other antioxidants are also available for foods, such as the synthetic phenolic antioxidants BHA and BHT. In recent years, there has been

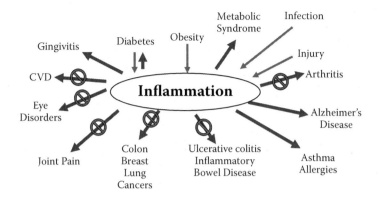

FIGURE 13.2 Antioxidants are reported to decrease inflammatory responses in many of these conditions. Thin arrows indicate causative factors of inflammation; thick arrows indicate inflammatory responses, and circles with a slash indicate responses reduced by antioxidants. CVD, cerebrovascular disease.

a trend away from synthetic antioxidants toward natural plant extracts. Rosemary extract has gained significant use as a natural antioxidant.

While lipid oxidation is well recognized in foods, similar reactions occur in vivo. Lipid oxidation in vivo has both positive and negative effects. Beneficial effects occur in vivo when the products of controlled lipid oxidation are used to kill unwanted bacterial cells in the body. If we have an infection from a wound, lymphocytes are recruited to the area, and they produce enzymes that oxidize arachidonic acid to produce radicals that can kill the invading bacteria. This is what occurs when the body protects itself from infection and is generally regarded as an acute inflammation. When tissue becomes invaded by foreign substances, such as between a bacteria and a splinter, the area becomes acutely inflamed. Such inflammations generally clear up in a few days. We are also prone to chronic inflammation when the production of oxidized materials continues for a prolonged period of time, generally at a lower level. Many diseases and adverse conditions have been associated with chronic inflammation. Many of these conditions can be mediated by antioxidants, as shown in Figure 13.2.

ANTIOXIDANT VITAMINS

The "antioxidant" vitamins, vitamins A, C, and E, can be found in a variety of functional foods. Supplementation with vitamin A has been under close scrutiny since adverse effects of high vitamin A were found in studies with smokers. Recent studies have demonstrated the benefits of the antioxidant vitamins, particularly as delivered from fruits and vegetables (Davis, 2008). In the review, Davis discussed the mechanisms by which the antioxidant vitamins can help prevent cancer. The review also emphasized the importance of interactions with other components in foods that appear to produce synergistic effects.

It has been speculated for some time that antioxidant vitamins would be particularly effective against prostrate cancer because it has a high prevalence, long latency, endocrine dependency, and reliable serum markers (prostate-specific antigen). Oxidative stress over prolonged periods has been associated with several types of cancer, such as prostrate cancer. Antioxidants can reduce oxidative stress if and when they reach the target tissues. Therefore, diets rich in antioxidants may offer some protection from cancers that are initiated by long-term oxidative stress. Many dietary factors have been related to prostrate cancer, such as dietary fat, cooked meat, micronutrients and vitamins (carotenoids, retinoids, and vitamins C, D, and E), and fruit and vegetable intake. There are clear trends that suggest positive effects, but more data are needed according to Schmid et al. (2007).

Ruano-Ravina et al. (2006) reviewed the relationship between lung cancer and antioxidant vitamins and addressed the mechanisms by which these nutrients might be exercising their activity. While tobacco use was a leading cause of lung cancer, the authors concluded that overall diet was also an important factor, but antioxidant vitamins demonstrated no clear protective effect.

CAROTENOIDS

Vitamin A conferred no protective effect; however, the provitamin A carotenoids lutein, zeaxanthin, lycopene, and α-carotene displayed limited protection. β-Cryptoxanthin showed a more consistent protective effect. Ruano-Ravine et al. (2006) pointed out that dose response studies are needed, and it is also clear that bioavailability is a question with the carotenoids.

Lycopene has been purported to have many health benefits. There are hundreds of carotenoids in the diet, and while lycopene is the most abundant, other carotenoids may have interesting activities. Supplementation of lycopene in foods to produce "functional foods" may not be necessary, particularly if appropriate levels of fruits and vegetables are consumed. Lycopene's potential benefits, its availability, and the role of carotenoids in health and disease states, focusing on the metabolism of the carotenoids, have been extensively reviewed (Bruno and Wildman, 2001; Faulks and Southon, 2001; Bruno et al., 2007).

PHENOLIC ANTIOXIDANTS

Currently, there is enormous interest in the many functional properties of plant-based phenolic compounds. As shown in Figure 13.1, phenolics encompass a wide range of phytochemicals, ranging from relatively simple phenolic acids to flavonoids to complex anthocyanins and tannins.

Epidemiology studies claimed a range of health benefits associated with tea consumption, and consumption of red wine, which is rich in phenolics, has been associated with the French paradox. The French diet is rich in fat, particularly saturated fat, yet the incidence of heart disease is much lower than seen in other countries where similar diets are consumed. Because the French consume higher amounts of red wine, it has been proposed that the red wine phenolic compounds prevent some of the adverse effects associated with high-fat diets. Dubick and Omaye (2007) reviewed

TABLE 13.3

Polyphenolic Content of Red and White Wine

	Average Red Wine (mg/L)	Average White Wine (mg/L)
Gallic acid	95	7
Catechin	191	35
Caffeic acid	7.1	2.8
Epicatechin	82	21
Cyanidin	2.8	0
Malvidin-3-glucoside	23.5	1.0
Rutin	9.1	0
Myrcetin	8.5	0
Quercetin	7.7	0
Resveratrol	1.527	0.027

Source: Frankel, 1995.

the literature on the influence of grape wine and tea polyphenols on atherosclerosis; they concluded that the wine and tea polyphenols possess antioxidant activity and may reduce total cholesterol, lower lipoprotein oxidation concentrations, and help reduce chronic inflammation. Table 13.3 summarizes the bioactive polyphenolics in wine as reported by Frankel et al. (1995). It is clear that the bulk of the phenolic compounds are associated with red wine, most likely coming from the skins. Falchi et al. (2006) demonstrated that flesh and skin of grapes had equivalent levels of tartaric acid, malic acid, shikimic acid, and *trans* caftaric acid; the flesh had slightly lower levels of caffeic, p-coumaric, cinnamics, catechin, and epicatechin and were approximately equal in providing cardioprotection.

Catechin in tea has been shown to strongly correlate with the antioxidant capacity. Henning et al. (2003) showed a correlation between total catechin content and antioxidant capacity. The authors presented a list of commercial tea products that included a range of catechin contents from 204.7 mg/g tea to 0 catechin for some ice tea products. Green tea ranged from 59.7 mg/g to 105.7 mg/g green tea. This clearly points to a major issue with functional foods. Many functional foods are marketed as containing green tea and other herbal supplements, but the labeling requirements do not require a clear statement of the bioactives and how those might equate to a cup of tea.

Many foods contain anthocyanins, which are likely beneficial at some level. Wu et al. (2004) compiled an extensive list of anthocyanin content of fruits, vegetables, nuts, dried fruit, spices, and other foods. This provides a useful baseline in determining how much of these components is consumed. One measure of the antioxidant potential of a food is the oxygen radical absorbance capacity (ORAC). The method provides a useful means to compare the potential of a food or ingredient to remove oxygen radicals from a model system. The result is that foods can be compared for their potential antioxidant value. Wu et al. compared the antioxidant capacities of many common foods. Putting the two efforts by Wu et al. together, we can establish that the anthocyanins correlate with the increased antioxidant potential of foods. Many of these foods could be considered as functional because of the high potential to deliver antioxidant protection. As a population, we have to either convince the consumer to consume more of the highly protective foods or deliver formulated foods that deliver the benefits with low levels of consumption of fruits and vegetables.

Because the components must be absorbed to become effective, the relationship between chemical analyses and the ultimate health benefit remains controversial. Cao et al. (1998) demonstrated in a clinical trial that increased consumption of fruits and vegetables resulted in a lower level of oxidative stress, while a high-fat diet increased oxidative stress.

If it can be clearly demonstrated that the improved antioxidant status prevents chronic diseases and that the antioxidants from foods are available, it raises new opportunities to deliver convenient foods with high antioxidant capacity.

FLAVONOIDS

Flavonoids represent a large family (over 5000 identified) of bioactive compounds derived from fruits and vegetables that deliver high antioxidant activity. Table 13.1 includes examples of flavonoids found in some common foods. These materials can be concentrated and used as food additives in functional foods or used as supplements. A more extensive report of flavonoids in common U.S. foods has been provided by Harnly et al. (2006). Williams and Grayer (2004) presented a detailed review of the chemistry of the anthocyanins and other flavonoids.

DETERMINATION OF BIOACTIVITY

Many bioactive ingredients have been discovered by assessing traditional folk medicine and remedies. Rich sources of potential bioactive ingredients have come from traditional Chinese medicine, Ayurveda, and Native American traditional medicines. The breadth of treatments also means a variety of ways have been developed to assess the potential health benefits of functional foods. When we consider the thousands of ingredients or components in traditional foods, alternative medicine formulations, and the potential combinations of materials, the amount of testing to validate efficacy is staggering. As a result, short-term model systems have been developed as screening tools for various health conditions. For example, lung tissue can be used to assess antioxidant regimens that protect the lung, and liver cell models can be used for hepatic function studies.

Frequently, the cell culture models are employed to screen large numbers of candidate materials and fractions of materials to identify which component is most bioactive in the product. The next level of testing is usually small animals (usually rodents), which serve to assess if the material identified by cell culture is absorbed in the animals and carried to the target site for the bioactivity. The small animal screening also serves as an excellent test to ensure that the component tested does not cause any adverse effects either through high dose or chronic exposure. Finally, clinical trials can be applied to assess the efficacy in targeted human populations.

Model systems generally are isolated target cells and as a result usually do not test metabolites or secondary compounds. Many studies expose target cells at relatively high concentrations of test materials and therefore do not reflect physiological concentrations normally seen in vivo. The growth conditions and cell lines (frequently tumor cell lines) often do not reflect the same selectivity of action that are seen in normal cells. The in vitro conditions often do not replicate actual pH, redox, and other conditions that exist in vivo. As a result, it must be kept in mind that much of the current data on antioxidants and anti-inflammatory activities of phytochemicals is based on in vitro testing, and the purported health benefits are extrapolations. Much more research is needed to generate direct evidence for extranutritional health benefits to humans for the vast majority of compounds discussed in this chapter and for the foods that carry them. Epidemiological studies are frequently used to support the health benefits of various food ingredients. These projections are difficult to interpret because they are frequently confounded by multiple factors. For example, the Mediterranean diet is proposed to be healthier because it includes olive oil, fish, and wine. When compared to American diets, the dietary profile is without a doubt advantageous; however, most studies do not account for lifetime exercise and other life style differences.

NUTRIGENOMICS

The New Frontier in Functional Food Understanding

Bioactive materials in functional foods frequently act by altering gene expression. With the sequencing of the genome and the tools of high-throughput screening, we can now monitor the effects of specific components on specific genes. Food ingredients have an impact on metabolism at many stages. Figure 13.1 shows the various points where bioactive ingredients in food can influence gene expression, enzyme activity, metabolic function, and ultimately the impact on the system as a whole. Bioactive ingredients can influence transcription factors. For example, fatty acids influence the transcription factors PPARs and SREPBs. The flavonoids can influence the expression of NFκB. Amino acids, particularly leucine, regulate mTOR signaling and as a result reverse inhibition of translation in model systems (Proud, 2002). Many food components act as enzyme inhibitors, for example, the Bowman-Birk and Kunitz protease inhibitors in soy. Amylase inhibitors have been proposed as a means to reduce caloric availability of starches. When nondigestible carbohydrates such as sugars and starches reach the colon, they are fermented by the colonic bacteria. Short-chain fatty acids such as butyric and propionic are considered beneficial to the colon bacteria. The other products produced by these bacteria are methane and carbon dioxide. These gases can cause considerable discomfort and flatulence.

The ability to monitor direct effects on genes and pathways will eventually lead us toward more individualized nutrition. Currently, we know that groups of individuals can react differently to different food ingredients. The application of nutritional genomics will help us understand these individual differences and provide a basis for more targeted functional foods in the future.

FUNCTIONAL FOODS IN THE MARKET

Currently, there are numerous foods on the market that can be considered functional foods. Products include margarines fortified with plant sterols to help reduce cholesterol, orange juice fortified with calcium, yogurts with probiotic bacteria or prebiotics, products with added green tea, and foods with added N-3 fatty acids. One of the largest segments is the fortified beverage market, with products ranging from energy drinks to vitamin-fortified water available.

The products leading the beverage category are enhanced water drinks, sports drinks, functional tea products, smoothies, and soy-based beverages.

CRITERIA

- When a nutrient or ingredient is added to the product, the presence must be indicated. Currently, many products on the market do not report the amount of bioactive ingredient in the product. In the future, manufacturers should be encouraged to do this. For example, the addition of green tea or ginseng should be related to a cup of tea or an actual concentration of the bioactive ingredient.
- Structure/function claims can be used to describe the effect of a food component on structure or function in the body, but such claims require Food and Drug Administration (FDA) approval.
- Dietary guidance claims can be used to describe the benefits of broad categories of foods, for example, the benefits of whole-grain products.
- Qualified health claims must be reviewed by the FDA. They must demonstrate a relationship between the food component and risk of disease or health condition. Substantial data are required to support these claims.
- Health claims must reflect confirmed relationships between food components and the disease or health condition addressed. There must be significant scientific agreement.

STUDY TOPICS

- Functional foods are foods that provide nutritional benefit beyond basic nutrition.
- Basic nutrition includes the essential nutrients provided from protein, fat, carbohydrate, vitamins, and minerals in the diet.
- Bioactive ingredients or components are the materials in functional foods that provide the principal benefit. Bioactive compounds can come from many sources, including peptides from milk, soy protein, lipids rich in N-3 fatty acids, and a broad range of phytochemicals from plants.
- Soy protein and the soy phytosterols are significant functional foods because they may prevent or reduce coronary heart disease and relieve menopausal symptoms.

- Lipids can provide significant health benefits, particularly the N-3-rich fatty acids found in marine products. These fatty acids help prevent coronary heart disease, reduce chronic inflammation, and may help prevent Alzheimer's disease.
- Antioxidants represent a very broad class of functional ingredients. They include the antioxidant vitamins (vitamins A, C, and E) as well as complex families of polyphenolic and flavonoid compounds found primarily in plant-based foods. These antioxidants have a broad spectrum of benefits, including prevention of oxidation of lipids in the food as well as in vivo, scavenging free radicals in vivo, and acting as inhibitors of chronic inflammation.
- The inhibition of chronic inflammation occurs in many ways; one of the most significant is the signaling of various genes. The phenolic compounds can either up- or downregulate genes in the inflammatory pathway. This regulation of gene expression opens the door to understanding individual variation and will eventually lead to personalized nutrition.

REFERENCES

Bruno, R.S. and Wildman, R.E.C. (2001) Lycopene: source, properties and nutraceutical potential. In *Handbook of Nutraceuticals and Functional Foods*, R.E.C. Wildman, Ed. CRC Series in Modern Nutrition. Boca Raton, FL: CRC Press, pp. 157–168.

Bruno, R.S., Wildman, R.E.C., and Schwartz, S.J., (2007) Lycopene: food sources, properties, and health. In *Handbook of Nutraceuticals and Functional Foods,* 2nd ed., Boca Raton, FL: CRC Press, pp. 55–72.

Cao, G., Booth, S.L., Sadowski, J.A., and Prior, R.L. (1998) Increases in human plasma antioxidant capacity after consumption of controlled diets high in fruits and vegetables. *Am. J. Clin. Nutr.* 68:1081–1087.

Carroll, K.K. and Kurowska, E.M. (1995) Soy consumption and cholesterol reduction: review of animal and human studies. *J. Nutr.* 125:594S.

Davis, C.D. (2008) Mechanisms for cancer-protective effects of bioactive dietary components in fruits and vegetables. In *Handbook of Nutrition and Food,* 2nd ed., C.D. Berdanier, J. Dwyer, and E.B. Feldman. Boca Raton, FL: CRC Press, pp. 1187–1210.

De la Lastra, C.A., Barrranco, M.D., Mottilva, V., and Herrerias, V. (2001) Mediterranean diet and health: biological importance of olive oil. *Curr. Pharma. Design* 7:933–950.

Dubick, M.A. and Omaye, S.T. (2007) Grape wine and tea polyphenols in the modulation of atherosclerosis and heart disease. In *Handbook of Nutraceuticals and Functional Foods* 2nd ed., Boca Raton, FL: CRC Press, pp. 101–130.

Duffy, C., Perez, K., and Partridge, A. (2007) Implications of phytoestrogen intake for breast cancer. *CA Cancer J. Clin.* 57:260–277.

Falchi, M., Bertelli, A., Lo Scalzo, R., Morassut, M., Morelli, R., Das, S., Cui, J., and Das, D.K. (2006) Comparison of cardioprotective abilities between the flesh and skin of grapes. *Journal of Agricultural and Food Chemistry* 54(18): 6613–6622.

Faulks, R.M. and Southon, S. (2001) Carotenoids, metabolism and disease. In *Handbook of Nutraceuticals and Functional Foods*, R.E.C. Wildman, Ed. CRC Series in Modern Nutrition. Boca Raton, FL: CRC Press, pp. 143–156.

Frankel, E.N., Waterhouse, J.L., and Teissedre, P.L. (1995) Principal phenolic phytochemicals in selected California wines and their antioxidant activity in inhibiting oxidation in human low-density lipoprotein. *J. Agric. Food Chem.* 43:890–894.

Friedman, M. and Brandon, D.L. (2001) Nutritional and health benefits of soy protein. *J. Agric. Food Chem.* 49(3):1069–1086.

Harnly, J.M., Doherty, R.F., Beecher, G.R., Holden, J.M., Haytowitz, D.B., Bhagwat, S., and Gebhardt, S. (2006) Flavonoid content of U.S. fruits, vegetables, and nuts. *J. Agric. Food Chem.* 54(26):9966–9977.

Hendrich, S. and Murphy, P.A. (2001) Isoflavones: source and metabolism. In *Handbook of Nutraceuticals and Functional Foods*, R.E.C. Wildman, Ed. CRC Series in Modern Nutrition. Boca Raton, FL: CRC Press, pp. 22–76.

Henning, S.M., Fajardo-Lira, C., Lee, H.W., Youssefian, A.A., Go, V.L.W., and Heber, D. (2003) Catechin content of 18 teas and a green tea extract supplement correlates with the antioxidant capacity. *Nutrition and Cancer* 45(2): 226–235.

IOM/NAS. (1994) *Opportunities in the Nutrition and Food Sciences*, P.R. Thomas and R. Earl, Eds. Washington, DC: Institute of Medicine/National Academy of Sciences, National Academy Press, pp. 1–109.

Jessup, W., Kritharides, L., and Stocker, R. (2004) Lipid oxidation in atherogenesis: an overview. *Biochem. Soc. Trans.* 32:134–138.

Kris-Etherton, P., Pearson, T.A., Way, Y., Hargroave, R.L., Moriarity, K., Fishell, V., and Etherton, T.D. (1999) High-monounsaturated fatty acid diets lower both plasma cholesterol and triacylglycerol concentrations. *Am. J. Clin. Nutr.* 70:10009–10015.

Kubow, S. (1993) Lipid oxidation products in food and atherogenesis. *Nutrition Reviews* 51(2): 33–40.

McCue, P. and Shetty, K. (2004) Health benefits of soy isoflavonoids and strategies for enhancement: a review. *Crit. Rev. Food Sci. Nutr.* 44(5):361–367.

Meeker, D.R. and Kesten, H.D. (1940) Experimental atherosclerosis and high protein diets. *Proc. Soc. Exp. Biol. Med.* 45:543.

Messina, M., Barnes, S., and Setchell, K.D. (1997) Phyto-estrogens and breast cancer. *Lancet* 350:971.

Messina, M., Gugger, E.T., and Alekel, D.L (2001) Soy protein, soybean isoflavones, and bone health: a review of the animal and human data. In *Handbook of Nutraceuticals and Functional Foods*, R.E.C. Wildman, Ed. CRC Series in Modern Nutrition. Boca Raton, FL: CRC Press, pp. 77–78.

Potter, S.M., Jimenez, F.R., Pollack, J., Lone, T.A., and Berber, J. (1993) Protein-saponin interaction and its influence on blood lipids. *J. Agric. Food Chem.* 41:1287.

Proud, C.G. (2002) Regulation of mammalian translation factors by nutrients. *Eur. J. Biochem.* 269:5338–5349.

Ruano-Ravina, A., Figueiras, A., Freire-Garabal, M., and Barros-Dios, J.M. (2006) Antioxidant vitamins and risk of lung cancer. *Curr. Pharm. Design* 12(5):599–613.

Schmid, H.-P., Engeler, D.S., Pummer, K., and Schmitz-Draeger, B.J. (2007) Prevention of prostate cancer: more questions than data. *Recent Results Cancer Res.* 174:101–107.

Setchell, K.D.R. (1985) Naturally occurring non-steroidal estrogens of dietary origin. In *Estrogen in the Environment II*, J.A. McLaachlan, Ed. New York: Elsevier, p. 69.

Simopoulos, A.P. (1999). Essential fatty acids in health and chronic disease. *J. Clin. Nutr.* 70:560S–569S.

Williams, C.A. and Grayer, R.J. (2004) Anthocyanins and other flavonoids. *Nat. Prod. Rep.* 21:539–573.

Wu, X., Beecher, G.R., Holden, J.M., Haytowitz, D.B., Gebhardt, S.E., and Prior, R.L. (2004) Lipophilic and hydrophilic antioxidant capacities of common foods in the United States. *J. Agric. Food Chem.* 52:4026–4037.

Wu, X., Beecher, G.R., Holden, J.M., Haytowitz, D.B., Gebhardt, S.E., and Prior, R.L. (2006) Concentrations of antocyanins in common foods in the United States and estimation of normal consumption. *J. Agric. Food Chem.* 54:4069–4075.

Zhan, S. and Ho, S.C. (2005) Meta-analysis of the effects of soy protein containing isoflavones on the lipid profile. *Am. J. Clin. Nutr.* 81:397–408.

14 Global Food Safety Issues

Mindy Brashears and Tyler Stephens

CONTENTS

ABSTRACT

According to the World Health Organization (WHO), food-borne illness results in 1.5 billion cases of diarrhea in children each year and 3 million deaths. In the United States, each year there are an estimated 76 million cases of food-borne illness and 5000 deaths. In contrast, more than 1 million children under the age of 5 die annually from food- or water-borne diseases in Southeast Asia. The condition of the food supply, as well as public health efforts to protect the food supply, vary from country to country and have a direct impact on the number of food-borne illnesses and deaths that occur. In addition to loss of human life, the economic costs are tremendous. In the United States, it is estimated that the cost of food-borne illness is up to $6.9 billion, which is attributed to such things as medical expenses, litigation fees, and lost time from work. Economic data are less accurate for developing countries. Many challenges arise when addressing global food safety issues. In developing countries, much of the food is grown locally or may be traded in local "farmer's markets." The issues faced in developing countries include a lack of refrigeration or other means of preservation, food contaminated by sewage due to inadequate sewage treatment systems, and food contaminated by nonpotable water. In addition, food may be handled improperly by field workers because of inadequate hand washing or restroom facilities in the field. In the United States and other developed countries, there are different challenges. Food is typically processed in centralized facilities and shipped to supermarkets where it is purchased by the consumer. Food must be processed in sanitary facilities and maintained at proper temperatures during processing and distribution. Targeted interventions to improve food safety are generally implemented in these food systems as well as strict regulatory control of the safety of many of the products. This chapter discusses details of the challenges faced around the world as well as the varying health and economic statistics from country to country.

INTRODUCTION

According to popular opinion, the United States has the "safest, cheapest, and most abundant" food supply in the world. While this popular conception may be true, the food safety needs around the world are vastly different from those encountered in the United States. According to the Centers for Disease Control and Prevention (CDC), there are an estimated 76 million cases of food-borne illness that occur annually, with 325,000 hospitalizations and 5,000 deaths (Mead et al., 1999). In 2007, ground beef recalls due to contamination with *Escherichia coli* O157:H7 were again on the rise after subsiding in the previous years. In addition, in recent years recalls due to spinach and other leafy greens contaminated with *E. coli* O157:H7 and *Salmonella* have been initiated. The Economic Research Service (ERS) estimated annual cost of food-borne illness in the United States is $6.9 billion (Mead et al., 1999). Most of the costs are due to recalls, loss of product, litigation, time lost from work, and medical bills. Other developed countries face similar food safety dilemmas.

The challenges faced by developing countries are dramatically different from those faced in the United States and other developed countries. The World Health Organization (WHO) reported (1999) that around the world there are 1.5 billion cases of diarrhea annually that occur in children alone; these result in 3 million deaths. In this chapter, a comparison of the different food safety needs encountered around the world is presented.

THE "COST" OF FOOD-BORNE ILLNESSES

HUMAN COSTS

The human costs associated with food-borne illnesses are high. Generally, the most vulnerable groups are children, elderly, and individuals with compromised immune systems; indeed, the highest morbidity and mortality rates occur among these groups. However, practically anyone can contract a food-borne illness and succumb to illness or death under the right circumstances. Food-borne pathogens can also carry resistance to various antibiotics, which can potentially result in an increase in morbidity and mortality from infection. Typical food-borne illnesses manifest themselves as diarrhea. In combination with diarrheal disease, fever, vomiting, and nausea can occur. Dehydration or malnutrition can occur as a result of typical food-borne illnesses. Other problems can develop depending on the pathogen and range from kidney failure to miscarriages.

COMMON FOOD-BORNE PATHOGENS

Salmonella

Salmonella can originate from a number of animal and environmental sources. It is commonly found in the intestinal tract of animals and is most often associated with foods of animal origin. Typical food-borne illness symptoms usually develop within 6 to 48 hours (Food and Drug Administration [FDA], 2008) and last 1 to 2 days. Cases of salmonellosis have been on the rise in the United States and in other industrialized nations.

Campylobacter

Campylobacter is an enteric pathogen originating primarily from the intestinal tracts of warm-blooded animals. In the United States, *Campylobacter* is commonly identified as the leading bacterial cause of food-borne illness. Infection with this organism usually results in diarrhea, sometimes with nausea, abdominal cramps, vomiting, and muscle aches. Typical symptoms develop after 2 or 3 days and can last up to 7 to 10 days.

Enterohemorrhagic *Escherichia coli*

Escherichia coli O157:H7 is the most common enterohemorrhagic *E. coli* (EHEC) pathogen that produces potent toxins (verotoxin or shiga-like toxin) and has been identified as one of the causal agents of hemorrhagic colitis, hemolytic uremic syndrome, and thrombotic thrombocytopenic purpura. Cattle have been recognized as the primary reservoir for this pathogen. Undercooked ground beef, lettuce, spinach, milk, alfalfa sprouts, and unpasteurized fruit juices have all been implicated as vehicles of infection for this pathogen.

Listeria monocytogenes

Listeria monocytogenes is a gram-positive bacterium that causes listeriosis. The manifestation of this disease includes intrauterine or cervical infections in pregnant women (spontaneous abortion or stillbirth), meningitis, encephalitis, and septicemia. Raw milk, cheese, raw vegetables, ice cream, all types of raw meats, and ready-to-eat meat products have been implicated as vehicles for this pathogen.

Staphylococcus aureus

Staphylococcus aureus is a gram-positive coccus bacterium that has specific strains capable of producing a potent heat-stable protein enterotoxin that causes staphylococcal food poisoning. Sewage, dust, water, air, and many food products could be vehicles for this pathogen to spread. Nausea, vomiting, and abdominal cramping are common symptoms of staphylococcal food poisoning. The onset of these symptoms are usually rapid, and most cases are acute.

Clostridium botulinum

Clostridium botulinum is a gram-negative, spore-forming, anaerobic bacterium that produces a potent neurotoxin. Food-borne botulism is contracted by humans who consume foods containing this neurotoxin. Foods commonly associated with botulism vary according to eating habits in different regions and food preservation but are most commonly associated with canned foods due to the anaerobic environment. Symptoms (lassitude, vertigo, double vision, etc.) can occur as quickly as 4 to 8 hours after ingestion, but in most cases occur in 18 to 36 hours. Infant botulism consists of symptoms such as poor feeding, lethargy, weakness, and loss of head control.

Vibrio cholerae

Vibrio cholerae is commonly found in marine and estuarine environments of the United States. Diarrhea, abdominal cramping, and fever are common symptoms associated with *V. cholerae* gastroenteritis. Shellfish harvested from U.S. coastal waters are mostly implicated in this disease.

Shigella spp.

Shigella spp. are gram-negative, non-spore-forming bacteria that cause shigellosis (bacillary dysentery). The most common symptoms of shigellosis are abdominal cramping and pain, diarrhea, fever, vomiting, abnormal stools (blood, pus, or mucus), and tenesmus. Salads that include potato, tuna, macaroni, shrimp, and chicken are commonly associated food vehicles. Some strains produce Shiga toxins similar to the verotoxins produced by *E. coli* O157:H7.

Clostridium perfringens

Clostridium perfringens is a gram-positive, spore-forming anaerobe that causes food poisoning and enteritis necroticans (pig-belly) disease. Intense abdominal cramps and diarrhea within 8 to 22 hours after consumer foods contaminated with *C. perfringens* are symptoms of the common form of perfringens poisoning. Necrotic enteritis, a rare disease in the United States, is characterized by infection and necrosis of the intestines that causes septicemia and can lead to death. Temperature abuse of prepared foods (meats and gravy) is the most common cause of perfringens poisoning.

ECONOMIC COSTS

Specific economic costs are discussed in each section of this chapter. In industrialized nations, the economic costs include the cost of litigation/insurance, loss of product from recalls, cost of investigation of outbreaks, medical expenses, and time lost from work. Economic costs are also a burden when food trade is banned or prevented due to unsafe product or a lack of regulatory oversight.

FOOD SAFETY BY REGION

SOUTHEAST ASIA REGION

It is estimated that 1 million children under the age of 5 die annually in the Southeast Asia region due to diarrhea from consumption of contaminated food and water (WHO, 2002a). Cholera is a major concern and is spread not only through water, but also through food (WHO, 2000). Lack of education is a major contributing factor to food safety concerns in this region. A lack of regulatory authority also plays a major role in the spread of food-borne disease.

In this area, many are migrating from rural to urban environments and living in extreme poverty and in conditions of poor sanitation (WHO, 2002). Many are homeless and malnourished and obtain food from street vendors, who have been identified as a significant source of food-borne illnesses. Most of the government structures in Southeast Asia have no written policies on food safety. Food is most commonly purchased with little or no processing, and hazards associated with these foods need to be evaluated and controlled. There is a need for an organized government structure and a need for food-handling laboratories to assess the hazards associated with foods in question to identify hazards, develop regulations, and control them.

WESTERN PACIFIC REGION

In the western Pacific region, there is a contrast in food safety issues, depending on the country. New Zealand and Australia are industrialized and have food safety policies in place. However, in Australia, there are an estimated 11,500 illnesses daily, which costs the country AU $2.6 billion (ANZFA, 1999). Due to limited surveillance in many countries in this region, accurate estimates of costs and the number of illnesses is difficult.

A recent report indicated that the outbreaks associated with food-borne pathogens are increasing. EHEC, *Cyclospora*, *Listeria*, and *Campylobacter* are among the pathogens causing recent outbreaks (WHO, 2003a). There has also been an increase in antibiotic-resistant *Salmonella* isolated from outbreaks (OZfoodnetwork, 2003, Government of China, 2004).

In less-developed areas, there is less oversight of the food industry, and surveillance is limited. It is speculated that rapid urbanization, poor sanitation, lack of education, and lack of food labeling regulations contribute to food safety issues in the western Pacific region (Center for Science in the Public Interest [CSPI], 2005).

EASTERN MEDITERRANEAN REGION

The WHO is working in collaboration with countries in the eastern Mediterranean region to assist them in developing regulatory systems for control of food safety issues. At this time, there are few data available on the occurrence of food-borne illnesses in this area or on the leading causes of food-borne illnesses because they are lacking a centralized regulatory office to oversee this effort.

The WHO and Food and Agriculture Organization (FAO) have reported that food-borne illnesses/diarrheal diseases are perceived as a part of everyday life in this region (WHO/FAO, 2005). Many of the diseases are likely waterborne. Cultural means of food preparation contribute to the occurrence of diseases. Cheeses are commonly prepared from raw milk, and raw animal products are commonly consumed.

In non-oil-producing countries of the eastern Mediterranean, food exports are an important source of income (FAO, 2004). Therefore, the implementation of food safety policy is important in sustaining the economic balance in these countries. According to a report published by the WHO, in 2001 up to 27% of food exports from this area were rejected when they reached the United States because of potential food safety problems. An additional 58% were rejected due to label issues. In areas where food legislation laws do exist, there are several challenges. Most laws are out of date, and there are not enough inspection personnel to meet the needs of the industry. In addition, there is little opportunity for education and training and a substantial lack of technical expertise in these areas (WHO, 1999).

Despite the current lack of legislation and regulatory authority over food safety, in these areas, a regional "plan of action" was adopted in 1999 (WHO, 2005). This plan of action defines the needs in the region and has resulted in a growing acceptance of food safety programs commonly used in more industrialized countries. Good manufacturing practices (GMPs), good agricultural practices (GAPs), and HACCP (Hazard Analysis and Critical Control Points) are among the programs being adopted in this region.

There has been a concern about the misuse of chemicals in the region, and the implementation of organic farming practices has increased to alleviate this problem. However, with the implementation of organic practices come other potential food safety problems if manure is used for fertilizer and it is not properly composted.

AFRICAN REGION

Both food-borne and waterborne diarrheal diseases are tremendous problems in the African region. According to a press release from the WHO, a total of 700,000 deaths occur annually due to food and waterborne diarrheal diseases. On average, each child under the age of 5 in this region will experience five episodes of diarrheal disease annually (WHO, 2005). This problem, combined with malnutrition, is a serious health threat in this region.

The WHO regional office has developed a strategy for controlling these diseases (WHO, 2005):

- Development of food safety policies, programs, legislation, and regulations to ensure the safety of food from production to consumption
- Development and improvement of capacity to provide analytical skills for monitoring foods on the market
- Establishment of transparent health promotional systems and procedures to ensure that producers, processors, retailers, consumers, and other stakeholders are properly informed on safe food handling as well as food emergencies
- Development of systems to ensure national, regional, and international cooperation, collaboration, and coordination to ensure that stakeholders work in a concerted manner

While these efforts will positively have an impact on the health and safety of people in this region, there is a tremendous need for assistance to alleviate the continued health threat in this region.

Compounding the problem is the fact that a large portion of the population in this region suffers from acquired immunodeficiency syndrome (AIDS), which makes them more susceptible to the illness. Recent studies have reported that the prevalence of *Salmonella, Shigella, Campylobacter, S. aureus, E. coli,* rotavirus, *Bacillus cereus,* and *Brucella* is high in this region.

Severe poverty also contributes to the number of illnesses. According to a recent report by CSPI, poverty is a negative contributing factor in several ways (2005):

- Unsanitary conditions in rapidly growing urban centers
- Lack of access to clean water
- Unhygienic transportation and storage of foods
- Low education levels among consumers and food handlers, leading to reduced information on food safety
- Lack of government financial resources to enhance food-borne disease surveillance and monitoring capabilities
- Lack of implementation of food safety regulations through an efficient inspection system
- Lack of modern facilities and utilities
- Lack of development of food safety education programs
- No means of conducting disaster planning and relief

CSPI also emphasized that street vendors are "an important source of affordable food." However, most street vendors are not regulated and do not meet minimal standards for safe food. Most food purchased from street vendors may be prepared well in advance and held for hours or days prior to consumption. In addition, equipment used by the street vendors may not be adequate to hold the food at the proper temperatures for safety.

Natural disasters also contribute to food safety needs in this region. This area experiences a large number of natural disasters, including droughts, floods, earthquakes, and internal/civil strife. These disasters can disrupt the food distribution system in the country and lead to improper handling or unsafe storage times for product that ultimately will be consumed, thus exacerbating food safety issues.

WHO has recognized that food safety problems in this region are having a negative impact on the economy regarding exporting product. Some of the problems they have identified are the production of substandard products, shipment of spoiled product, improperly labeled products, products containing harmful levels of preservatives, and poor packaging of product. Because of the lack of food in this region, there is also a concern about importing food of low quality to meet the nutritional demands of the people.

Generally, there is a lack of any food law governing the safe production and distribution of food in this region. Regulatory issues are focused on simply providing food and not on necessarily providing safe food. Much progress needs to be made in this region to implement food laws and to provide safe and wholesome food to the vulnerable population.

European Region

In the European Union, bovine spongiform encephalopathy (BSE) (also known as "mad cow disease") has been a key food safety issue in recent years. *Salmonella* as well as *Campylobacter* are significant causes of food-borne illnesses in the European Union, with many outbreaks linked to poultry products (WHO, 2002b). Despite having well-organized regulatory agencies and oversight of processing in developed countries, food safety problems still occur.

The European Union consists of 25 European nations that approach food safety issues in an integrated fashion. The European Union has faced many challenges and has recently revised food laws and in general implements "cutting edge" food safety programs to control problems faced there. Despite these efforts, WHO reports that the numbers of food-borne illnesses in this region are on the rise. This increase could potentially be due to improvements made to the surveillance programs in the country and an increase in consumer awareness about food-borne illnesses and subsequent reporting of illnesses.

Recent surveillance reported an increase in the number of cases of *Campylobacter* and *Salmonella,* which are still the number one cause of death from food-borne illness, causing up to 75% of the outbreaks (WHO, 2003d). Other issues include botulism, which is a problem in eastern Europe because of the prevalence of home-prepared foods (CSPI, 2005). *Listeria monocytogenes* and *E. coli* O157:H7 are also reported, but to a lesser extent than other pathogens. This may be due to the fact that reporting of these particular pathogens is not mandatory in all countries.

Of primary concern in the European Union is BSE. So far, a total of 150 deaths due to variant Creutzfeldt-Jakob (vCJD) have been reported in the European Union (European Commission, 2002). This is closely linked to exposure to BSE-infected cattle. The European Union has responded to this problem by implementing control measures as follows (European Commission, 2002):

- Controls on animal feed, which includes banning the feeding of any mammalian meat and bone meal (MFM) to cattle, sheep, and goats
- A total suspension on the use of processed animal protein for feed used for animals destined for human consumption

- Heat-processing standards for the treatment of animal waste (133°F at 3 bars of pressure for 20 minutes) to reduce infectivity
- Implementation of surveillance measures for the detection, control, and eradication of BSE
- Specified risk materials (SRMs) removed from the food chain, including spinal cord, brain, eyes, tonsils, and portions of the intestinal tract
- Culling of animals from herds that may have been fed infected feed
- Testing of all cattle over 30 months of age destined for food consumption

Another food safety concern in the European Union is the use of genetically modified/genetically engineered (GE) foods (also called genetically modified organisms, GMOs). Some concerns reported recently in a report by CSPI (2008) include:

- The capability of GE plants and animals to introduce engineered genes into wild populations
- The impact of pesticidal trails on insects that are not pests
- The reduction in the spectrum of other plants and the loss of biodiversity
- The potential for allergic reaction and other adverse effects on human health
- The intellectual property right of the industry and the rights of the farmers to own their crops
- The lack of accountability in case of disaster
- The labeling and traceability of GE organisms

While none of these concerns have been proven as a real danger in the scientific community, the public concerns over the use of genetically modified foods has led to a ban in the European Union.

Another concern in the European Union is the use of antibiotics in food animals and the emergence of antibiotic-resistant microorganisms in the food chain. While there has been an increase in drug-resistant *Salmonella* and *Campylobacter* reported, there is no evidence that using antibiotics in the live animal results in increased amounts of resistance in pathogens isolated from the food supply. Despite this lack of correlation between resistance and drug use, the European Union in 1998 adopted legislation that banned the use of some antibiotics (Phillips, 2007). The use of virginiamycin, spiramycin, tylosin phosphate, and bacitracin was banned as feed additives. Despite this ban, which has been in effect for more than 10 years, a recent study reported that the amount of resistant bacteria in the food supply had not changed dramatically (Ageso et al, 2005).

As mentioned, the European Union takes an integrated approach to food safety from a regulatory perspective. The countries work together, with similar food safety legislation governing most. The regulatory authority is well established and functions effectively. The "food law" was adopted in 2002 by the European Parliament and the council to provide a comprehensive "farm-to-table" approach to food safety (Phillips, 2007). The food law states that no countries will be admitted into the European Union with food safety standards that do not meet minimum requirements. This law has led to the establishment of food safety laboratories and an improved surveillance system, which has led to more accurate reporting of food-borne illnesses.

In addition, training systems are in place for government officials and food industry employees to further ensure the safety of the food supply in the European Union.

An agency, the European Food Safety Authority (EFSA) was developed in 2002 (JEC, 2002). The goal of this agency is to provide "independent scientific advice" to those involved with food safety issues. The EFSA provides input on issues from production to consumption of the food.

The European Union is the most progressive region in the world regarding food traceability and labeling. *Traceability* is defined as the ability to trace and follow food, feed, and ingredients through all stages of production, processing and distribution. There is a labeling requirement in the European Union that ensures that the consumer receives all information possible about the history of the product. The goal is to provide as much information as possible to the consumer so that the consumer is not misled.

CENTRAL AND SOUTH AMERICAN REGION

The situation in Central and South America is similar to that seen in other less-developed countries. The incidence rate of diarrheal disease is high, especially among children, and there is a lack of an adequate surveillance system to assess the true risk of the situation. Poverty, lack of clean water, and a lack of government oversight in this region contribute to the problems experienced by the population. In addition, cultural influences on food preparation practices as well as the climate play a role in increasing the food safety risks.

A report by the Pan American Health Organization (PAHO) concluded that between 1993 and 2002, there were 10,400 outbreaks of food- and waterborne diseases in Latin American and Caribbean countries (Pan American Institute for Food Protection and Zoonoses. An estimated 400,000 illnesses occurred from these outbreaks, with more then 500 deaths reported. *Salmonella, S. aureus,* and *C. perfringens* were all associated with outbreaks (WHO, 2003b). This has had a negative impact on the economy, which thrives on tourism. It is reported that up to 20% of tourists become ill from something they ate when visiting Caribbean countries.

In addition to these reports, *E. coli* O157:H7 and other EHEC bacteria are becoming more prevalent in Central and South America. This pathogen, with the potential to cause kidney failure, is of special concern among children when the incidence of hemolytic uremic syndrome occurs at the highest levels (INPPAZ, 2003).

The regulatory and surveillance systems in Central and South America are starting to develop. SIRVETA, which is a regional food-borne disease surveillance network, is in place for reporting of food-borne disease outbreaks. Also, a network of laboratories is in place, known as INFAL. This network provides information among surveillance laboratories in the region and provides training. A PULSENET for Latin America system is in place, which is similar to PULSENET in the United States. This is a system of surveillance for food-borne illnesses and provides assistance for trace sources of illnesses. Despite these efforts, improvements are needed, and it is speculated that the reporting of food-borne illness is still too low.

NORTH AMERICAN REGION

As stated in the introduction to this chapter, the United States claims to have the safest, cheapest, and most abundant food supply in the world. Despite this claim, the countries in the North American region face food safety challenges. The processing and distribution systems differ radically from other areas of the world, which creates the unique food safety challenges faced in this region.

Mexico, the least-developed nation in the North American region, reported that the estimated number of food-borne illnesses is 6.8 million annually, with a mortality rate of 25 per 100,000 in children under 5, with many of these losses linked to either food- or waterborne illness (Luna et al., 2002). Canada reported 10,000 cases of food-borne illness annually, with the actual estimated number around 2 million (Health Canada, 2004). In the United States, there are an estimated 76 million cases of food-borne illness with 325,000 hospitalizations and 5,000 deaths per year (Mead et al, 1999). The CDC estimates that every year, one in four Americans will become ill from a food-borne illness (CDC, 2004).

A relatively new surveillance system in the United States has indicated that there is a decline in illnesses resulting from *Yersinia, Campylobacter, E. coli* O157:H7, and *Salmonella*. However, illnesses associated with *Listeria monocytogenes* have not declined significantly.

There are many educational efforts as well as regulatory programs in place to prevent food-borne illnesses in this region. However, as stated, food-borne illnesses are still occurring and result in significant monetary losses for the countries in this region. In the United States, many industries have been required by regulatory agencies to implement HACCP systems in their operations. To date, meat and poultry industries and seafood- and juice-processing industries must have HACCP systems in place. Many other industries follow HACCP systems but are not required by law to do so.

In addition to outbreaks of food-borne disease, a new threat, "bioterroism," is a concern in the food industry in this region. WHO recently reported that, due to the diversity of the food supply and global distribution, intentional contamination with toxic agents could occur and should be addressed (WHO, 2002b). In Canada, the CEPR (Centre for Emergency Preparedness and Response) was created in 2000 to coordinate food security (Canadian Food Inspection Agency, 2008). This organization is responsible for creating emergency plans for food safety emergencies. In 2002, the United States approved the Bioterrorism Act, which empowers the FDA to implement a system of record keeping and tracing procedures to protect the U.S. food supply.

While food-borne pathogens are still the primary concern in this region, there has been a small problem with BSE. In May 2003, a cow in Canada tested positive, and therefore measures were put into place to prevent the spread of this illness. Specific bans include excluding SRMs from the human food chain and enhanced surveillance. This enhanced surveillance resulted in identification of an additional two cases in January 2005.

In the United States, December 2003 marked the first identified case of BSE in the food chain. It was found in an adult cow in Washington State that been shipped to the United States from Canada. The identification of this animal increased surveillance in the United States and banned particular materials from entering the food supply. To date, there have been no cases of BSE reported in Mexico.

The approach to genetically modified foods in this region differs from the view held in the European Union. In Canada, any genetically modified food must be made known by notifying the Health Products and Food Branch. This branch will conduct a safety assessment of the product prior to marketing. In Mexico, the Comision Intersecretarial de Bioseguridad y Organismos Geneticamante Modificados oversees genetically modified foods.

In the United States, the FDA, the U.S. Department of Agriculture (USDA), and the Environmental Protection Agency (EPA) all play a role in oversight of genetically modified plants. Approximately 40% of field corn, 80% of soybeans, and 73% of cotton grown in the United States is genetically engineered (CSPI, 2005). The FDA ensures that plants are safe for human consumption.

In the United States and Canada, consumer education plays an important role in controlling food safety issues. In the United States, the Fight BAC! campaign is a partnership among government and consumer and industry organizations to provide consumers with safe food-handling information. In Canada, a governmental agency, the Canadian Partnership for Food Safety Education, was formed, and it launched a Fight BAC! in 1998 (Fight BAC!, 1998).

Mexico also has a limited amount of information available to consumers for educational efforts. The Office for Consumers' Communication has been formed to educate consumers on safe food-handling practices.

In North America, regulatory agencies and a wide array of food safety laws govern the way food is handled from farm to table. Health Canada administers food safety regulatory standards in Canada under the Food and Drugs Act. The minister of agriculture oversees the CFIA (Canadian Food Inspection Agency), which is responsible for inspections and oversight of food safety regulations in Canada.

In 2001, Mexico formed the National Service of Agro-food Safety and Quality, which provides oversight for food safety regulations. This is a relatively new group that was formed to improve food safety standards in Mexico.

The most complicated regulatory structure is in the United States. Both the USDA and the FDA play a role in oversight of food safety policy. The USDA primarily oversees meat and poultry production, while the FDA has oversight over other food products. This is somewhat of a generalization because 12 different government agencies are involved at some point with food safety policy.

In the United States, recalls of product are voluntary. The recalls are not initiated by either the USDA or the FDA, but by the food-processing company. This system allows the processing plant to identify their internal problems and take action when necessary.

CONCLUSION

There are food safety issues in every segment of the world. In very general terms, it appears that in developing countries, the problem could be solved with the establishment of government agencies and a set of standards to control food safety issues. However, the problem goes beyond this because even in industrialized countries with well-established programs to control food safety issues, there are still significant numbers of illnesses occurring annually, and a whole set of new/different problems arises. There are a number of needs that need to be addressed.

INDUSTRIALIZED NATIONS

- Coordination and communication among government agencies for consistent enforcement of food safety regulatory policies
- Research to study emerging issues that may be a problem in the future
- More information about the safety and long-term impact of genetically modified foods

DEVELOPING COUNTRIES

- Establishment of government agencies for development and oversight of food safety policy
- Assistance with providing the nutritional needs of the population and thus improvement of the health status for less susceptibility to foodborne pathogens
- Establish adequate food preparation and preservation facilities
- Establish educational programs that promote food safety

Food safety has become a topic of worldwide concern due to the ability of multiple countries to import and export food products. Providing safe food for a world that is increasing in population and decreasing in resources has become increasingly difficult.

REFERENCES

Ageso, J., H. Emborg, O.E. Heuer, V.F. Jensen, A.M Seyfarth, A.M. Hammerum, L. Bagger-Skjot, A.M. Rogues, C. Brandt, R.L. Skov, and D.L. Monnet. 2005. DANMAP 2005—use of antimicrobial agents and occurrence of antimicrobial resistance in bacteria from food animals, foods and humans in Denmark. Available at: http://www.danmap.org/pdf-Files/Danmap_2005.pdf.

Australian New Zealand Food Standards. 1999. Incidence of Food Borne Illnesses. Available at www.foodstandards.gov.au.

Canadian Food Inspection Agency. 2008. CFIA *emergency response plan*. Available at: http://www.inspection.gc.ca/english/anima/heasan/man/ahfppfsa/ahfppfsa_5_4e.shtml.

Center for Science in the Public Interest. 2005. Food safety around the world. Available at: http://safefoodinternational.org/local_global.pdf.

Center for Science in the Public Interest. 2008. Biotechnology project: frequently-asked questions. Available at: http://www.cspinet.org/biotech/faq.html.

Centers for Disease Control and Prevention. 2004. Preliminary FoodNet data on the incidence of infection with pathogens transmitted commonly through food—selected sites, United States, 2003. *Morbidity and Mortality Weekly Report* 53:338–343.

Centers for Disease Control and Prevention. 2006. Update on multi-state outbreak of E. coli O157:H7 infections from fresh spinach, October 6, 2006. Available at: http://www.cdc.gov/foodborne/ecolispinach/100606.htm.

Directive 2003/89/EC of the European Parliament and of the Council of 10 November 2003: amending 2000/13/EC as regards indication of the ingredients present in foodstuffs. 2003. *Official Journal of the European Communities* L30815-18. Available at: http://eur-lex.europa.eu/LexUriServ/LexUriServ.do?uri=OJ:L:2003:308:0015:0018:EN:PDF.

European Commission. 2002. Questions and answers on BSE. Memo/03/3. January 8, 2002. Available at: http://ec.europa.eu/dgs/health_consumer/library/press/m03_3_en.pdf.

FightBAC!™ 1998 campaign. Available at: http://www.fightbac.org.

Food and Agriculture Organization. 2004. Food safety and international trade in the Near East Region. Discussion paper for the Technical Consultation on Food Safety and International Trade in the Near East (Cairo, December 2003) in preparation for the 27th FAO Regional Conference for the Near East (Doha, Qatar, March 2004). Available at: ftp://ftp.fao.org/es/esn/food/nerc_report.pdf.

Health Canada. 1998. The Canadian Partnership for Consumer Food Safety Education. January 1, 1998. Available at: http://www.hc-sc.gc.ca/food-aliment/mh-dm/mhe-dme/e_fightbac. html.

Health Canada. 2004. Health Canada Policy—Food Program—Food Safety Assessment Program. November 1, 2004. Available at: http://www.hc-sc.gc.ca/fn-an/securit/eval/pol/index_e.html.

Huan, N.H. and D.T. Anh. 2001. Vietnam promotes solutions to pesticide risks. *Pesticides News* 53:6–7.

Luna, J.L.F. 2002. Communication and participation—the experience of Mexico. FAO/WHO Global Forum of Food Safety Regulators, Morocco, January 28–30. GF 01/6. Available at: http://www.fao.org/docrep/meeting/004/y2122e.htm.

Mead, P.S., L. Slutsker, V. Dietz, L.F. McCaig, J.S. Bresee, C. Shapiro, P.M. Griffin, and R.V. Tauxe. 1999. Food-related illness and death in the United States. *Emerging Infectious Diseases* 5:607–625.

OzFoodNet Working Group. 2003. Food borne disease in Australia: incidence, notifications and outbreaks, annual report of the OzFoodNet network, 2002. *Communicable Disease Intelligence* 27:209–243.

Phillips, I. 2007. Withdrawal of growth-promoting antibiotics in Europe and its effects in relation to human health. *International Journal of Antimicrobial Agents* 30:101–107.

United States Food and Drug Administration. 2003. FY2003 budget program narratives. Available at: http://www.fda.gov/oc/oms/ofm/budget/2003/Narratives.pdf.

United States Food and Drug Administration. 2008. Bad bug book. Available at: <http://vm.cfsan.fda.gov/~mow/intro.html.

World Health Organization. 1999. Food Safety Programme: Food safety—an essential public health issue for the new millennium. Available at: http://www.who.int/foodsafety/publications/general/brochure_1999/en/index.html.

World Health Organization. 2000. Food safety regional programme evaluation report prot II: country profiles. Available at: http://www.afro.who.int/des/fos/country_profiles/index.html.

World Health Organization. 2002a, Core health data selected indicators. Data updated to 2002; and the health situation analysis and trends summary. Available at: http://www.paho.org/English/DD/AIS/cp_484.htm.

World Health Organization. 2002b. Food safety issues: terrorist threats to food; guidance for establishing and strengthening prevention and response systems. Food Safety Department. Available at: http://www.who.int/foodsafety/publications/general/en/terrorist.pdf.

World Health Organization. 2002c. Health situation in the South East Asia Region 1998–2000. Office for the South East Asia Region. Available at: http://www.searo.who.int/EN/Section1243/Section1382/Section1386/Section1898_9352.htm.

World Health Organization. 2002d. Statistical information on food-borne disease in Europe—microbiological and chemical hazards. FAO/WHO Pan-European Conference on Food Safety and Quality, Budapest, Hungary, February 25–28. PEC 01/04. Available at: http://www.fao.org/docrep/meeting/004/x6865e.htm.

World Health Organization. 2003a. Food safety issues—public health significance of foodborne illnesses. Food Safety Program, Regional Office for the Western Pacific. Available at: http://www.wpro.who.int/NR/rdonlyres/F2F62765-B851–4982-BE63–5AC9B52E81A2/0/food_safety01.pdf.

World Health Organization. 2003b. Proposed plan of action of the Pan American Institute for Food Protection and Zoonoses (INPPAZ), 2004–2005. Pan American Health Association, 13th Inter American meeting, at the ministerial level, on health and agriculture, Washington, DC, April 24–25. RIMSA 13/5. Available at: http://www.ops-oms.org/English/AD/DPC/VP/rimsa13–05-e.pdf.

World Health Organization. 2003c. Several foodborne diseases are increasing in Europe. Press release EURO/16/03, Copenhagen, Rome, Berlin, December 16. Available at: http://www.euro.who.int/mediacentre/PR/2003/20031212_2.

World Health Organization. 2003d. WHO Surveillance Programme for Control of Foodborne Infections and Intoxications in Europe, 8th report—1999–2000. Available at: http://www.bfr.bund.de/internet/8threport/8threp_fr.htm.

World Health Organization. 2004a. Bangladesh country paper. FAO/WHO Regional Conference on Food Safety for Asia and Pacific, Seremban, Malaysia, May 24–27. CRD 15. Available at: http://www.fao.org/docrep/meeting/006/ad730e/ad730e00.htm.

World Health Organization. 2004b. Developing and Maintaining Food Safety Control Systems for Africa Current Status and Prospects For Change. Second FAO/WHO Global Forum of Food Safety Regulators, Bangkok, Thailand, October 12–14. CRD 32. Prepared by WHO Regional Office for Africa. Available at: http://www.fao.org/docrep/meeting/008/ae144e/ae144e00.htm.

World Health Organization. 2004c. Epidemio-surveillance of foodborne diseases and food safety rapid alert systems. Second FAO/WHO global forum of food safety regulators, Bangkok, Thailand, October 12–14. Prepared by Thailand. Available at: http://www.fao.org/docrep/meeting/008/ae192e.htm.

World Health Organization. 2004d. Fact sheet 4: food safety in emergencies. Regional Office for Africa, Division of Healthy Environments and Sustainable Development, Food Safety Unit. Available at: http://www.afro.who.int/des/fos/afro_codex-fact-sheets/fact4_latest-emergencies.pdf.

World Health Organization. 2004e. International cooperation on food contamination monitoring and foodborne disease surveillance. A case study in the AMRO region. Second FAO/WHO global forum of food safety regulators, Bangkok, Thailand, October 12–14. CRD 66. Available at: ftp://ftp.fao.org/docrep/fao/meeting/008/ae196e.pdf.

World Health Organization. 2004f. Prevention and management system for food poisoning in Korea. FAO/WHO Regional Conference on Food Safety for Asia and Pacific, Seremban, Malaysia, May 24–27. CRD 11, Republic of Korea. Available at: http://www.fao.org/docrep/MEETING/006/AD704E/AD704E00.HTM.

World Health Organization. 2004g. Prevention and response to intentional contamination. Second FAO/WHO global forum of food safety regulators, Bangkok, Thailand, October 12–. CRD 47. Available at: ftp://ftp.fao.org/docrep/fao/meeting/008/ae173e/ae173e00.pdf.

World Health Organization. 2007. Cholera: fact sheet no. 107. Available at: http://www.who.int/mediacentre/factsheets/fs107/en/.

World Health Organization/FAO. 2005. The impact of current food safety systems in the Near East/Eastern Mediterranean region on human health. FAO/WHO Regional Meeting on Food Safety for the Near East, Amman, Jordan, March 5–6. NEM 05/3. Available at: ftp://ftp.fao.org/es/esn/food/meetings/NE_wp2_en.pdf.

Section V

Food Production: Synergy of Science, Technology, and Human Ingenuity

15 Challenge and Threats to Sustainable Food Production

Larry W. Harrington and Peter R. Hobbs

CONTENTS

ABSTRACT

Previous chapters have focused on food needs. This chapter introduces the challenges and threats that must be addressed if food needs are to be met in a sustainable and equitable way. Challenges and threats introduced here include demographic change, water scarcity, land degradation, climate change, energy insecurity, and loss of ecosystem services. Demographic change, along with income growth and urbanization, determines the shape of future food needs. Food insecurity will increasingly be concentrated in low-income and food-insecure countries where population growth remains relatively high. Water scarcity is becoming a major threat to sustainable production where food demand is rising (especially for "water-intensive" products such as fresh fruits and vegetables); where competition is increasing between agricultural and nonagricultural uses of water; and where river basins are "closing," meaning that the demand for water has come to exceed its supply and rivers no longer reach the sea. Climate change has also come to be recognized as a threat to sustainable food production. Analysis of multiple scenarios suggests that climate change will have serious distributional impacts across countries, with poor countries in warm environments suffering the most. Questions of energy insecurity pose further threats. Higher energy prices will make manufacturing and applying fertilizer, running farm machinery, pumping water, and transporting inputs and marketing products more expensive. A final threat to sustainable food production is the loss of ecosystem services. Past successes in producing food may be part of a larger process that threatens future food production. Agriculture will need technical, institutional, and policy innovations to achieve sustainable improvements in productivity, equity, and resilience through advances in the management of energy, soils, water, livestock, and crops.

INTRODUCTION

Since the beginnings of agriculture, change has been the norm. An important part of human history has been the coevolution of food needs and preferences, food production systems, and resource management practices. Over time, as resource management has improved and food production systems have became more productive, food needs and preferences have evolved. Evolving needs and preferences in turn have placed further demands on production systems, pushing them in new directions. This process of coevolution continues, but at a much faster pace than ever before.

Previous chapters have focused on food needs. They have portrayed some of the relationships among food choice, nutrition, and human health and have discussed how diseases may develop when food habits (or food availability) are poorly matched with nutritional requirements. They have described how food choice has influenced and has been influenced by culture and religion; how the evolution of food choice has helped cultures define themselves; and how food traditions have contributed to human perceptions of identity.

Much of the rest of this book focuses on how these food needs can be met. An important challenge is to sustainably meet evolving and growing food needs in the face of unprecedented change in the global environment. Such changes pose threats but also create opportunities.

This chapter introduces the unfolding challenges and threats that must be addressed if food needs are to be met and the evolving supply-times-demand equation "solved" in a sustainable and equitable way. Challenges and threats introduced here include demographic change, water scarcity, land degradation, climate change, energy insecurity, and loss of ecosystem services. Agriculture will need technical, institutional, and policy innovations to achieve sustainable improvements in productivity, equity and resilience through advances in the management of energy, soils, water, livestock, and crops. How these improvements may be achieved is the topic of subsequent chapters.

DEMOGRAPHIC CHANGE

Just a few decades ago, the world had half as many people as it does today. They were less wealthy; consumed fewer calories; ate fewer fresh fruits, vegetables, and animal products; depleted less water; emitted fewer greenhouse gases; and in many other ways placed less stress on the environment.

Today, the world's population is around 6.5 billion. Demographers suggest that global population growth will slow and eventually peak at about 9.2 billion later this century (Alexandratos 2005). This 50% increase in the world's population will bring with it a more than proportional increase in food needs. This is because urbanization and income growth will lead to more diversified diets (more livestock and fish products and high-value crops) and higher per capita consumption of calories. The exact way in which diets will evolve in different societies will be influenced by history, culture, and religion. Food needs/demands are swiftly evolving even as the production environment itself is transformed beyond recognition.

In the future, the world's population will not only be larger and slower growing; it will also be older, more urban, and increasingly concentrated in developing countries. Some analysts see the present decade as a unique moment in the history of humanity: For the first time, old people will outnumber young people, urban people will outnumber rural people, and the median woman worldwide will barely have enough children to replace herself and the father (Cohen 2005).

However, the global level of analysis masks important differences between countries. Peak populations in some low-income and food-insecure countries are projected to be multiples of present ones (Alexandratos 2005). Many issues relating to food security, water depletion, land degradation, and environmental destruction will increasingly be concentrated in such countries.

In the meantime, much unfinished business remains. In 2003, 850 million people in the world were food insecure, most of them in South Asia and sub-Saharan Africa, where the average daily per capita food supplies in 2003 were only 2400 kcal and 2200 kcal, respectively, well below the world average of 2800 kcal (Comprehensive Assessment 2007). Fortunately, there is evidence to suggest that successful development and dissemination of productivity-improving agricultural technologies can help slow population growth by generating employment, raising incomes, and improving food availability (Vosti et al. 1994). (It is recognized that sociocultural and demographic variables such as female literacy and marriage rates are equally important in determining fertility change.)

The challenge for food production technology and resource management practices is considerable. Several goals must be met simultaneously. Food security must be achieved for rapidly rising populations in poor countries, but this must be done in ways that allow reallocation of scarce water to nonagricultural uses, that reverse or slow land degradation, and that preserve a wide range of ecosystem services. And, all of this must be done in the context of ongoing processes of climate change that are likely to have an impact most heavily on developing countries in tropical environments. Further information on population growth and food availability is provided in Chapters 21 through 24.

WATER SCARCITY

Water scarcity is becoming a major threat to sustainable food production in many areas (see also Chapter 19). This is particularly the case in areas where the following trends are converging:

- Demand for food is rising, driven by population growth. Without improvements in productivity, ever-increasing amounts of water will be needed to grow food.
- Demand is growing particularly fast for "water-intensive" products such as fresh fruits and vegetables. This demand growth, driven by rising incomes, will further increase water use in agriculture.
- Competition is increasing between agricultural and nonagricultural uses of water (direct consumption, hydropower, industry, ecosystem services, and the environment). Improvements in agricultural water productivity will be needed to maintain desired levels of food production—more food from less water.
- River basins are "closing," meaning that the demand for water has come to exceed its supply, and rivers at times no longer reach the sea. Examples of closed basins may be found in Africa (Limpopo), South Asia (Indus), China (Yellow), North America (Colorado), and Australia (Murray-Darling), among others.

At the global level, food production from rain-fed and irrigated agriculture currently uses around 7,130 cubic kilometers of water each year. By the year 2050, this is expected to increase to around 13,000 cubic kilometers. Annual freshwater withdrawals from lakes, rivers, and groundwater are presently estimated at about 3,800 cubic kilometers, with around 70% used for agriculture (Comprehensive Assessment 2007). Where such water is scarce, pressures to reduce agricultural uses in favor of other uses may become irresistible.

The global level, however, is not necessarily the best place to analyze water scarcity. At the global level, there appears to be adequate water for agriculture, industry, and domestic and other uses. Water scarcity is in fact highly concentrated. The most severe problems are found in such regions as southeastern Australia, northern and western China, central India, central and western Asia, North Africa, southern Africa and the Sahelian zone of West Africa, northeast Brazil, and northern Mexico and contiguous areas of the southwest of the United States (Comprehensive Assessment 2007).

In terms of human needs, a "water-stress threshold" may be said to have been reached when per capita water availability is less than 1700 cubic meters per year. At present, around 700 million people live below this threshold, and by 2025 this number may reach 3 billion. Even when water is not physically scarce, many people may not have access to adequate water supplies. People may be excluded by their poverty, their lack of legal rights to water, or simply by a lack of water collection and distribution infrastructure. Some analysts have concluded that water scarcity is in many instances an artifact of political processes and institutions that discriminate against the poor (UNDP 2006).

Water scarcity is not the only water-related problem that threatens sustainable food production. Floods are an extreme form of water insecurity. So are droughts, which can occur even where annual rainfall is on average adequate to support agriculture.

As water becomes scarce, further problems emerge. Food production may suffer as water is diverted from agricultural to nonagricultural uses. Irrigated lands may become salinized, groundwater overexploited and depleted, and freshwater fisheries damaged. In addition, pollution may increase, lakes may shrink, ecosystems may become degraded, biodiversity may be reduced, and habitats may be threatened. Competition and conflict over water also becomes more likely—between nations in a river basin, between pastoralists and farmers, between rural and urban areas, and in general, between upstream and downstream populations (Harrington et al. 2006). Further analysis of water issues is provided in Chapters 18 and 19.

LAND DEGRADATION

Land degradation in its many forms is another threat to sustainable food production. *Land degradation* is understood to mean a loss in intrinsic quality that leads to a decline in the capacity of land to satisfy particular uses (Blaikie and Brookfield 1987). Serious degradation can result in land losing its capacity to satisfy a wide range of productive uses.

Land degradation and rehabilitation are influenced by natural forces as well as by humans. *Net degradation* is the difference between naturally degrading processes plus human-induced degradation on the one hand and natural reproduction plus restorative management on the other. Land degradation is understood to include wind and water erosion; loss of organic matter and soil nutrients; soil salinization, acidification, compaction, and crusting; and reduced biological activity in soils.

At the global level, land availability appears adequate to meet future growth in demand for food. Around 11% of the world's land surface is thought to be generally suitable for food and fiber production, while another 24% is used for grazing and 31% for forests (Eswaran et al. 1999). Land degradation, however, whittles away at these totals, effectively reducing the quantity and quality of land available for agricultural uses. Unfortunately, there are few good estimates of the global extent of land degradation and its effect on productivity (World Bank 2007). Some analysts continue to rely on the GLASOD (Global Assessment of Human Induced Soil Degradation) study, which estimated that at the global level, forest areas, grazing areas, and agricultural areas are 18, 21, and 38% degraded, respectively (Oldeman 1992). These estimates, based on consensus judgment, are somewhat contentious and in any event say little about the effect of degradation on productivity.

Land degradation is considered a threat to sustainable food production because it reduces present and undermines future productivity. In point of fact, however, the effect of land degradation on productivity is notoriously difficult to measure. Farmers may mask the effects of degradation by adjusting land use or applying extra inputs. In addition, there may be "threshold effects." Deep, highly fertile soils may undergo erosion for extended periods of time with little loss in productivity—until at some point they become so thin that rapid yield losses begin to unfold (Scherr and Yadav 1996).

Land degradation has an additional dimension: Its effects are often felt beyond the field or farm where it occurs. Erosion on hillsides on upper catchments can result in siltation of downstream irrigation infrastructure, degradation of water quality for downstream urban consumers, and reduced productivity of downstream fisheries.

It has been estimated that the global cost of replacing water storage capacity lost to sedimentation might be as much as $13 billion per year (World Bank 2007).

Because land management in upper catchments affects everyone else in a river basin, policy makers tend to emphasize conservation goals for these areas. But, catchments may also be home to large numbers of poor people. In some parts of the world, upper catchment communities are often economically, politically, and culturally marginalized, and their limited livelihood options center on the exploitation of land, water, and forest resources. Equity considerations suggest that resource management in upper catchments should allow for continued productive use of land, water, and other resources by local communities, while also keeping in mind the interests of downstream populations (Harrington et al. 2006).

Like water scarcity, land degradation demonstrates substantial spatial variability. Land degradation in the form of erosion is usually more serious on sloping lands such as are found in the foothills of the Himalayas; sloping areas in the Andes, southern China, and Southeast Asia; rangelands in Africa and Central and West Asia; and the semiarid lands of the Sahel. Pressure to farm hillsides and other marginal areas is reduced when agricultural productivity and employment are increased in favored production areas.

Favored, intensively farmed agricultural areas are not immune from land degradation. Here, however, it is not a result of erosion, but rather detrimental soil chemical, physical, and biological processes associated with unwise use of fertilizers and pesticides, excessive tillage, unsuitable rotations, and inappropriate water management practices. These factors threaten sustainable food production just as much as soil loss from a hillside field—perhaps moreso given the heavy reliance placed on these favored areas for maintaining global food security. Further information on soil management and conservation is provided in Chapter 18.

CLIMATE CHANGE

Climate change has recently come to be recognized as a serious threat to sustainable food production. There are widespread concerns that climate change may undermine the capacity of agroecosystems to meet food needs. In addition, it may trigger severe water shortages. Higher temperatures and the associated changes in hydrological regimes may shorten growing seasons, increase the frequency of extreme and destructive weather events, and shift the incidence of pests and diseases (Harrington et al. 2006).

Analysis of multiple scenarios suggests that climate change will have serious distributional impacts across countries, with poor countries in warm environments suffering the most. Countries in low latitudes start with relatively high temperatures. Further warming pushes these countries farther and farther away from optimal temperatures for the production of climate-sensitive crops (Mendelsohn et al. 2006). Even moderate warming in tropical countries (1 to 2°C) can significantly reduce yields of crops (such as maize) that are already near the limit of their heat tolerance. For temperature increases above 3°C, yield losses are likely to be widespread and severe. Maize and wheat yields could decline by as much as 20 to 40% in parts of sub-Saharan Africa, Central America, and tropical Asia (World Bank 2007).

Some adverse effects of climate change are water related. Many of the world's most water-stressed areas will get even less water and in less-predictable patterns. Water availability is expected to fall in East Africa, the Sahel, and southern Africa, with large productivity losses in food production. Some projections for East Africa suggest that productivity losses may be as much as 33, 20, and 18% for maize, sorghum, and millet, respectively (UNDP 2006). Climate change may also accelerate glacial melt, reducing water availability for particular areas in East Asia, South Asia, and Latin America and may even disrupt monsoon patterns in South Asia, with enormous consequences for food security.

At the global level, the world is expected to be able to continue to feed itself for the next century, even in the presence of anticipated climate change. However, this outcome will be achieved through increased production in developed countries located in temperate climates, which mostly benefit from climate change, especially in high-rainfall temperate areas where moderate increases in CO_2 levels may actually stimulate plant growth. Increased production in temperate climates compensates for reduced production in tropical and subtropical climates (Parry et al. 2004). This begs the question of food access and affordability—how will the poor be able to purchase food they can no longer produce for themselves? Globalization of food distribution is unlikely to be the answer: With few exceptions, food flows from surplus to deficit areas only when the latter can pay for it.

As climate change continues to unfold, however, it is possible that agriculture may become less productive even in temperate climates. In these areas, it appears that there initially may be a parabolic relationship between temperature and impacts (benefits at lower temperature increases, damages at higher increases). Beyond an approximate 3 to 4°C increase in GMT (global mean temperature), however, most studies show increasing adverse impacts. It appears that, beyond several degrees of GMT increase, damages are likely to be adverse, increasing, and ubiquitous (Hitz and Smith 2004).

The future impacts of climate change need to be incorporated into development planning. They need to be taken into account when selecting priorities for agricultural and resource management research and in the design of water storage and control investments. Research and investment can help some farm communities adapt to the adverse effects of climate change through advances in the management of energy, soils, water, livestock, and crops. For some communities, however, eventual out-migration appears to be the only solution. Issues of climate change are further analyzed in Chapter 16.

ENERGY INSECURITY

Questions of energy insecurity pose further threats to sustainable food production. Higher energy prices make manufacturing and applying fertilizer, running farm machinery, pumping water, and transporting inputs and marketing products more expensive. Other things being equal, this is likely to result in reduced input use, less irrigation, lower yields, higher marketing margins, and higher food prices for consumers.

There can be little doubt that energy prices will be higher in the future than they have been for the past several decades. Higher energy prices will particularly affect developed country agriculture, where food production is energy intensive. For example, it takes the fossil fuel equivalent of 160 L of oil to produce a ton of maize in the United States, versus 4.8 L in Mexico (using traditional methods). In 2005, energy costs were estimated to account for about 16% of agricultural production costs in the United States (World Bank 2007).

Developing country agriculture is not immune from these effects. Food security in China and India depends on maintaining high crop yields in favored food-producing regions like the Punjab. Fertilizer is important in maintaining these yield levels. In sub-Saharan Africa, increased fertilizer use is seen as key to improving farm system productivity. Many developing country farmers have been able to escape from poverty by producing high-value crops and exporting them, often via air cargo, to developed countries. Higher energy prices will have an impact on all of these.

The key role of energy in producing food and in other ways driving development is receiving increased recognition. The government of India, for example, considers energy security second only to food security in national priorities. Reliable access to gas and oil resources is considered a vital national interest, one so important that it has a major influence on foreign policy stances such as relationships with Iran and Russia (Blank 2005).

Among the most controversial topics regarding energy in food production is that of biofuels. These include wood, charcoal, ethanol, biodiesel, and biogas. Proponents see them as environmentally friendly sources of energy that will help address problems of climate change as well as being a boon to farmers. Skeptics argue that biofuel production will threaten food supplies for the poor while producing few environmental benefits.

The balance of costs and benefits of using bioenergy very much depends on the source of the biofuel and how it is processed. Here are a few of many factors that affect the cost/benefit equation:

- A biofuel industry could provide developing country farmers a use for crop residues (von Braun and Pachauri 2005). This, however, may hinder efforts to introduce resource-conserving practices such as conservation agriculture that require the use of residues for soil cover and are needed to feed soil microbes, an important component of biological soil health.
- Raw material for biofuel such as the oil-bearing plant *Jatropha curcus* can be grown on marginal areas unsuitable for crop production (von Braun and Pachauri 2005). This strategy, however, may have unknown consequences for dryland ecosystems. Yields of oil-bearing biofuel crops are also low, and their production may not be profitable on marginal lands with poor soil moisture.
- The use of maize and other cereals to produce ethanol increases their price—good for farmers, but less good for food consumers. Biofuels are most likely to impinge on food security (by raising food prices) when they are processed from cereals. However, if growing crops for biofuel production generates employment and raises incomes for poor farmers, food security might still be increased.

- Greater reliance on bioenergy will further intensify competition for water, soil nutrients, and land and will increase the amount of water used in agriculture. Some estimates suggest that by 2050, the amount of water depleted for biofuel production may be similar to what is used today for all of agriculture (Comprehensive Assessment 2007). Nutrient depletion by biofuel crops will also need to be balanced by nutrient additions as otherwise crop productivity is likely to decline over time.

There are numerous areas for research on biofuels: how to best produce them from sources other than cereals, their consequences for water use, their role in slowing climate change, and their relative advantage when compared to alternative sources of energy like solar, wind, geothermal, and tidal. The question of biofuels is further analyzed in Chapter 17.

LOSS OF ECOSYSTEM SERVICES

A final threat to sustainable food production is the loss of ecosystem services. An *ecosystem* is defined as a dynamic complex of plant, animal, and microorganism communities and the nonliving environment interacting as a functional unit (Millennium Ecosystem Assessment [MEA] 2005). *Ecosystem services* are the benefits people obtain from ecosystems. These benefits include food, water, disease management, climate regulation, spiritual fulfillment, and aesthetic enjoyment. The concept of ecosystem services, then, brings together the discussion in other sections on food needs, water scarcity, land degradation, and climate change, while integrating these with many other factors. Humans are fundamentally dependent on the flow of ecosystem services.

The MEA (2005) found that over the past few decades, humans have had a greater impact on ecosystems than at any other time in history. People have been successful in meeting rapidly growing demands for food, freshwater, timber, fiber, and fuel but at the cost of many other ecosystem services. A full 60% of ecosystem services examined were found to be subject to degradation or unsustainable use. Of special concern are services relating to freshwater, capture fisheries, air and water purification, and the regulation of regional and local climate, natural hazards, and pests. There has also been a substantial and largely irreversible loss in the diversity of life on Earth. Among other things, the MEA found that during the past few decades:

- More land was converted to cropland than was done in the preceding 150 years.
- Approximately 20% of the world's coral reefs and 35% of its mangrove area were lost.
- More water is held in reservoirs than in natural rivers because of dam construction.
- Flows of nitrogen in terrestrial ecosystems have doubled.
- Across taxonomic groups, the population size or range of most species is declining.

- The spatial distribution of species on Earth is becoming more homogeneous.
- Some 10 to 30% of mammal, bird, and amphibian species are threatened with extinction.
- Genetic diversity has declined globally, particularly among cultivated species.

A basic conclusion of the MEA is that:

The changes that have been made to ecosystems have contributed to substantial net gains in human well-being and economic development, but these gains have been achieved at growing costs in the form of the degradation of many ecosystem services, increased risks of nonlinear changes, and the exacerbation of poverty for some groups of people. These problems, unless addressed, will substantially diminish the benefits that future generations obtain from ecosystems. (p. 5)

This lesson, when applied to sustainable food production, means that past successes in producing food may be part of a larger process that threatens future food production. One purpose of this book is to present information that may be helpful in amending the negative impacts of increased food production on ecosystem services.

INTERRELATIONSHIPS

In the sections presented, several challenges or threats to sustainable food production were introduced. These include demographic change, water scarcity, land degradation, climate change, energy insecurity, and loss of ecosystem services. There are many interrelationships among these challenges. A surprisingly large number of interrelationships feature coevolution and trade-offs.

Clearly, a major driver is demographic change. Along with changes in incomes, this determines the magnitude and structure of food needs, with some shaping and filtering by cultural norms and religious beliefs. But, success in meeting food needs as well as other development needs is part of the process by which population growth itself slows and eventually peaks. There is coevolution among demographic change, food production, and development.

Farming practices, including land and water management, influence resource quality. Water scarcity and land quality in turn influence the productivity of agriculture. Coevolution exists between resource management, farming practices, and resource quantity and quality.

There is adequate evidence that demographic change, combined with urbanization, modern transport, and other forms of technical change, is driving climate change. Climate change in turn is having a large effect on water availability, crop yields, farm system productivity, and the production of other ecosystem services. People seek to mitigate climate change while maintaining increasing food production and maintaining energy security, critically important for modern lifestyles. An important element in achieving this may be an increased reliance on biofuels, which, however, may exacerbate water shortages, accelerate land degradation, and have other unknown consequences for the environment. Thus, there are subtle but important kinds of trade-offs among energy use, climate change, and food and biofuel production.

Agriculture will need technical, institutional, and policy innovations to achieve sustainable improvements in productivity, equity, and resilience through advances in the management of energy, soils, water, livestock, and crops. The evaluation of alternative interventions, however, should be sufficiently broad. It is clear that interventions to address one challenge or threat to sustainable food production must take account of interrelationships, coevolution, and trade-offs with other challenges and threats.

MOVING FORWARD

Science continues to provide innovations capable of making food production more productive and sustainable. These include innovations in crop improvement to increase yield potential, improve biotic and abiotic stress tolerance, and make crops more competitive with weeds. They also include crop and farm system management innovations such as no till, conservation agriculture, residue management, precision agriculture, innovative rotations, advanced crop-livestock integration, and on-farm water harvesting. Some innovations go beyond the farm level and are implemented at the community level, such as the design and construction of multiple-use water systems, catchment-level water harvesting and sharing, and even community-level aquaculture in areas prone to monsoon season flooding. There are also water management innovations such as conjunctive water use, drip irrigation, treadle pumps, and so on.

Not all innovations take the form of new technologies. Changes in property rights, so that farmers have the right to use their own crop residues, are sometimes needed before conservation agriculture is feasible at the farm level. Collective action in water system management by water users' groups is usually essential to improving water productivity in irrigated systems. Effective programs of "payment for environmental services" begin with institutional innovations for payment targeting and verification.

Finally, policy reform is almost always needed to support and foster institutional and technical innovation. A good example comes from groundwater depletion— only water pricing or top-down control of tubewell use can reverse this classic case of private overuse of a common property resource. Policies can encourage or discourage conservation agriculture, make water availability more or less equitable, encourage or discourage the user of fertilizers, and increase or decrease marketing margins through infrastructure development.

The next several chapters describe in more detail the potential of new technical, institutional, and policy innovations to ensure adequate food for all in the twenty-first century.

CONCLUSIONS AND STUDY TOPICS

Several key points have emerged from this chapter:

- Food needs are influenced by demographic change and income growth as well as by culture and religion. Population increase and food insecurity are increasingly concentrated in particular countries in the developing world.
- Research that seeks to sustainably increase food production should take account of threats and challenges, among them water scarcity, land degradation, climate change, energy insecurity, and loss of ecosystem services. Most of these challenges are relatively more serious in developing countries. Even climate change is expected to negatively impact poor countries in the tropics more than richer countries in temperate areas.
- Agriculture will need technical, institutional, and policy innovations to achieve sustainable improvements in productivity, equity, and resilience through advances in the management of energy, soils, water, livestock, and crops. Some of these innovations are discussed in the following chapters.
- A basic question is whether the innovations described in this book will be adequate to meet the challenges of sustainable food production for the future.

REFERENCES

Alexandratos, N. 2005. Countries with rapid population growth and resource constraints: issues of food, agriculture, and development. *Population and Development Review* 31(2):237–258.

Blaikie, P. and H. Brookfield. 1987. *Land Degradation and Society*. London: Methuen.

Blank, S. 2005. India's energy offensive in Central Asia. *Central Asia-Caucasus Institute Analyst*, March 9, pp. 3–5.

Cohen, J.E. 2005. Human population grows up. *Scientific American* 293(3):48–55.

Comprehensive Assessment of Water Management in Agriculture. 2007. *Water for Food, Water for Life: A Comprehensive Assessment of Water Management in Agriculture*, David Molden, Ed. London and Colombo, Sri Lanka: Earthscan and IWMI.

Eswaran, H.F., F. Beinroth, and P. Reich. 1999. Global land resources and population supporting capacity. *American Journal of Alternative Agriculture* 14:129–136.

Harrington, L., F. Gichuki, B. Bouman, N. Johnson, V. Sugunan, K. Geheb, and J. Woolley. 2006. *Synthesis 2005*. Colombo, Sri Lanka: CGIAR Challenge Program on Water and Food.

Hitz, S. and J. Smith. 2004. Estimating global impacts from climate change. *Global Environmental Change* 14(3):201–218.

Mendelsohn, R., A. Dinar, and L. Williams. 2006. The distributional impact of climate change on rich and poor countries. *Environment and Development Economics* 11:159–178.

Millennium Ecosystem Assessment. 2005. *Ecosystems and Human Well-Being: Synthesis*. Washington, DC: Island Press.

Oldeman, L. 1992. *Global Extent of Soil Degradation*. Biannual report. Wageningen, The Netherlands: International Soil Reference and Information Center.

Parry, M.L., C. Rosenzweig, A. Iglesias, M. Livermore, and G. Fischer. 2004. Effects of climate change on global food production under SRES emissions and socio-economic scenarios. *Global Environmental Change* 14(1):53–67.

Scherr, S. and S. Yadav. 1996. *Land Degradation in the Developing World: Implications for Food, Agriculture, and the Environment to 2020*. Washington, DC: IFPRI. 2020 Vision Food, Agriculture, and the Environment Discussion Paper 14.

UNDP. 2006. *Human Development Report 2006—Beyond Scarcity: Power, Poverty and the Global Water Crisis*. New York: Palgrave Macmillan and UNDP.

von Braun, J. and R.K. Pachauri. 2005. *The Promises and Challenges of Biofuels for the Poor in Developing Countries.* Washington, DC: IFPRI.

Vosti, S.A., J. Witcover, and M. Lipton. 1994. *The Impact of Technical Change in Agriculture on Human Fertility: District-Level Evidence from India.* Washington, DC: IFPRI. EPTD Discussion Paper 5.

World Bank. 2007. *World Development Report 2008—Agriculture for Development.* Washington, DC: International Bank for Reconstruction and Development/World Bank.

16 Global Climate Change and Agriculture

Roberto César Izaurralde

CONTENTS

ABSTRACT

The Fourth Assessment Report of the Intergovernmental Panel on Climate Change released in 2007 significantly increased our confidence about the role that humans play in forcing climate change. There is now a high degree of confidence that the (1) current atmospheric concentrations of carbon dioxide (CO_2), methane (CH_4), and nitrous oxide (N_2O) far exceed those of the preindustrial era; (2) global increases in CO_2 arise mainly from fossil fuel use and land use change, while those of CH_4 and N_2O originate primarily from agricultural activities; and (3) net effect of human activities since 1750 has led to a warming of the lower layers of the atmosphere, with an increased radiative forcing of 1.6 W m^{-2}. Depending on the scenario of human population growth and global development, mean global temperatures could rise between 1.8 and 4.0°C by the end of the twenty-first century. The agricultural sector is likely to be significantly affected by regional changes in temperature and precipitation. Overall changes in crop productivity and global food production could be positive in the lower range of temperature rise but negative if the upper range of temperature increases were to occur. While crops may benefit from the so-called CO_2 fertilization effect, yield losses may be expected if crops were to operate under high temperature, drought, and other environmental stressors such as rising tropospheric ozone (O_3) concentrations. One challenge will be to develop cultivars and production systems that are adapted to the changing climatic conditions expected to prevail during this century and productive enough to satisfy the dietary needs of increasing populations. There will also be other challenges and opportunities for the development and application of agricultural technologies capable of (1) mitigating greenhouse gas emissions and (2) producing bioenergy crops in a sustainable manner.

INTRODUCTION

Climate, the long-term average condition of weather and its variations over a region, has changed significantly during the evolution of Earth. Climate has been relatively stable during the last 10,000 years, a period known as the Holocene. Coincidentally, or perhaps because of this climate stability, it was at the beginning of this geologic epoch when humans learned how to domesticate wild plants and convert them into food and fiber crops. This revolutionary change marked the beginning of the first culture, agriculture, that allowed for the progression of human civilizations. Major cultures emerged and developed around a relatively small number of crops such as rice (*Oryza sativa* L.) in Asia, wheat (*Triticum aestivum* L.) in the Middle East, and maize (*Zea mays* L.) in the Americas. Today, these crops still account for a significant fraction of the caloric intake and nutritional diet of humans. To meet this demand, these and other important crops have been adapted to grow under a wide range of climatic and edaphic conditions prevailing around the world.

The development of the steam engine in Great Britain in the middle of the eighteenth century started a second revolution of transformational changes that affected agriculture, manufacturing, and transportation. These changes in turn had profound effects on socioeconomic and cultural conditions worldwide. Use of coal to power engines and expansion of agriculture grew quickly to satisfy energy and food demands of burgeoning populations. The development of the internal combustion engine and electrical power generation toward the end of the nineteenth century spurred even more the need for fossil fuels and agricultural land.

Accompanying all this and as a result of these activities, however, many invisible changes started to alter the chemical composition and dynamic properties of the atmosphere. Several heat-trapping trace gases (carbon dioxide [CO_2], methane [CH_4], and nitrous oxide [N_2O]) started to accumulate beyond their normal levels as a result of combustion of fossil fuels and conversion of land under native vegetation to agricultural use. And gradually, scientists began to study, document, and understand the impacts of these changes. There were very many important contributions, such as de Saussure's demonstration of the greenhouse effect, Tyndall's experiments on the absorption of thermal radiation by complex molecules (such as CO_2), and Callendar's equations linking greenhouse gases and climate change (Le Treut et al. 2007).

Two examples are given here to illustrate the evolution of our understanding of the effect of the accumulation (or variation) of trace gases in the atmosphere. The first example relates to Svante Arrhenius (1859–1927), a Swedish physical chemist who in 1895 theorized about the influence of carbonic acid in air on surface temperatures as a way to explain past climate changes. His work described a model of energy balance that included the effects of the absorptive power of CO_2 and water vapor for long-wave, infrared radiation and its possible effects on the temperature of terrestrial surfaces. The second is C. David Keeling (1928–2005), an American chemist who in 1958 established a laboratory in Mauna Loa, Hawaii, for accurate measurements of CO_2 concentrations in air samples and demonstrated throughout the years the connection between human activities and the chemical composition of the atmosphere (Keeling, 1960; Le Treut et al., 2007).

Today, and as a consequence of decades of research and observations, we know that the Earth is warming, and that this warming could bring undesired consequences on life as we know it. In May 2007, the Intergovernmental Panel on Climate Change (IPCC) (IPCC, 2007) announced the latest advancements in the state of knowledge of anthropogenic climate change (www.ipcc.ch). This report, the fourth in a series that started back in 1990, represented the culmination of almost 20 years of continuous activity of the international community assembled under the IPCC geared toward the comprehensive, objective, and transparent evaluation of the scientific, technical, and socioeconomic basis of anthropogenic climate change, its impacts, and options for its mitigation and adaptation.

One of the outstanding conclusions of Working Group I (WG I)—in charge of the studies related to the origin and causes of climate change—was to assert with a "high degree of confidence" that human activities since 1750 have induced a net warming effect over the Earth that translates into a "radiative power" of 1.6 (range 0.6 to 2.4) W m^{-2} (Solomon et al., 2007). This increase in radiative forcing[1] is due primarily to the continuous increase in emissions of three "trace" greenhouse gases since the beginning of the industrial period: carbon dioxide (CO_2), methane (CH_4), and nitrous oxide (N_2O).

The high degree of confidence about the occurrence of global warming of the climate system is based on observations of temperature increases in the atmosphere and oceans, ample evidence of sea ice and glacier melting, and sea-level rise. This global warming is also expressed in changes in the spatial distribution and magnitude of precipitation, salinity of ocean water, distribution and intensity of winds, and in general an increase in extreme events such as droughts, floods, and heat waves and in the intensity of tropical cyclones (Rosenzweig et al., 2007). Figure 16.1, from IPCC AR4 (Solomon et al., 2007), portrays a sophisticated analysis with respect to the evolution of greenhouse gases throughout hundreds of thousands of years before the present. The figure reveals the genuine increase in atmospheric concentrations of three greenhouse gases during recent years due mainly to the incessant increase in fossil fuel combustion and the conversion of tropical forests into pasturelands or croplands.

Agriculture is one human activity that should be responsive to climate change. In its most basic essence, agriculture consists of the guided capture of solar energy through domesticated plants (crops) and in the transformation of this energy into plant- or animal-based food, fiber, or bioenergy products. Thus, the changes in temperature, precipitation, cloudiness, windiness, and so on anticipated to occur during

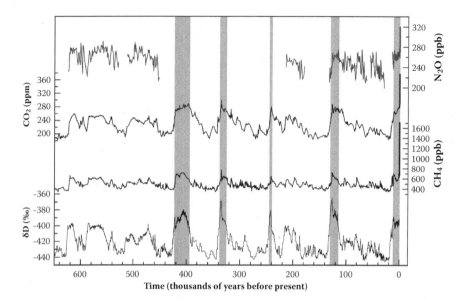

FIGURE 16.1 Deuterium (δD) variations in Antarctic ice (an estimator of local temperatures) and atmospheric concentrations of the greenhouse gases carbon dioxide (CO_2), methane (CH_4), and nitrous oxide (N_2O) in air trapped inside ice cylinders and in recent atmospheric samples. The data span 650,000 years. The shaded bands indicate previous and current interglacial warm periods. (From Intergovernmental Panel on Climate Change (IPCC), 2007, In: *Climate Change 2007: The Physical Science Basis. Contribution of Working Group I to the Fourth Assessment Report of the Intergovernmental Panel on Climate Change,* S. Solomon, D. Qin, M. Manning, Z. Chen, M. Marquis, K.B. Averyt, M. Tignor, and H.L. Miller, Eds. New York: Cambridge University Press, with permission from IPCC).

this century are projected to induce profound changes in agriculture production systems worldwide. The purpose of this chapter is to review and discuss the impacts of climate change on agriculture, the needs for adaptation to climate change, and the opportunities offered by agriculture to mitigate climate change. The review of the literature is not exhaustive; rather, it is restricted to the subjects needed to facilitate or illustrate the discussion of the topics presented in the chapter.

CLIMATE CHANGE IMPACTS ON AGRICULTURE

PLANT GROWTH UNDER ELEVATED CO_2

As photoautotrophic organisms, plants are able to use solar energy to convert CO_2 into sugars and utilize this chemically stored energy to fulfill processes of growth and development. Plants use this energy for maintenance and growth, and during this process (known as *respiration*) they consume O_2 and generate CO_2, which is emitted back to the atmosphere. The difference between photosynthesis and respiration is known as *net primary productivity* (NPP), of which, in the case of domesticated

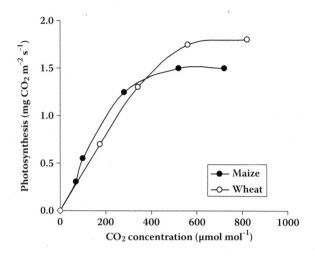

FIGURE 16.2 Photosynthetic response of maize and wheat to increasing levels of CO_2 concentration. (Redrawn from Akita, S. and D.N. Moss, 1973, *Crop Science* 13:234–237.)

plants or crops, a portion (e.g., seeds, fruits, tubers) is harvested for human uses. Heterotrophic terrestrial organisms, which include humans, depend directly or indirectly on plants for their growth and survival.

Plants exchange gases with the atmosphere through pores called *stomata* located underneath their leaves. The main gases exchanged through stomata are O_2, CO_2, and water vapor. When plants are exposed to elevated levels of CO_2, the stomata partially close to regulate the entry of CO_2 and indirectly reduce the rate of water vapor loss, which leads to an increase in *water use efficiency* (WUE), which is defined as the amount of plant biomass produced per unit of water used. As the CO_2 concentration of air increases, plants respond by increasing the rate of photosynthesis, more or less linearly at sub- and above-ambient concentration levels, up to a maximum (Figure 16.2) (Akita and Moss, 1973). Notably, maize plants reach a saturation of photosynthetic activity at lower levels of CO_2 concentration than wheat plants. This photosynthetic response to elevated CO_2 is species specific or, more precisely, dependent on the plant's photosynthetic pathway. Plants, in general, and crops, in particular, are classified as C_3 or C_4 according to the photosynthetic pathway used to fix CO_2. A plant is said to have a C_3-type metabolism because the first organic compound formed during photosynthesis is a three-carbon (C) molecule (3-phosphoglycerate). Conversely, a plant is of the C_4 type when the first compounds formed contain four-carbon atoms (oxaloacetate, malate). There are many evolutionary, morphological, and biochemical differences between C_3 and C_4 plants, so refer to any good text of plant physiology to learn about these mechanisms, including a third one (Crassulacean acid metabolism or CAM) not discussed here. Wheat, rice, alfalfa (*Medicago sativa* L.), cotton (*Gossypium hirsutum* L.), and potato (*Solanum tuberosum* L.) are examples of C_3 crops. Maize and sugarcane (*Saccharum officinarum* L.) are representatives of the C_4-type metabolism.

Understanding the Effects of Elevated CO_2 and Other Factors on Plant Growth

Many experiments were conducted during the 1960s, 1970s, and early 1980s to learn more about the effects of elevated CO_2 on plant growth and development. Kimball (1983) reviewed more than 70 greenhouse and growth chamber studies of the effects of CO_2 enrichment on the yield of 24 agricultural crops and 14 other species. In spite of data limitations, the meta-analysis concluded that crop yields could increase 33 $\pm$ 9% if the atmospheric CO_2 concentration were to double from 330 to 660 ppmv (parts per million by volume).

In 1990, a new technology was implemented in Arizona to study the response of crops to elevated CO_2 under field conditions and, eventually, the terrestrial feedback of CO_2 exchange with the atmosphere (Hendrey and Kimball, 1994). The technology, called FACE (free-air carbon dioxide enrichment), allowed plants to grow in the field in microclimates with a CO_2 level of 550 ppmv, which was expected to prevail around the middle of this century. Yields of irrigated cotton grown under elevated CO_2 were 43% greater than those grown under current ambient conditions (Mauney et al., 1994). The FACE treatment also increased WUE. Quickly, the FACE technology was applied to other crops and vegetation types, including grasslands, forests, and arid shrublands. Furthermore, the studies not only covered aspects of yield and biomass accumulation but also a range of ecosystem properties (e.g., water balance, carbon cycling).

Conditions of elevated CO_2 are expected to interact with many other factors, such as temperature, availability of water and essential nutrients. Thus, Amthor (2001) reviewed 50 studies on the effects of elevated CO_2 on wheat. The studies reviewed revealed that wheat yields increased by 31% when CO_2 concentrations doubled from 350 to 700 ppmv, with increased yields due mainly to an increase in ear density. Relative to ambient conditions, the *harvest index* (HI; defined as the fraction of plant biomass harvested due to its alimentary or economic value) of wheat under elevated CO_2 tended to increase under water stress conditions and decrease with nutrient limitations. Increased yields due to elevated CO_2, however, may reduce grain quality due to lower protein content and inferior bread-making qualities. These are very relevant findings since, as Amthor (2001) pointed out, wheat contributes on average about 20% of the calories and about 22% of the proteins consumed daily by the world population.

In addition to the negative impacts caused by warmer temperatures and water stress, crops exposed to elevated levels of tropospheric ozone (O_3) may not realize the beneficial effects of CO_2 enrichment due to the counteracting effect of this phytotoxic molecule formed at ground levels by industrial activities. Wheat and maize plants responded differently when grown in open-top chambers under ambient and elevated CO_2 (500 ppmv) with and without 40 ppbv (parts per billion by volume) of O_3 (Rudorff et al., 1996). Under ambient CO_2 conditions, wheat grown under elevated O_3 yielded 23% less than wheat grown under ambient levels of O_3. Correspondingly, under elevated CO_2, wheat yielded only 9% less when exposed to elevated concentrations of O_3. Similar results were obtained with maize, suggesting that the stoma-closing effect of elevated CO_2 could ameliorate the inhibitory effects of elevated O_3 on photosynthesis.

PREDICTING THE EFFECTS OF CLIMATE CHANGE ON AGRICULTURAL PRODUCTIVITY

Simulation models driven by historical, current, and future climate scenarios have been essential tools for testing hypotheses concerning the possible impacts of climate change on agricultural production and water resources (Rosenberg, 1992). As an example, Izaurralde et al. (2003) applied results of the HadCM2 GCM (Hadley Climate Model 2 General Circulation Model) and the EPIC (Environmental Policy Integrated Climate) agroecosystem model to evaluate climate change impacts on crop yields and ecosystem processes. EPIC was implemented to run the main crops (with and without irrigation) grown in the United States under current (1961–1990) and two future climates (2025–2034, 2090–2095). The simulation runs were implemented at two CO_2 concentrations (365 and 560 ppmv). The simulation results revealed a high spatial dependence driven mainly by regional changes in temperature and precipitation. For example, wheat yields in the Northern Plains region are expected to remain unchanged in the two future periods but benefit by the presence of the CO_2 fertilization effect (Figure 16.3). In the warmer Southern Plains, the projected increases in temperature would reduce yields, but losses would be partially compensated by the CO_2 effect. This effect appears less expressed in maize than in wheat crops. Water use efficiency is reduced with warming in both crops, but again, the CO_2 fertilization effect, if present, would attenuate this decrease.

Despite many advances during the last four decades toward an improved understanding of the possible impacts of climate change on agricultural productivity and ecosystem function, much remains to be done to reduce the uncertainties arising from experimental and simulation results. For example, a recent meta-analysis of FACE experiments (Ainsworth and Long, 2005; Long et al., 2006) suggested that grain yield increases under elevated CO_2 were lower than those of previous enclosure (non-FACE) studies. However, a reinterpretation of crop yields reported across FACE and non-FACE data sets found the responses to be rather similar (Tubiello et al., 2007). The relevant aspect of this discussion in relation to this chapter resides in our ability to predict the influence of climate change on global food supply. As this section has shown, future crop production will depend on the dynamics and interaction of many climatic factors, such as precipitation, temperature, CO_2, and tropospheric O_3, among others.

THE ROLE OF AGRICULTURE IN CLIMATE CHANGE MITIGATION AND ADAPTATION

DEFINITIONS OF MITIGATION AND ADAPTATION

What can be done to prepare agriculture for climate change? One action could be that of *mitigation*, that is, use current or future agricultural technologies to counteract emissions of greenhouse gases and thus contribute to their stabilization in the atmosphere. Another action is that of *adaptation*, which is designed to lessen adverse impacts of climate change on human and natural systems. In this context, the main objective is to reduce the vulnerability of agriculture to the harm that may be caused by climate change.

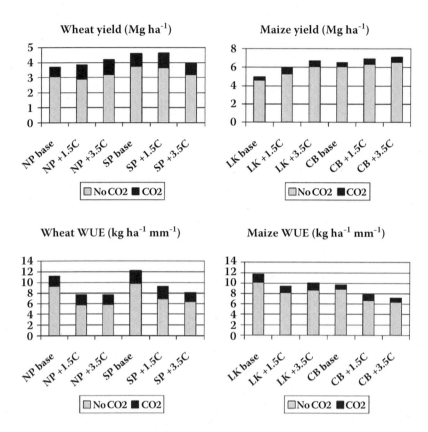

FIGURE 16.3 Modeled effects of climate change and CO_2 concentration on crop productivity and water use efficiency (WUE) in four U.S. regions: Northern Plains (NP), Southern Plains (SP), Lakes (LK), and Corn Belt (CB). The term *base* refers to current climate. The numerals 1.5 and 3.5 refer to increases in global mean temperature of 1.5 and 3.5°C, respectively. Top panel: wheat and maize yields. Bottom panel: water use efficiency in wheat and maize. (From Izaurralde, R.C., N.J. Rosenberg, R.A. Brown, and A.M. Thomson, 2003, *Agriculture and Forest Meteorology* 117:97–122.)

As discussed, not all changes brought about by climate change may have a negative effect on agricultural production. From what we know, the so-called CO_2 fertilization effect or some benign warming may benefit crop production, at least in regions where the changes in temperature and precipitation are favorable. It is also anticipated, however, that climate change will not only change the mean conditions of climate properties but also their distribution and frequencies likely will increase the occurrence of extreme events such as droughts and floods. For the last decade and a half, research and technical efforts have concentrated on climate mitigation and, in the case of agriculture, deploying technologies such as soil carbon sequestration and non-CO_2 greenhouse gas reductions (N_2O and CH_4). Many believe, however, that the time to start adapting to climate change is now. In November 2006, the U.N. Climate Change Conference held in Nairobi, Kenya (UNFCC, 2007) sent a clear message on the need for adaptation, stating that even if emissions were to be stopped now, greenhouse gases already in the atmosphere will continue to induce global warming. Despite agriculture being a highly adaptive sector, there has been little effort geared toward adapting crops and agricultural systems to upcoming changes in climate, except for plant-breeding efforts to develop cultivars with increased resistance to high temperatures and droughts.

MITIGATION OF CLIMATE CHANGE THROUGH SOIL CARBON SEQUESTRATION AND REDUCTIONS IN GREENHOUSE GAS EMISSIONS

Agricultural activities are very important from the point of view of climate change because of their role in the emission and absorption of greenhouse gases. Agricultural lands (cropland, grasslands, and permanent crops) occupy about 40% of the Earth's land surface, estimated at 13.4 billion hectares. Historically, it is estimated that agricultural soils have released to the atmosphere approximately 55 Pg (Petagram; one billion metric tons) carbon (Cole et al., 1997) as a result of conversion of forests and grasslands into croplands and application of management practices unable to maintain adequate levels of soil organic matter (e.g., excessive tillage, insufficient nutrient replacement). The agricultural sector is also the largest emitter of CH_4, due mainly to paddy rice cultivation and livestock activities (enteric fermentation), and N_2O, mainly due to excessive or untimely application of nitrogenous fertilizers. Smith et al. (2008) estimated that agricultural activities currently emit about 6.1 Pg CO_2-eq yr^{-1} (10 to 12% of total global anthropogenic emissions of greenhouse gases). Methane emissions from agricultural activities represent about 3.3 Pg CO_2-eq yr^{-1} (50% of the world total), while N_2O emissions represent 2.8 Pg CO_2-eq yr^{-1} (60% of world total). Carbon dioxide emissions are estimated to contribute about 0.04 Pg CO_2-eq yr^{-1} (~0% of world total).

Cole et al. (1997) estimated that agriculture could, during a period of 50 years, contribute to the mitigation of CO_2 emissions by applying soil carbon sequestration practices aimed at the recovery of two thirds of the historical losses of soil organic carbon. Figure 16.4 shows a simple representation of soil carbon sequestration in an agroecosystem. Organic carbon stabilizes in soil due to biochemical and physicochemical mechanisms that operate at different spatial and temporal scales (Jastrow et al., 2007). Soil carbon sequestration can be effected by such practices as direct seeding (zero tillage, no till), efficient use of nutrients, crop intensification, and residue management. For example, the adoption of direct-seeding systems in several South American countries (Brazil, Argentina, Paraguay, and Uruguay) has been very important recently. Izaurralde and Rice (2006) used global average soil carbon sequestration rates of 0.57 Mg (Megagram; one metric ton) C ha^{-1} yr^{-1} (West and Post, 2002) under no till and the global area under no till (~70 million hectares) to estimate a soil carbon sink of about 40 Tg (Teragram; one million metric tons) C yr^{-1}. Much work remains to be done to foster the adoption of this practice at global scales and to design mechanisms to monitor soil carbon changes (Izaurralde and Rice, 2006).

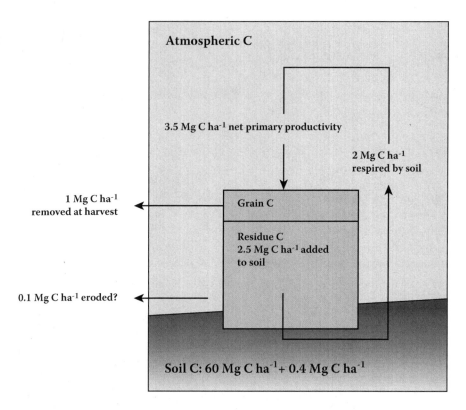

FIGURE 16.4 Schematics of annual carbon flows and soil carbon stocks in an agroecosystem. Most of the carbon captured by plants from the atmosphere via net primary production (photosynthesis-respiration) is respired back to the atmosphere by microbes, removed from the ecosystem as harvestable products, or transferred to other places via erosion. Soils gain (sequester) carbon due to a positive balance between inputs and outputs of carbon. Conversely, soils may lose (emit) carbon when the balance is negative. Some other terms of the carbon balance (gains of carbon via soil deposition, leaching of dissolved carbon) are not represented here.

Soil carbon sequestration is a unique mitigation option because of the many ancillary benefits and environmental services associated with its practice (see Chapter 18). Carbon accumulation in soil has a direct effect on ecosystem quality and function. Carbon is the main constituent of soil organic matter, which in turn is a universally accepted reference for quantifying soil fertility, productivity, and resistance to erosion (Carter, 2002). For example, sequestration of 1 Mg ha^{-1} of atmospheric carbon in soil leads to the retention of about 100 kg ha^{-1} of nitrogen as well as significant amounts of phosphorus, sulfur, and other essential plant nutrients in soil organic matter. Sequestration of 1 Mg ha^{-1} of atmospheric carbon in soil can increase crop productivity in the range of 20–70 kg ha^{-1} for wheat, 10–50 kg ha^{-1} for rice, and 30–300 kg ha^{-1} for maize (Lal, 2005). Thus, soil carbon sequestration can contribute not only to climate change mitigation but also to global food security, especially in developing countries.

ADAPTING AGRICULTURE TO CLIMATE CHANGE

Whether it would be possible to apply current and future technologies to mitigate climate change or stop the emissions of greenhouse gases altogether, there is agreement that the world has already committed to a certain degree of global warming given the magnitude of historical emissions already added to the atmosphere. This means that the agricultural industry must be ready to adapt crops and practices for the environmental conditions anticipated to prevail during this century. Agriculture is an activity with a high capacity for adaptation, something that is easily demonstrated when we consider the geographic distribution of crops such as wheat, maize, and soybean. Advances in biotechnology, such as cultivars with pest resistance or herbicide tolerance, are examples of optimization and efficiency of the agricultural enterprise. However, it is not known how vulnerable the main agricultural systems are to the extreme changes in climate, such as prolonged droughts, floods, or heat waves.

From a historical point of view, one of the first regional studies that considered adaptation to climate change was that of Easterling et al. (1992a, 1992b). The simulation study known as MINK (Missouri, Iowa, Nebraska, Kansas) used meteorological data of the 1930s (known as the Dust Bowl era for the persistent drought and erosion of that period) as analogue climate to evaluate the impacts of these climatic conditions on agricultural production as well as possible adaptations to these change. Both known adaptations (e.g., early planting, long-season cultivars, and planting density) and advanced adaptations (cultivars with improved radiation use efficiency or stress tolerance) were able to partly alleviate yield losses induced by climate change.

For an adaptation technology to be effective, it is important to know the patterns of adoption or patterns of substitution at the regional scale. Easterling et al. (2003) suggested that, facing climate change, not all farmers would adopt changes in their practices in response to climate change at the same time. In many adaptation studies, the convention is to simulate perfect adoption or perfect ignorance of the adaptation practice. Easterling et al. (2003) simulated the adoption of adaptive technologies by farmers as a logistic (S-shaped) diffusion process and recommended this approach as a sound method to test adaptations that rely on technological innovation and substitution.

In general, the agricultural sector has exhibited good adaptive capacity to environmental and economic conditions. However, there are significant differences in adaptive capacity throughout the world. Because of this and the anticipated nature of future climate change, there is a need intensify adaptation studies to guide the development of adaptation practices.

TOWARD AGRICULTURAL SUSTAINABILITY IN THE TWENTY-FIRST CENTURY

Climate change will not occur in isolation during this century. Instead, it will occur amid many other concurrent socioecological events and trends, many of them hard to predict. In this context, land use and land use competition will play fundamental roles in the design and implementation of mitigation and adaptation practices. One of the topics related to agriculture and climate change is the production of biofuels (see

Chapter 17). Whether for climate change or energy security reasons, many countries have high interest in the production of biofuels. What is not known, however, is how production of biomass crops at global scales could affect the functioning and quality of both managed and unmanaged ecosystems. Water demand, competition for land for food crops, and losses of biodiversity are among the challenges to be dealt with before recommending the expansion of these crops worldwide.

In this context, the design and implementation of sustainable management practices will play a key role in the development of effective solutions for climate change. The challenge will be to harmonize the management of food, fiber, and energy systems with the unmanagement of natural systems. Increasingly, this will require a higher degree of interdisciplinary activities to ensure a comprehensive evaluation of mitigation and adaptation practices to address climate change.

CONCLUSIONS

The most recent scientific evidence synthesized by the IPCC in 2007 appears to confirm that the observed changes in climate occurring during the past century are due in part to human activities, and that these change will persist or even more likely increase still further during this century. Because of its global dimension, agriculture has an important role in climate change, not only for the possibility of suffering its impacts, but also for the need to adapt to or mitigate these changes.

SUGGESTIONS FOR STUDY TOPICS

- *Climate change:* Review and discuss the nonlinearity of the climate system and the possibility of "abrupt climate change."
- *Impacts:* Review and discuss observed and projected changes in agricultural practices as a result of climate change.
- *Mitigation:* Review and discuss the potential of agricultural soils to act as sinks of atmospheric carbon dioxide.
- *Adaptation:* Review and discuss adaptation measures in agriculture (crops, livestock, soils) in relation to ongoing and future climate change.

NOTE

1. As defined by the IPCC, radiative forcing is the "radiative imbalance ($W\ m^{-2}$) in the climate system at the top of the atmosphere caused by the addition of a greenhouse gas (or other change)" (Le Treut et al., 2007).

REFERENCES

Ainsworth, E.A. and S.P. Long. 2005. What have we learned from 15 years of free-air CO_2 enrichment (FACE)? A meta-analytic review of the responses of photosynthesis, canopy properties and plant production to rising CO_2. *New Phytologist* 165:351–372.
Akita, S. and D.N. Moss. 1973. Photosynthetic responses to CO_2 and light by maize and wheat leaves adjusted for constant stomatal apertures. *Crop Science* 13:234–237.

Amthor, J.S. 2001. Effects of atmospheric CO_2 concentration on wheat yield: review of results from experiments using various approaches to control CO_2 concentration. *Fields Crops Research* 73:1–34.

Carter, M.R. 2002. Soil quality for sustainable land management: Organic matter and aggregation interactions that maintain soil functions. *Agronomy Journal* 94:38–47.

Cole, C.V., J. Duxbury, J. Freney, O. Heinemeyer, K. Minami, A. Mosier, K. Paustian, N. Rosenberg, N. Sampson, D. Sauerbeck, and Q. Zhao. 1997. Global estimates of potential mitigation of greenhouse gas emissions by agriculture. *Nutrient Cycling in Agroecosystems* 49:221–228.

Easterling, W.E., N. Chhetri, and X. Niu. 2003. Improving the realism of modeling agronomic adaptation to climate change: simulating technological substitution. *Climatic Change* 60:149–173.

Easterling, W.E., M.S. McKenney, N.J. Rosenberg, and K.M. Lemmon. 1992a. Simulations of crop responses to climate change—effects with present technology and no adjustments (the dumb farmer scenario). *Agriculture and Forest Meteorology* 59:53–73.

Easterling, W.E., N.J. Rosenberg, K.M. Lemmon, and M.S. McKenney. 1992b. Simulations of crop responses to climate change—effects with present technology and currently available adjustments (the smart farmer scenario). *Agriculture and Forest Meteorology* 59:75–102.

Hendrey, G.R. and B.A. Kimball. 1994. The FACE program. *Agriculture and Forest Meteorology* 70:3–14.

Intergovernmental Panel on Climate Change (IPCC). 2007. Summary for policymakers. In: *Climate Change 2007: The Physical Science Basis. Contribution of Working Group I to the Fourth Assessment Report of the Intergovernmental Panel on Climate Change,* S. Solomon, D. Qin, M. Manning, Z. Chen, M. Marquis, K.B. Averyt, M. Tignor, and H.L. Miller, Eds. New York: Cambridge University Press, pp. 1–18.

Izaurralde, R.C. and C.W. Rice. 2006. Methods and tools for designing pilot soil carbon sequestration projects. In: *Carbon Sequestration in Soils of Latin America*, R. Lal, C.C. Cerri, M. Bernoux, J. Etchevers, and C.E. Cerri, Eds. Boca Raton, FL: CRC Press, pp. 457–476.

Izaurralde, R.C., N.J. Rosenberg, R.A. Brown, and A.M. Thomson. 2003. Integrated assessment of Hadley Centre climate change projections on water resources and agricultural productivity in the conterminous United States. II. Regional agricultural productivity in 2030 and 2095. *Agriculture and Forest Meteorology* 117:97–122.

Jastrow, J.D., J.E. Amonette, and V.L. Bailey. 2007. Mechanisms controlling soil carbon turnover and their potential application for enhancing carbon sequestration. *Climatic Change* 80:5–23.

Keeling, C.D. 1960. The concentration and isotopic abundances of carbon dioxide in the atmosphere. *Tellus* 12:200–203.

Kimball, B.A. 1983. Carbon dioxide and agricultural yield: an assemblage and analysis of 430 prior observations. *Agronomy Journal* 75:779–788.

Lal, R. 2005. Enhancing crop yields in the developing countries through restoration of the soil organic carbon pool in agricultural lands. *Land Degradation and Development* 17: 197–209.

Le Treut, H., R. Somerville, U. Cubasch, Y. Ding, C. Mauritzen, A. Mokssit, T. Peterson, and M. Prather. 2007. Historical overview of climate change. In: *Climate Change 2007: The Physical Science Basis. Contribution of Working Group I to the Fourth Assessment Report of the Intergovernmental Panel on Climate Change,* S. Solomon, D. Qin, M. Manning, Z. Chen, M. Marquis, K.B. Averyt, M. Tignor, and H.L. Miller, Eds. New York: Cambridge University Press, pp. 94–127.

Long, S.P., E.A. Ainsworth, A.D.B. Leakey, J. Nosberger, and D.R. Ort. 2006. Food for thought: lower-than-expected crop yield stimulation with rising CO_2 concentrations. *Science* 312:1918–1921.

Mauney, J.R., B.A. Kimball, P.J. Pinter Jr., R.L. LaMorte, K.F. Lewin, J. Nagy, and G.R. Hendrey. 1994. Growth and yield of cotton in response to a free-air carbon dioxide enrichment (FACE) environment. *Agriculture and Forest Meteorology* 70:49–67.

Rosenberg, N.J. 1992. Adaptation of agriculture to climate change. *Climatic Change* 21:385–405.

Rosenzweig, C., G. Casassa, D.J. Karoly, A. Imeson, C. Liu, A. Menzel, S. Rawlins, T.L. Root, B. Seguin, and P. Tryjanowski. 2007. Assessment of observed changes and responses in natural and managed systems. In: *Climate Change 2007: Impacts, Adaptation and Vulnerability. Contribution of Working Group II to the Fourth Assessment Report of the Intergovernmental Panel on Climate Change*, M.L. Parry, O.F. Canziani, J.P. Palutikof, P.J. van der Linden, and C.E. Hanson, Eds. Cambridge, U.K.: Cambridge University Press, pp. 79–131.

Rudorff, B.F.T., C.L. Mulchi, C.S.T. Daughtry, and E.H. Lee. 1996. Growth, radiation use efficiency, and canopy reflectance of wheat and corn grown under elevated ozone and carbon dioxide atmospheres. *Remote Sensing of the Environment* 55:163–173.

Smith P. et al. 2008. Greenhouse gas mitigation in agriculture. *Phil. Trans. Royal Soc. B. Biol. Sci.* 363:789–813.

Solomon, S. et al. 2007. Technical summary. In *Climate Change 2007. The Physical Science Basis*, Solomon, S. et al. (Eds.), Cambridge, UK: Cambridge University Press.

Tubiello, F.N., J.S. Amthor, K.J. Boote, M. Donatelli, W. Easterling, G. Fischer, R.M. Gifford, M. Howden, J. Reilly, and C. Rosenzweig. 2007. Crop response to elevated CO2 and world food supply. A comment on "Food for thought ..." by Long et al., Science 312:1918–1921, 2006. European Journal of Agronomy 26:215–223.

United Nations Framework Convention on Climate Change (UNFCC). 2007. Report of the Conference of the Parties on its 12th session, held at Nairobi from 6 to 17 November 2006. Available at http://unfccc.int/resource/docs/2006/cop12/eng/05.pdf (accessed March 6, 2008).

West, T.O. and W.M. Post. 2002. Soil organic carbon sequestration rates by tillage and crop rotation: a global data analysis. Soil Science Society America Journal 66:1930–1946.

17 Bioenergy: *Energy Sources and Costs for Agriculture*

Donald C. Erbach and Wallace W. Wilhelm

CONTENTS

ABSTRACT

The way in which the world's population currently meets its energy needs is not sustainable. Fossil petroleum, natural gas, and coal reserves currently relied on are finite and nonrenewable, and their use has serious security, environmental, and economic consequences. Therefore, it is critical to develop renewable and sustainable alternative energy sources. Solar radiation is the only significant renewable and sustainable source of energy for the earth. This energy can be captured in plant material via photosynthesis and converted to liquid transportation fuel alternatives for petroleum-based gasoline and diesel. However, the resources required for producing biomass for energy are the same as those required to produce biomass for food, and this poses a fuel-versus-food conflict. Some feel the world's land and water resources are sufficient to meet energy needs and also satisfy food and feed requirements. Others are concerned that use of food crops and of agricultural resources for energy are causing food shortages and starvation. It is critical that both energy and food be adequately provided, and fortunately, the world has great capacity to meet both needs. The challenge is to sustainably and efficiently manage that capacity, including agricultural and forestry resources, to effectively meet human needs. Agriculture has the potential to be a significant energy provider and in doing so bring vitality to rural communities. For a successful bioenergy program to be implemented, effective energy policy must be enacted, policy that encourages producers and users of energy and food to make choices and take actions that are in the best interest of humanity.

INTRODUCTION

The world's population consumes a very large amount of energy, about 500 quadrillion BTU (Energy Information Administration [EIA], 2008) annually, and demand is growing. Energy is needed for heat, power, electricity, and transportation fuel. Currently, energy is supplied mainly from nonsustainable geologic hydrocarbon sources. In addition to supplies being finite and nonrenewable, there are serious security, environmental, and economic concerns with continued reliance on fossil energy. Bioenergy is a potentially sustainable carbohydrate alternative that has been used throughout human existence and in some locations continues to be the predominant energy source. Biomass currently accounts for about 15% of energy used worldwide and about 3% of energy currently used in the United States. For bioenergy to become a major sustainable source of energy, an expanded energy-from-agriculture infrastructure must be developed. This will require agriculture to undergo significant changes, changes that will have an impact on agriculture and everything associated with agriculture, including food and its availability and cost. An efficient bioenergy infrastructure must be developed such that sustainable, renewable energy is maximized while negative impacts on food, feed, fiber, the environment, ecosystems, and the economy are minimized.

THE PROBLEM: FOSSIL ENERGY

Fossil feedstocks provide most of the energy consumed by the world's population and have for more than a century. Petroleum, natural gas, and coal are burned to produce over 80% of the current world energy use. All industrialized economies are dependent on petroleum. However, because it is a nonrenewable, nonreplenishable resource, reliance on fossil energy for the long-term is obviously not sustainable. As people's energy expectations and demands grow, as developing countries become increasingly industrialized and adopt more mobile lifestyles with greater demand for electrical and liquid fuel-powered devices, and as populations increase, the competition for fossil energy, in particular petroleum and natural gas, will become more intense and contentious. Although there is considerable uncertainty associated with estimates of recoverable reserves, it is certain that fossil energy is finite. Even with worldwide energy use growing, some estimates of existing recoverable reserves suggest there is sufficient fossil energy in conventional and unconventional reserves to meet overall demand for many years. Other projections suggest the capacity to produce conventional petroleum is reaching a peak, and the world is nearing, if not already past, the point at which one half of the petroleum resources have been consumed; this has occurred in only a century. Oil production in the United States reached a peak in 1970 and has since steadily declined. There are of course areas that contain potentially significant amounts of oil but are off limits to oil recovery primarily for environmental reasons. If worldwide oil production and discovery are waning and worldwide demand is increasing, supplies may soon be insufficient to meet demand. Rising oil prices are an indication that this is already occurring.

Fossil fuels have attributes that make them great energy sources. These fuels are generally easily mined or extracted and are relatively low cost, transportable, and energy dense. Because fossil fuels have been such good energy sources and have been in use for some time, efficient and effective infrastructures have been developed for their use. As a result, most people in the world have come to rely on coal, petroleum, and natural gas for heating, cooling, electricity, power, and transportation.

Unfortunately, dependency and continued reliance on fossil energy has associated collateral damage. Three serious negative side effects are associated with energy security, the environment, and the economy.

ENERGY SECURITY

Securing an adequate supply of energy involves maintaining and deploying a military with the capacity for intervention to provide political stability for regions of the world with petroleum and natural gas resources and to protect supply lines that deliver these products. As supplies become scarce and more expensive and as demands increase, nations will become more aggressive in acting to ensure that their supplies are secure and adequate for current and future needs. These actions have serious political, military, and economic consequences, as demonstrated by Japan's invasion of Southeast Asia in 1941, Iraq's invasion of Kuwait in 1990, and the U.S.

attack on Iraq in 1991. The U.S. invasion of Iraq in 2003 certainly had energy security implications.

ENVIRONMENTAL IMPACT

The second serious concern is that greenhouse gases (GHGs) emitted when fossil fuels are consumed will alter the environment. An estimated 80% of GHG emissions, discharges linked with global warming, are associated with consumption of fossil fuels. One GHG of much concern is the carbon dioxide formed during burning of fossil fuels when carbon in the fuel combines with oxygen from the air. In 2004, an estimated 8 billion metric tons of carbon, previously sequestered in coal, petroleum, or natural gas, were emitted into the atmosphere as a result of fossil fuel use. The atmospheric concentration of carbon dioxide, which had historically been fairly constant, has risen nearly 20%, from about 315 parts per million by volume (ppmv) in 1959 to over 377 ppmv in less than 50 years. Research has found that much of the increase is due to use of fossil energy and has raised concern that elevated levels of atmospheric carbon dioxide are increasing global temperature and driving other serious environmental changes. Changes that scientists project, if not averted, may well have costly consequences and potentially calamitous effects on Earth's climate and, in turn, on its inhabitants. Changes of concern include sea-level rise from melting ice that could flood large inhabited areas, loss of glaciers that may cause some major rivers to cease flowing during portions of the year, and changes in ocean currents that could significantly alter climates. Some experts feel strongly that unless carbon emissions are stabilized within a few decades, irreversible climatic changes will begin to occur.

To reduce carbon emissions, clean technologies that capture and sequester carbon from fossil fuels, particularly from coal, have been proposed. Unfortunately, carbon capture is difficult, costly, and unproven. Capture is especially difficult from distributed uses and users, including from mobile engines such as cars, planes, trucks, trains, tractors, recreational vehicles, and outdoor power equipment. Capture from large point sources, such as power plants, is potentially feasible but has not been commercialized. Sequestration is also a challenge. Because the quantities of carbon are large, finding locations and procedures for safe, sustainable, and economical sequestration is difficult. Sequestration sites include soil, where the carbon could increase organic matter and enhance soil quality and productivity, and oil fields, where injecting carbon dioxide could enhance oil recovery. Other sites proposed are aquifers, caverns, and the ocean; however, the sustainability of sequestration in these sites is not proven. Certainly, planting more trees would also remove and store considerable carbon. Possibly the best method, however, to ensure safe sequestration of the carbon in unrecovered oil, natural gas, and coal deposits is to leave those materials where they are—in the earth unused. This also conserves these fossil natural resources.

In addition to carbon dioxide, other emissions associated with use of fossil energy, especially coal, are problematic. These include particulate matter, sulfurous compounds, nitrous oxides, and mercury. Fortunately, most of these can be controlled by use of technology that cleans the fuel either before it is used or with aftertreatment of combustion gases.

Recovery and processing of fossil energy also have environmental impacts. As supplies dwindle, increasingly large amounts of energy will be required to extract the remaining usable energy, causing the extraction to be less energy efficient and more costly. As conventional oil supplies become more difficult to obtain, pressures to extract energy from alternative fossil sources grow. These sources, such as tar sands, extra-heavy oil, and oil shale, contain large amounts of energy but require more energy to extract, transport, and process than more desirable petroleum sources. In addition, expanded drilling, mining, and extraction activity will likely be more environmentally damaging and use larger amounts of process water, leading to additional potentially damaging ecological impacts.

ECONOMIC IMPACT

The third major issue with fossil energy, and especially with oil, is the economic impact of depending on it as a primary energy source. This impact is manifest in the negative effects on economies of countries that must purchase large quantities of energy. For the United States, oil import costs have exceeded $1.5 billion dollars per day, an annual cost of over $0.5 trillion. Petroleum import costs account for about two thirds of the negative U.S. trade balance. Of course, as the price of oil increases, these expenditures, with their significant national and international security implications, will also increase.

If collateral damage concerns associated with fossil energy use and dependence could be alleviated such that the security, environmental, and economic issues associated with their use would vanish, it is still important that this source of energy be used wisely. Fossil fuels have many positive energy attributes, but because supplies are finite, what remains should be considered precious and be prudently conserved for the long-term benefit of humanity.

In fact, conservation is an energy management strategy vastly undervalued in current energy discussions. For example, in spite of technological advances, the U.S. corporate average fuel economy (CAFE) standard for its motor vehicle fleet had remained unchanged for over three decades (Public Citizen, 2008) until recently passed legislation mandated that fuel economy of cars and light trucks be increased by about 30%. But, it will take several years for the policy to fully take effect in spite of the fact that technology is currently available that could bring about much greater increases in vehicle fuel economy and could significantly reduce transportation fuel demand.

The need to use energy, whether fossil or renewable, more efficiently is essential. Economics has driven development of some technology, such as combined heat and power facilities, to use energy more efficiently. Generally, however, fossil energy has been cheap or at least sufficiently low cost to discourage development of new, or commercialization of existing, technology that would improve energy-use efficiency. As a result, inefficient practices and dependence on fossil energy have become firmly entrenched. Improved energy use efficiency would lower energy demand, lessen the challenge of providing adequate amounts of energy, reduce the environmental impact of producing and consuming fuels, and improve the environment while enhancing the quality of human life.

THE SOLUTION: RENEWABLE ENERGY

To avoid the serious negative consequences of fossil energy dependence, systems that provide adequate amounts of economical, reliable, and sustainable renewable energy must be developed and commercialized. A truly adequate and sustainable energy system must be renewable and provide sufficient energy to meet demand. Renewable energy must come from renewable sources, must not depend on nonrenewable inputs, and must have the potential for its production to be sustained in perpetuity. Currently, most alternative energy relies to some extent (directly or indirectly) on nonrenewable fossil energy. As a result, indices such as fossil energy ratio and net energy are used to assess the relative sustainability of renewable energy feedstock and fuel production practices and processes (Farrell et al., 2006).

When evaluating potential renewable energy sources, it must be considered that the earth's only real source of renewable energy is solar energy, an energy source humans have relied on as long as they have inhabited the earth. It is also the energy that drives the climate and ecology that make the earth the place that it is. Fortunately, it is a very large energy source (Crabtree and Lewis, 2007). In only a matter of minutes, the energy equivalent to that used by humans each year arrives from the sun at the earth's surface.

Solar radiation is low intensity and widely distributed. These characteristics are ideal for warming the earth and for the safety of plants and animals inhabiting the planet. Wind, solar, hydro, and biomass energies are all solar radiation driven. Much of the energy from these sources has traditionally been used at or near where it is captured. Solar forms of energy are well suited for distributed, small-scale uses such as pumping water for animals and for home heating of water. However, economically capturing and making use of solar radiation for large-scale generation of heat, power, and electricity has proven difficult. The challenge is to efficiently and effectively convert solar energy into useful forms and to make that energy available in adequate amounts when and where needed. Possibly the most difficult form of energy to provide is an energy-dense and transportable transportation fuel that can replace currently used gasoline and diesel fuel.

Because energy use is large, and increasing, it is unlikely that any single source of renewable energy will replace fossil fuel. However, collectively solar, wind, hydro, and biomass could provide all of the transportation fuel, electricity, heat, and power for the world's industrial and domestic needs.

RENEWABLE BIOENERGY

Green plants, through the chlorophyll molecule and the process of photosynthesis, capture and use the energy in sunlight to form carbohydrates from carbon dioxide and water. The energy in these carbohydrates is in the form of reduced carbon or carbon-hydrogen bonds. The solar energy captured in plant biomass provides energy and nutrition for the animal kingdom. Biomass can also be used directly as a solid fuel or converted into other forms, including liquid or gaseous fuels. These fuels are renewable bioenergy that can be converted to heat, power, and electricity or used for transportation.

A downside of using plants to capture energy from solar radiation is the low efficiency of the photosynthetic process. Only about 0.2% of the energy falling on a growing plant is captured in the biomass produced (Long et al., 2006). Because photosynthetic solar energy capture is low, biomass production, on the scale needed to significantly contribute to the transportation fuel supply, will require large land areas.

As with fossil fuel, combustion of biobased fuel forms carbon dioxide, but use of biofuels is carbon neutral because the carbon emitted when biomass is consumed is carbon removed from the atmosphere as the plants, which produced the biomass, were growing. The process will repeat as the carbon is again removed from the atmosphere as plants grow to produce biomass in subsequent seasons. Thus, the carbon cycle for bioenergy is a closed loop with no net addition of carbon to the atmosphere. This is in contrast with carbon flow from use of fossil fuel, in which carbon sequestered in hydrocarbons formed millions of years ago is released when the fuel is burned. Unlike the case with bioenergy, there is no process to again recapture the carbon as fossil fuel, and as a result atmospheric carbon content increases. Therefore, bioenergy is renewable while fossil energy is not.

Bioenergy is carbon neutral only if it is produced completely without use of fossil energy. At this time, however, most bioenergy produced is not carbon neutral. That is because fossil energy (e.g., diesel fuel for tractors and natural gas for nitrogen fertilizer) is used in production of crops grown to produce biobased feedstock. In addition, conversion facilities often rely on fossil fuel (natural gas or coal) for energy to process biomaterial into biofuels. To be renewable, sustainable, and carbon neutral, agriculture must eliminate its dependency on fossil fuels.

Biomass for use as energy will likely come from agricultural and forestry enterprises. Agricultural biomass useful for energy includes herbaceous crops grown specifically for energy and residues of herbaceous crops grown for other purposes. Plant residues include stover from corn and straw from wheat when these crops are harvested for grain. Forestry biomass includes wood, branches, and other organic material from trees and shrubs. Additional biomass feedstocks include wastes from the production, processing, and use of agricultural and forestry materials. These include animal wastes, wastes from processing of food and nonfood bio-based products, and municipal wastes, including household, lawn and garden, construction, and demolition materials.

Critical Questions for Renewable Bioenergy

What infrastructure is available, and what is needed for agriculture to become a significant source of energy? What are the economic, natural resource, and social implications, both positive and negative, that energy production would have for agriculture and rural areas and for society in general? What are the barriers that must be overcome to commercialize, on a large scale, bioenergy production and use? These and other questions must be answered for bioenergy to become viable, commercial, and sustainable.

BIOENERGY: POTENTIAL FROM AGRICULTURE AND RURAL AMERICA

Agriculture is primarily the business of producing crops and livestock. Crop production is the growing of plants to capture solar energy in plant material for use as food, feed, fiber, and other useful products. In the same way, forestry is primarily the growing of woody plants for production of wood and fiber. Therefore, agriculture and forestry are logical industries to produce biomass for energy.

Although biomass currently provides only 3% of the energy consumed in the United States, it is the largest source of renewable domestic energy and the only source of renewable liquid transportation fuel. Even though other renewable alternatives are needed to completely address the earth's overall energy needs, replacing liquid petroleum-based transportation fuel (gasoline and diesel) is probably the most pressing need and one with few alternatives. The most likely source of renewable liquid transportation fuel is biofuel.

No single bioenergy feedstock will provide all or even a major portion of energy needs, and one geographical area will not be able to produce an adequate amount of biomass. Rather, the overall bioenergy feedstock supply will consist of a combination of plant materials that can be efficiently produced economically, sustainably, and in a viable quantity in each of the many agroecosystems that make up the range of environments found in rural areas of the United States and the world. For energy production to be sustainable, only biomass that is renewed on a recurring basis can be used, and only that portion of the material not otherwise needed to sustain soil productivity may be removed. That excludes removal of plant material needed on the soil surface to control erosion and that needed to replenish nutrients and retain soil organic matter. Organic matter, although in most cases representing 5% or less of the soil volume, is critical in giving soil many of the attributes critical to successful and sustained plant growth; aggregate stability, water and air exchange dynamics, nutrient cycling, and energy for soil flora and fauna at the micro and macro levels, to mention a few. Soil organic matter content (or the change in organic matter content with change in management practice) is the current best single parameter for assessing soil quality (Shukla et al., 2006). Soil quality must be enhanced to improve productivity or at the minimum be maintained so that productivity is not degraded. Similarly, other production inputs, including water, nutrients, crop protection materials, and energy, must be used sustainably. If the agricultural system is not managed properly, neither energy nor food, feed, and fiber production will be sustainable.

Given these constraints, how much feedstock or energy can agriculture and forestry sustainably provide? Are the land resources of the United States and of the world adequate to produce the quantity of biomass needed to meet demands for food, feed, and fiber and to provide energy?

U.S. ethanol production has been, and continues to be, almost totally from corn grain. However, the quantities of corn and other grain available for use as an energy feedstock will probably provide no more than 18 billion gallons of ethanol annually (NCGA, 2007). In fact, the U.S. Congress-approved Energy Independence and Security Act of 2007 caps corn ethanol production at 15 billion gallons per year.

The amount of oil and fat projected to be available for conversion to energy will produce possibly 2 billion gallons of biodiesel. These are significant quantities of fuel, and their production is economically important. However, to replace the energy in the 140 billion gallons of gasoline and the 40 billion gallons of diesel fuel currently used for ground transportation in the United States with biofuel would require nearly 200 billion gallons of ethanol and 50 billion gallons of biodiesel. Therefore, for biofuel to replace fossil transportation fuel, very large quantities of biomass feedstock must be made available, and technologies for conversion of these feedstocks must be commercialized.

A joint USDA-DOE (U.S. Department of Agriculture/Department of Energy) study concluded that, "The land resources of the United States are capable of producing a sustainable supply of biomass sufficient to displace 30% or more of the country's present petroleum consumption" (Perlack et al., 2005). The study estimated that agriculture could provide nearly a billion dry tons of cellulosic biomass annually, and that forestry resources could add in excess of 0.3 billion additional dry tons. This is less than 1% of the biomass of the earth's biosphere produced for human use and consumption as estimated by Whittaker and Likens (1975) (Table 17.1). At a conversion efficiency of 100 gallons of ethanol per dry ton, 1.3 billion tons of biomass would yield in excess of 130 billion gallons of liquid fuel. Based on energy content, approximately 100 billion of the current 180 billion gallons of liquid petroleum transportation fuel now consumed as gasoline and diesel fuel in the United States could be replaced. Obviously, future demand for transportation fuel will depend on changes in miles driven and in vehicle fuel use efficiency. If the fuel economy of all vehicles was increased by 30%, an amount similar to the mandate for cars and light trucks recently approved by the U.S. Congress, it is reasonable to project that biofuels could provide all of the highway transportation fuel now used. To reach this point, the biofuel industry must grow significantly, a growth that will require enacting governmental policies to provide incentives and mandates and will also require development of technology to improve biofuel production.

Opinions differ on changes in land use and in production practices that would be required for agriculture and forestry to provide significant amounts of bioenergy. Changes will by necessity occur. The challenge is to ensure that land use changes made and production practices adopted are sustainable such that the country and the world move toward a sustainable, renewable energy future.

On a global basis, the energy content of biomass produced by photosynthesis is greater than the energy that human activities consume. Therefore, from purely an energy content perspective, bioenergy can meet the world's energy needs. However, biomass must also provide adequate food, feed, and fiber as well as the ecological needs of the planet's forests, grasslands, marshes, and so on. For this to be realistic, it must be possible to efficiently and sustainably collect and make this energy available economically in the form and quantities and at the times needed. In addition, all other needs for biomass must be sustainably met. This no simple task.

TABLE 17.1

Biosphere Productivity: Biomass Production for Human Use and Consumption, Not Including that Not Harvested or Utilized

Biome Ecosystem Type	Area (million km²)	Mean Net Primary Production (gram dry C/m²/year)	World Primary Production (billion tonnes/year)	Mean Biomass (kg dry C/m²)	World Biomass (billion tonnes)	Minimum Replacement Rate (years)
Tropical rain forest	17.0	2200	37.40	45.00	765.00	20.50
Tropical monsoon forest	7.5	1600	12.00	35.00	262.50	21.88
Temperate evergreen forest	5.0	1320	6.60	35.00	175.00	26.52
Temperate deciduous forest	7.0	1200	8.40	30.00	210.00	25.00
Boreal forest	12.0	800	9.60	20.00	240.00	25.00
Mediterranean open forest	2.8	750	2.10	18.00	50.40	24.00
Desert and semidesert scrub	18.0	90	1.62	0.70	12.60	7.78
Extreme desert, rock, sand, or ice sheets	24.0	3	0.07	0.02	0.48	6.67
Cultivated land	14.0	650	9.10	1.00	14.00	1.54
Swamp and marsh	2.0	2000	4.00	15.00	30.00	7.50
Lakes and streams	2.0	250	0.50	0.02	0.04	0.08
Total continental	**149.00**	**774.51**	**115.40**	**12.57**	**1873.42**	**16.23**
Open ocean	332.00	125.00	41.50	0.003	1.00	0.02
Upwelling zones	0.40	500.00	0.20	0.020	0.01	0.04
Continental shelf	26.60	360.00	9.58	0.010	0.27	0.03
Algal beds and reefs	0.60	2500.00	1.50	2.000	1.20	0.80
Estuaries and mangroves	1.40	1500.00	2.10	1.000	1.40	0.67
Total marine	361.00	152.01	54.88	0.01	3.87	0.07
Grand total	510.00	333.87	170.28	3.68	1877.29	11.02

Source: Whittaker, R.H. and Likens, G.E. 1975. In H. Leith and R.H. Whittaker, *Primary Productivity of the Biosphere*. Ecological Studies, Vol. 14 (Berlin). Berlin: Springer-Verlag, pp. 305–328. Available at http://en.wikipedia.org/wiki/Biomass.

RURAL ECONOMIC BENEFITS

The potential impact that energy production can have on the economies of agriculture and of rural communities in developed and developing countries is like nothing previously seen in the history of agriculture. *Industrialization of agriculture*, that is, the need to expand to farm larger areas to survive economically, has led to consolidation of farms and therefore to fewer farms and to the general reduction in rural opportunities; fewer farms require fewer businesses to provide products and services for their support. The whole process has reduced job opportunities in rural areas and has reduced the quality of rural life. This has led to a reduced rural population, which has furthered the reduction in the desirable aspects of life in rural areas and added to the decline in economic opportunity. Many rural areas have experienced this spiral of decline. Bioenergy production can provide an increased economic vitality and quality of life for the farmers and communities involved. This increased vitality has occurred in communities in which corn-to-ethanol refineries have been built.

The impact of cellulosic biomass production and conversion to liquid fuel will be much larger than the impact of ethanol from corn grain. In the United States, the total potential farm gate income for production of cellulosic biomass for energy could well be larger than that from traditional production of corn and soybean, the country's two major agricultural crops, combined. Prices received for production of commodity crops vary, as will prices for sale of bioenergy feedstocks. A couple of years ago, it was projected that biomass would be worth $40 per dry ton or a potential farm gate income of $40 billion for 1 billion dry tons. At the time, the farm gate income for corn and soybean combined was less than $40 billion, and oil sold for $60 per barrel. Currently, oil is over $130 per barrel, and prices for corn and soybean have also increased, as has the price for bioenergy feedstocks, some now selling for $110 per ton. Although prices cannot be predicted with certainty, if the $500 plus billion per year currently spent to import oil were spent to obtain domestically produced bioenergy, the impact on not only the rural economy but also the nation's economy would be very positive. An infusion of income of that magnitude would reverse the spiral of rural decline, initiate restoration of job opportunities, and restore vitality to agriculture and to agricultural communities.

In addition to biomass, rural agricultural areas are often locations appropriate for collection of other forms of renewable energy, including solar and wind.

BIOENERGY: CONCERNS

There are also many barriers and concerns that must be overcome and addressed if agriculture and forestry are to become major energy producers. Although many promote bioenergy as the preferred option to fossil energy, especially for liquid transportation fuel, others have doubts. Energy efficiency, food, climate, land use, and water are among the concerns raised. As ethanol and biodiesel production has increased, biofuels have been pointed to as the cause of, as having contributed to, or as likely to exacerbate a wide range of concerns. The issues are already being raised, generally in opposition, with regard to the production of ethanol from corn and, to a lesser degree, with the production of biodiesel from soybean. There is concern that food should not be used for energy. There is no doubt that a balance must be maintained. Corn producers, who have pushed for production of ethanol from corn grain in order to raise prices, have been successful. However, it must be recognized that corn production must remain great enough to meet food, feed, and export demands as well as to produce ethanol.

Net Energy

At this time, some amount of fossil energy is used in all bioenergy production. However, the systems used to produce the two most common biofuels, ethanol and biodiesel, produce more energy in the fuels and coproducts than that required for their production (Durante and Miltenberger, 2004).

Net energy has been discussed and debated in the literature extensively. Authors of these articles have used somewhat different assumptions and different domain boundaries and as a result have delivered an array of conclusions about the sustainability of biofuels. An extremely thorough analysis, attempting to level the differences in assumptions and domains of several previous reports, was published in *Science* (Farrell et al., 2006). The conclusions, although not above criticism, suggest that biofuels are energy positive; that is, energy contained in the fuel is greater than the energy (other than solar radiation) that is consumed in creating and delivering the fuel to the point of use. They wisely pointed out that gasoline has a net energy value of only 0.80, meaning that 20% of the energy in a gallon of gasoline is used to extract, transport, and refine the crude and deliver the resulting fuel to the point of use. Compare this to the net energy value of ethanol from grain of 1.30 and from switchgrass (cellulosic ethanol) of 11.2.

Fuel versus Food

Production of bioenergy feedstocks has much the same requirements and makes use of much the same resources as does production of human food and animal feed. This has given rise to a controversy of fuel versus food and feed. It has been estimated that there are sufficient land and water resources, and that sufficient infrastructure can be developed to produce, transport, and process sufficient bio-based feedstock to supply required bioenergy as well as to meet food and feed demands. However, the justification for these estimates has not been adequately documented. And, even if resources are sufficient, as the bioenergy production system develops there will be a transition phase consisting of changes that will have perceived and real implications for food and feed producers and consumers.

The rise in price of corn grain attributed to the amount being converted to ethanol has been linked to increases in prices of other commodity grains and to products made from them. There is concern that demand and price paid for biomass for energy purposes will cause increased production of energy feedstock and decreased production of crops for food and feed. Perlack et al. (2005) suggested production of 1 billion tons of feedstock for ethanol production will not cause significant changes in commodity use and prices. Recent changes in Board of Trade values for corn and other grains appear to contradict that suggestion—at least in the short run. The ultimate impact of high grain prices on food has yet to be seen. Almost certainly, feed prices three times their historical levels will elevate food prices. The demand created for biomass feedstock to produce energy will no doubt increase feedstock cost. This will negatively affect current consumers of feedstock, those that use it for food, feed, and fiber. Changes in supply-demand dynamics will cause at least short-term disruptions that will create winners and losers.

WATER

Water is essential for biological life. Although the Earth's surface is largely covered with water, most contains levels of dissolved salts that make it generally unusable by plants and animals. Growing the plants to produce the large quantities (billion or more dry tons) of biomass feedstock that will be necessary to produce sufficient quantities of biofuels will require large amounts of water. In addition, conversion of biomass feedstock to liquid fuel and products may require additional significant amounts of water.

Water is a finite natural resource that exists in a constant quantity on the earth. But, is that quantity sufficient? The global supply of water is estimated to be 1,386,000,000 cubic kilometers or 206 billion liters for each of the world's nearly 6.7 billion humans. Unfortunately, 97% of this water contains excessive dissolved solids and without treatment is unusable by most plants and animals. This leaves 5 billion liters of freshwater per person. Of that, over two thirds is frozen in ice caps and glaciers and not readily available. That still leaves about 1.6 billion liters of fresh surface and groundwater for each person on the planet. Assuming that per capita usage of freshwater is 2 million liters per year (an estimate for personal, agricultural, and industrial use in industrialized countries), human needs would require only one eighth of 1% of the freshwater available.

If desalinated, seawater could easily provide all water needs. The major obstacle is the energy required to remove the dissolved solids. The minimum energy to desalinate seawater is about 0.66 kcal/L. That is true regardless of the technology used. Therefore, it would take 2 million L * 0.66 kcal/L = 1.32 million kcal (5544 million J) to desalinate the water needed by one person in an industrial lifestyle. That equals the energy in about 70 gallons of ethanol.

This suggests that there is plenty of water; unfortunately, it is not well distributed, and therein lies the problem (see Chapter 19). In many cases, local water supplies are simply insufficient. In other situations, the water is not effectively or efficiently used. The challenge then is to engineer systems for the sustainable use of a water resource that is sufficient in overall quantity but is lacking in quality and distribution. This challenge must be met to reduce human misery, to enhance the environment, and to eliminate a major cause of strife.

LAND USE CHANGE

There are concerns that cropping changes to produce bioenergy feedstocks in one location will result is land use changes in other locations, changes that could result in significant oxidation of carbon in the existing vegetation and in the soil, with a resulting increase in carbon dioxide emissions into the atmosphere. These carbon dioxide emissions could be greater than those associated with fossil energy use. It is important that policies be established to manage land use change so that practices are sustainable.

GREENHOUSE GAS EMISSIONS

Increased use of corn and other grains is based on use of large quantities of nitrogen fertilizer. Higher prices for grains may result in application of more fertilizer to existing and newly cultivated marginal lands to spur crop yield. Excessive application of nitrogen fertilizer can result in release of nitrous oxide, one of the most potent GHGs, on a molecule per molecule basis, 300 times more potent than carbon dioxide.

Although biofuel will address carbon dioxide emissions to the atmosphere, increased production of corn and other crops requiring high rates of nitrogen fertilizer to achieve the high yields necessary to supply the massive amount of feedstock needed to make biofuels a significant player in the energy game will increase the potential for some of this nitrogen to enter the ecosystem as leachate or nitrous oxide (Kim and Dale, 2005).

YIELDS

Historically, average yields of commodity crops have steadily increased as a result of genetic modification through conventional breeding and more recently through biotechnology. Expected future yield increases will allow even greater amounts of both food and fuel to be grown on each unit of land area. In fact, the Billion Ton Vision (Perlack et al., 2005) uses the assumption that grain crop yields will double in the 25-year time frame of their analysis (through 2030). They also assumed that harvest index or grain-to-stover ratio would remain constant for all grains except soybean. The outcome of these assumptions is that crop biomass production will increase linearly with grain yield. Others question the validity of the assumption that the rate of yield increase we have experienced over the past 40 years can be maintained over the next 25 years (Cassman, 2007; Council for Agricultural Science and Technology [CAST] 2006).

RURAL ECONOMIC DEVELOPMENT

Because photosynthetic solar energy capture is low, biomass production, on the scale needed to significantly contribute to the transportation fuel supply, will require large land areas. This in turn means that its production and gathering of the biomass feedstock will be distributed and will provide jobs and income for residents over large rural areas. This activity has potential to revitalize many economically depressed regions and will as a result provide significant social benefits. In addition, because cellulosic biomass tends to have low physical density as well as low energy density, transportation of the raw feedstock will be expensive. On-farm and local community preprocessing may be economically beneficial and even essential. These pretreatment facilities will add value to the biomass feedstock and in so doing will create additional jobs and enhanced rural community vitality.

Policy

Because policy changes are needed to internalize the external costs of energy security, of negative environmental changes, and of economic effects of our fossil energy economies, there will be political implications associated with large-scale production and conversion of biomass. Those political implications are already being felt with regard to ethanol production from corn and biodiesel from soybean. Issues include government subsidies, how much and for whom; increasing feedstock cost and its effects on price of food and feed; effects on imports and exports; and short-term adjustments versus long-term goals.

Sustainability

Biomass production and harvest, as with any other production practice, must be done sustainably. It must conserve and enhance natural resources, including soil and water. Also of concern are undesirable effects of rural ecosystems and ecology. If soil quality is degraded, sustainable production is not possible. Water has already become seriously deficient in many places. Further, as demands increase, extraction of nonrenewable resources will also increase and reduce our capacity to provide adequate amounts of energy (renewable or not) in the future.

There is need for sustainability in land use. This is as true for biofuel production as for any other purpose. Consideration must be given for changes in land use in one location caused by land use changes to produce bioenergy in another location. There is a strong interdependency among nations and their policies and practices. Sustainable practices and policies must extend across political and geographical boundaries.

Invasive Species

There is concern that bioenergy crops could become invasive. An ideal crop for production of biomass energy feedstock is one that is easily established, grows rapidly, is robust, and is resistant to pests. That is basically the definition of an invasive plant. Therefore, any acceptable energy crop plant must have characteristics that allow it to be managed such that it does not become invasive. Species invasiveness, as related to biofuel feedstock production, has been discussed in detail in a recent CAST commentary (Council for Agricultural Science and Technology [CAST] 2007a).

THE TRANSITION TO RENEWABLE BIOENERGY: MAKING IT HAPPEN

The transition from nonrenewable and nonsustainable dependence on fossil energy to renewable and sustainable bioenergy will require addressing the barriers to change with technology development and policy action.

TECHNOLOGY

There is a need for technological improvements that will improve water use efficiency, increase plant productivity (make plants more productive in the environments in which they must be grown), improve the efficiency with which plants capture solar energy, improve biomass collection and handling, improve the efficiency with which energy in biomass is converted into useful forms, and in general improve the energy efficiency in all aspects of bioenergy production and use.

A challenge is to make the production of energy feedstock as great as possible from each unit of land and to do this as sustainably, efficiently, and at as low a cost as possible. One problem is that photosynthesis, while effective, is not a very efficient method of energy capture. Design and development of energy crop plant varieties that capture solar energy with improved efficiency would increase yields of biomass per unit land area and decrease the land needed for energy production. Efficient technology must be developed to convert biomass feedstocks to desired energy forms. Feedstocks with the characteristics that match the capabilities of conversion processes to convert them to desired forms of fuel, whether liquid or solid, need to be developed. Design of crops to be used as energy feedstocks and of the conversion technology for processing these crops must be done collaboratively such that the overall system is efficient and economical.

The good news is that the technology we need to begin the transformation to a low-carbon economy exists, and the investment dollars are available if the policy ground rules are properly established. A great deal of investment and effort will be needed to make this vision real, but the hard work of ushering it in can become a powerful engine for growth, competitive advantage, and jobs.

POLICY

Under current energy policies, bioenergy is generally unable to economically compete at the point of sale with fossil energy. Therefore, policy changes are needed to aid the purchaser of energy to make choices that are in society's best interest— policies that incorporate the externalities of security, environment, and economics associated with fossil energy, along with benefits and negatives of bioenergy—into the prices of energy at the point of sale. Expansion of biomass energy production must be addressed as part of an overall energy policy, an overall agricultural policy, and an overall economic development policy. These policy actions will require reconciling the interests of the existing fossil industry with those of the developing bioenergy industry and aligning both with the interests of society in general. This is not an easy task and, despite energy supply and use concerns, a task that policy makers have chosen not to deal with seriously. Political action has not been taken because there is no sense of urgency. Unfortunately, it appears that a sense of urgency will not likely occur without a crisis.

Humans must develop a new way, a plan for sustainable use of the world's resources, including energy, water, air, and soil. These issues are also addressed in other chapters (see Chapters 15, 16, 18, 19, 23). The plan must be developed and implemented collectively among the entire Earth's population. It is obviously an idealistic and optimistic outlook considering the difficulty in reaching agreement over and getting along on the "simple" things. For this to be possible, it is necessary to view from a high, long-term level. The size of the human population and the level of technology that has been achieved make it possible for humans to change Earth's environment. There are many examples of effects on air, water, and soil quality as a result of land clearing and agricultural practices, of contamination associated with meeting energy needs and with industrial production, and as a result of use of consumer products. It is not clear that sustainability/survivability at a desired level is or can be by itself market driven. It is not clear or intuitively obvious that if it makes a profit it will be in the long term desirable or sustainable. There is need for long-term goals and policies that steer market forces toward the desired outcome.

From a national perspective, a policy that promotes bioenergy, or renewable energy in general, will have significant economic benefits.

CONCLUSIONS

The world's population has become dependent on petroleum, coal, and natural gas for most of its energy needs. These fossil energy sources are finite and therefore not sustainable. They also come with significant security, environmental, and economic side effects. Therefore, renewable energy sources and systems must be developed and implemented. The sun is the only significant source of renewable energy for the earth. The challenge is to effectively and efficiently capture and make use of that solar energy. One method is through photosynthesis, by which plants gather and store in their tissues energy from solar radiation. The energy, *bioenergy*, stored in plant biomass can, by use of biochemical or thermochemical processes, be used directly for heat or power or can be converted into liquid or gaseous fuels. Bioenergy has the potential to provide a significant portion of the world energy needs, to improve energy security, to improve environmental quality, and to enhance both rural and national economies and to do so sustainably. Bioenergy production systems must overcome barriers to commercialization and must avoid potential side effects that could render these practices nonsustainable. If managed properly, bioenergy production will provide significant benefits to agriculture and rural areas and will do so while at the same time adequately meeting the world's food, feed, and fiber needs. *Renewable energy must be developed such that it is sustainable.* The need for sustainable energy is not whether, but how and when. Although the need is urgent, a crisis may have to occur before the sense of urgency is sufficient for significant action.

Bioenergy cannot fulfill the total energy needs of the United States or of the world. Sustainable bioenergy can, however, provide a significant portion of the liquid fuels needed for transportation and a significant amount of renewable energy for other purposes.

The good news for the earth is that the sun provides a huge amount of energy, many times what human activities consume. Therefore, the world can operate on an energy-sustainable basis. It is not necessary to rely on nonrenewable, finite energy sources. The bad news is that incident solar radiation is distributed and reasonably low intensity, or from the reverse perspective, it is not concentrated and as energy dense as fossil fuel resources tend to be. The challenge then is to efficiently and sustainably collect this energy such that it can be concentrated and economically used.

When establishing energy policy, it is important to consider the production and use of energy as part of human involvement in the global system, a system that includes the global climate and its stability as well as the production, distribution, availability, and cost of food. The short-term implications of policy actions must be considered, but the long-term effects on sustainability are critical to the continued welfare of the earth's inhabitants and to survival of humans on the earth. At this time, it is not yet certain what the correct energy system will be. What is important is to have the debate and move forward with a rational plan. Global effects must be considered in that the actions of anyone have implications for everyone.

The consequences of not dealing with the collateral damage of fossil fuel use are significant enough to seriously consider alternatives. However, that concern has not yet become sufficient to create a sense of urgency.

Solving the energy problem and avoiding the associated collateral damage requires policy action, but that will not happen without a sense of urgency. Both technology development and policy action are necessary to avoid an energy crisis and to ensure that individual nations and the world move forward on an appropriate path toward a sustainable energy future, a future that holds great promise for agriculture, for rural economies, and for the world.

REFERENCES

Cassman, K.G. 2007. Climate change, biofuels, and global food security. *Environment Resources Letters* 2:011002.

Council for Agricultural Science and Technology (CAST). 2006. *Convergence of Agriculture and Energy: Implications for Research and Policy.* CAST Commentary QTA 2006-3. Ames, IA: CAST.

Council for Agricultural Science and Technology (CAST). 2007a. *Biofuel Feedstocks: The Risk of Future Invasions.* CAST Commentary QTA 2007-1. Ames, IA: CAST. (http://www.cast-science.org/displayProductDetails.asp?idProduct=146)

Council for Agricultural Science and Technology (CAST). 2007b. *Convergence of Agriculture and Energy: II. Producing Cellulosic Biomass for Biofuels.* CAST Commentary QTA2007-2. Ames, IA: CAST.

Crabtree, G.W., and N.S. Lewis. 2007. Solar energy conversion. *Physics Today* 60:37–42.

Durante, D. and M. Miltenberger. 2004. The net energy balance of ethanol production. Available at: http://www.ethanol.org/pdf/contentmgmt/Issue_Brief_Ethanols_Energy_Balance.pdf.

Energy Information Administration. 2008. International energy outlook 2008. Available at: http://www.eia.doe.gov/oiaf/ieo/world.html.

Farrell, A.E., R.J. Plevin, B.T. Turner, A.D. Jones, M. O'Hare, and D.M. Kammen. 2006. Ethanol can contribute to energy and environmental goals. *Science* 311:506–508.

Kim, S. and B.E. Dale. 2005. Life cycle assessment of various cropping systems utilized for producing biofuels: bioethanol and biodiesel. *Biomass and Bioenergy* 29:426–439.

Long, S.P., X.-G. Zhu, S.L. Naidu, and D.R. Ort. 2006. Can improvement in photosynthesis increase crop yields? *Plant, Cell and Environment* 29:315–330.

Perlack, R.D., L.L. Wright, A.F. Turhollow, R.L. Graham, B.J. Stokes, and D.C. Erbach. 2005. *Biomass as Feedstock for a Bioenergy and Bioproducts Industry: The Technical Feasibility of a Billion-Ton Annual Supply.* DOE/GO-102005–2135 and ORNL/TM-2005/66. Oak Ridge, TN: Oak Ridge National Laboratory.

Public Citizen. 2008. Corporate average fuel economy. Available at: http://www.citizen.org/ autosafety/fuelecon/.

Shukla, M.K., R. Lal, and M. Ebinger. 2006. Determining soil quality indicators by factor analysis. *Soil and Tillage Research* 87:194–204.

Whittaker, R.H. and G.E. Likens. (1975). The biosphere and man. In H. Leith and R.H. Whittaker, *Primary Productivity of the Biosphere*. Ecological Studies, Vol. 14 (Berlin). Berlin: Springer-Verlag, pp. 305–328. Available at: http://en.wikipedia.org/wiki/ Biomass.

18 Soil Science: *Management and Conservation*

R. Lal

CONTENTS

ABSTRACT

Conversion of natural to agricultural ecosystems, involving 1070 million hectare (Mha) of forests and woodland/shrubland and 660 Mha of grassland, and subsequent soil cultivation caused depletion of the soil organic carbon pool, disruption in elemental cycling, and increase in susceptibility to degradation by erosion, salinization, and fertility depletion. The severe problem of soil degradation, caused by extractive farming practices and fragile soils, is likely to be exacerbated by the projected climate change. Despite the severe problem of soil degradation, cereal yields in developing countries must be increased by 35 to 63% by 2025 and 58 to 121% by 2050 without and with any dietary change, respectively. The global fertilizer use is projected to increase from 136 million nutrient tons in 2000 to 165 million tons in 2015 and 188 million tons by 2030, and most of the projected increase is to occur in India, China, and other Asian and Latin American countries. Similarly, global irrigated land area is projected to increase, with most expansion in South and East Asia, but not in sub-Saharan Africa. Strategies of improving soil quality include increasing soil organic matter pool, improving soil structure, and strengthening elemental cycling. Conservation and sustainable management of soil resources involve adaptation of no-till farming, using crop residue mulch, growing cover crops, creating positive nutrient balance, using soil-specific or precision farming, conserving and recycling water, and using drip or subirrigation. Biofuel must be produced from lingocellulosic biomass grown on energy plantations of short-rotation woody perennials or warm-season grasses.

INTRODUCTION

Of the six prehistoric natural global ecosystems, 36.0% of Earth's land area was occupied by forests, 26.3% by grassland, 12.5% by shrubland, 12.2% by desert, 7.5% by woodland, and 5.7% by tundra (Table 18.1). The land area under different ecosystems progressively changed since about 10 to 12 thousand years ago when the settled agriculture and domestication of plants and animals began. Increase in food supply with expansion of agriculture led to increase in human population, from about 4 million in 10,000 B.C. to 170 million in 0 B.C., 190 million in 500 A.D. to 6.5 billion in 2006 A.D. (Table 18.2). The world population is presently increasing at the rate of 1.3% per year and is projected to be 7.5 billion by 2020, 9.4 billion by 2050, and 10 billion by 2100 (Fischer and Helig 1997; Cohen 2003). Almost the entire increase in future population will occur in developing countries, where soil resources are in short supply and are prone to degradation because of the harsh climate and extractive farming practices.

Increase in agricultural land area was caused by conversion of natural to managed ecosystems. The latter involved reduction in area by 750 million hectare (Mha) of the forests, 660 Mha of grasslands, 180 Mha of woodland, 140 Mha of shrubland, and 90 Mha of deserts. Of these changes, 1550 Mha were used for cultivation of food crops (Table 18.1). Furthermore, anthropogenic transformation of ecosystems had a strong impact on soil processes and properties, energy balance, water balance, and elemental cycling. The soil organic carbon (SOC) pool declined by 66–90 Pg (Petagram, one billion metric tons) (Lal 1999), with the attendant adverse impact on soil quality and reduction in ecosystem services. Decline in soil quality was exacerbated by other degradative processes (Lal 2004), such as erosion by water and wind, nutrient depletion, elemental imbalance, acidification, salinization, compaction, crusting, and decline in soil structures. The objective of this chapter is to deliberate processes, factors, and causes of soil degradation; the attendant impact on the environment and ecosystem services; and the conservation and management options to restore degraded soils and achieve sustainable management.

SOIL AREA UNDER NATURAL AND CULTIVATED ECOSYSTEMS

Of the world cropland area of 1550 Mha, 19.4% is on Oxisols, 18.7% on each of Alfisols and Mollisols, 8.4% on Ultisols, and 5.2% on Entisols (Table 18.3).

Growing crops, especially by the use of extractive farming practices, leads to decline in soil quality by accelerated erosion and other degradation processes. The problem is more severe in developing countries of the tropics and subtropics (Table 18.4) because the climate is harsh, soils are fragile, and farmers are resource

TABLE 18.1

Estimates of the Pre-Agricultural and Present Land Area in Major Ecosystems

Ecosystem	Prehistoric Area		Present Area		Changes	
	10^9 ha	% of Total	10^9 ha	% of Total	10^6 ha	% of Total
1. Forest	4.68	36.0	3.93	30.2	750	20.6
(a) Tropical Rainforest	1.28	9.8	1.20	9.2	80	2.2
(b) Other Forests	3.40	26.2	2.73	21.0	670	18.4
2. Woodland	0.97	7.5	0.79	6.1	180	4.9
3. Shrubland	1.62	12.5	1.48	11.4	140	3.8
4. Grassland	3.40	26.3	2.74	21.0	660	18.1
5. Tundra	0.74	5.7	0.74	5.7	0	0
6. Desert	1.59	12.2	1.50	11.5	90	2.5
7. Cultivated	0.0	0.0	1.55	11.9	1550	42.6
8. Pasture/ Plantations	0.0	0.0	0.27	2.2	270	7.5
Total	13.0	100	13.0	100	3640	100

Source: Adapted from Williams, 1994; NRC, 1993.

TABLE 18.2

Increase in World Population

Year	Population (millions)
10,000 BC	4
5,000 BC	5
1,000 BC	50
0 BC	170
500 AD	190
1000 AD	310
1500 AD	425
1700 AD	600
1800 AD	980
1900 AD	1650
1950 AD	2400
2000 AD	6070
2006 AD	6500

Source: http://en.wikipedia.org/wiki/Image:Population_curve.svg

TABLE 18.3
Predominant Soils of the World and Their Use in Crop Production

	Total Historic Area		Cropland Area	
Order	10^6 ha	% of Total	10^6 ha	% of Total
Alfisols	1330.3	10.2	290	18.7
Andisols	106.0	0.8	50	3.2
Aridisols	1555.5	11.9	40	2.6
Entisols	2167.8	16.6	80	5.2
Histosols	161.0	1.2	0	0
Inceptisols	946.1	7.3	150	9.7
Mollisols	924.6	7.1	290	18.7
Oxisols	1011.7	1.8	300	19.4
Spodosols	347.9	2.7	50	3.2
Untisols	1174.6	9.0	130	8.4
Vertisols	320.0	2.5	60	3.9
Gellisols	1119.9	8.6	0	0
Others	1870.0	14.3	110	7.0
Total	13035.4	100.0	1550	100

Source: Adapted from Buringh, 1984; Eswaran et al., 1995.

poor and either cannot or do not adopt recommended practices of soil and crop management. Of the total global land area prone to degradation, area affected in developing countries is 77% by water erosion, 83% by wind erosion, 97% by loss of nutrients, 94% by salinization, 100% by pollution, 83% by acidification, and 91% by waterlogging (Table 18.4). Next to erosion, secondary salinization is the most serious problem. Of the 270 Mha of irrigated cropland in the year 2003, there were 72 Mha or 26.5% salinized due to excessive irrigation, lack of adequate drainage, or use of poor-quality water (Table 18.5). The problem of secondary salinization of irrigated area is especially severe in Egypt, India, Iran, Pakistan, and China (Table 18.5).

FUTURE RISKS OF SOIL AND ENVIRONMENTAL DEGRADATION

The problem of soil and environmental degradation may be exacerbated during the twenty-first century because of the confounding effects of many interactive factors. Important among these are the following:

1. Future increase in the world population will occur in developing countries where soil resources are already under great stress and the per capita cropland area is low and decreasing.
2. The per capita food supply in the world, compared with that of the United States and other developed countries, is low (Table 18.6) and decreasing in resource-

TABLE 18.4
Estimates of Land Area Affected by Soil Degradation in the Developing Countries

Degradation process	Area in developing countries (Mha)	% of the world's degraded area in the tropics
Water erosion	837	77
Wind erosion	457	83
Loss of nutrients	132	97
Salinization (L1)	72	94
Pollution	21	100
Acidification	5	83
Compaction	32	47
Waterlogging	10	91

L1 from Pritchard and Amthor (2005), see Table 18.5.

Source: Recalculated from Oldeman, 1994.

TABLE 18.5
Estimates of Cropland Area Irrigated and Salinized in 2000

Country	Cropland area (10^6ha) Total	Irrigated	Salinized Area (10^6ha)	% Salanized
Argentina	25.0	1.6	0.6	37.5
Australia	50.3	2.4	0.2	8.3
China	124.0	54.4	7.5	11.9
Egypt	2.8	2.3	1.0	43.4
India	161.8	54.8	20.0	36.4
Iran	14.3	7.5	2.0	26.6
Pakistan	21.3	18.1	4.5	24.8
South Africa	14.8	1.5	0.2	13.3
Thailand	14.7	5.0	0.45	0.9
United States	177.0	22.4	5.2	23.2
USSR (former)	204.1	19.9	3.5	17.5
World total	1364.2	271.7	72	26.5

Source: Adapted from Pritchard and Amthor, 2005.

TABLE 18.6
Per Capita Food Supply in the World and the U. S.

Parameter	World		USA	
	1980	2000	1980	2000
Energy (cal d^{-1})	2535	2805	3166	3772
Protein (g d^{-1})	66.9	75.6	97.9	114.0
Fat (g d^{-1})	59.4	75.2	127.5	151.3

Source: Adapted from Pritchard and Amthar, 2005.

poor countries where the growth rate of the population is high (Rosegrant and Cline 2003). Consequently, the required average yield of cereals in developing countries must be increased by 35 and 58% by 2025 and 2050, respectively, to maintain the same dietary intake. With possible change in dietary habits of people in rapidly developing economies (e.g., China, India), increase in crop/cereal yield must be 63% by 2025 and 121% by 2050 (Table 18.7). The desperateness of bringing about a quantum jump in agricultural production may exacerbate elemental imbalance, soil compaction, secondary salinization, and accelerated erosion.

3. The problem of soil degradation may also be accentuated by the projected climate change. Increase in soil temperature, decrease in SOC pool because of cultivation and increase in rate of mineralization, increase in losses by

TABLE 18.7
Present and Expected Average Yield of Cereals to Meet the Future Food Needs in Developing Countries[a]

Parameter	Cereal grain yield (t/ha)	Total production (Mt/yr)
Present – 2000	2.34	1267
Required 2025		
+35%[b]	3.60	1706
+63%[b]	4.40	2045
2050		
+58%[b]	4.30	1995
+121%[b]	6.00	2786

[a] Africa, Asia, South America (excluding Japan)

[b] Estimated increase above 2000 level to account for increase in population and possible change in food habits.

Source: Adapted from Wild, 2003.

TABLE 18.8
World Primary Energy Supply in 2004

Source	% of Total Energy
Oil	34
Coal	24
Gas	21
Combustible renewable & waste	11
Nuclear	7
Hydro	2
Others	1

Source: Adapted from EIA, 2004; EPA 2005.

erosion (Lal 2005a) and leaching, and increase in frequency and intensity of extreme events may exacerbate the problem of soil degradation.

4. Increase in the future energy demand may accentuate soil degradation directly and indirectly. Presently, only 11% of the world's primary energy supply is met through combustible renewable and other biosolids (Table 18.8). There is a strong emphasis on increasing the use of biofuels. Toward this strategy, the goal is to procure 1 billion Mg (Megagram, one metric ton) of cellulosic biomass in the United States and 4 to 5 billion Mg in the world (Lal 2005b; Somerville 2006). Harvesting crop residues for biofuel production (ethanol) may increase the risk of erosion, compaction, depletion of the SOC pool, and so on.

Indirectly, the severity and extent of soil degradation may be affected by the rapid change in land use, especially conversion of forest/woodlands and grasslands to cropland and pastures. Presently, 19 to 20% of the total annual emissions (CO_2 equivalent) are attributed to land use change and deforestation (Table 18.9). The effect of land use, land use change, and soil cultivation on the radiative forcing may

TABLE 18.9
Sources of Global Emissions in 2000

Source	% of Total PgC Equivalent
Fossil fuel combustion	53.0
Cement manufacture	2.0
Land use change and forestry	19.0
Methane emission	16.0
Nitrous oxide	9.0
High GWP gases	1.0

Source: DOE, 2006.

TABLE 18.10

Estimates of U. S. and Global CH$_4$ and N$_2$O Emissions from Agriculture

Source	USA		World	
	Total (Tg CO2 Eq.)	% of Non- CO2	Total (Tg CO2 Eq.)	% of Non- CO2
1 N$_2$O from Agriculture	282	26	2875	32
2 CH$_4$ emission				
a. Enteric	116	11	1712	19
b. Manure	38	3	199	2
c. Rice paddies	8	1	643	7
3 Total	443	40	5429	60

Source: DOE, 2006; IPCC 2001.

increase due to increase in the demand to enhance food production. Furthermore, emissions of N$_2$O (from increase in fertilizer use) and of CH$_4$ (from increase in area under rice paddies, livestock raising, and use of manure) may increase (Table 18.10). These trace gasses have a high global warming potential relative to CO$_2$ (21 for CH$_4$ and 310 for N$_2$O) and a positive feedback on soil degradation risks. Emission of CH$_4$ and N$_2$O from agricultural activities is estimated at 444 Tg (Teragram, one million metric tons) per year of CO$_2$ equivalent (44% of total non-CO$_2$ gases) for the United States compared with 5428 Tg per year of CO$_2$ equivalent (60% of total non-CO$_2$ gases) for the world (Department of Energy [DOE] 2006).

BASIC PRINCIPLES OF MANAGEMENT AND CONSERVATION OF SOILS

Strategies of sustainable management of soil to increase food production while restoring degraded/desertified soils include the following:

1. Minimize risks of soil erosion caused by water and wind through conversion of plow tillage to no-till farming and afforestation of steep terrain and highly erodible land
2. Enhance SOC pool by liberal use of crop residue mulch, growing cover crops, and use of complex crop rotations and farming systems (e.g., agroforestry, ley farming)
3. Maintain a positive nutrient balance through integrated nutrient management systems, including the use of manure and other biosolids, biological nitrogen fixation, and strengthening processes of nutrient recycling
4. Improve soil structure by enhancing bioturbation through increase in activity of earthworms and microbial processes

5. Conserve water in the root zone by reducing losses due to runoff and evaporation
6. Adopt improved irrigation/fertilization techniques that deliver water and nutrients directly to plant roots at the most critical stage of plant growth
7. Use improved/genetically modified (GM) plants/cultivars that are efficient in resource utilization and are adapted to current and future biotic and abiotic stresses
8. Synchronize crop requirements (rooting depth, pH, water and nutrient needs) with soil properties through use of soil survey data and land use capability classification
9. Minimize losses by adopting soil-specific management that increases productivity
10. Adopt techniques of soil/crop management that increase productivity per unit area of land, time, and input (energy, nutrients, and water)

Basic principles of management and conservation of soil are outlined in Figure 18.1. The strategy is to adopt recommended management practices (RMPs) that increase soil organic matter (SOM), improve soil structure, and strengthen elemental cycling. Adoption of RMPs would decrease losses of water and nutrients out of the ecosystem, reverse degradative trends, and restore degraded/desertified soils.

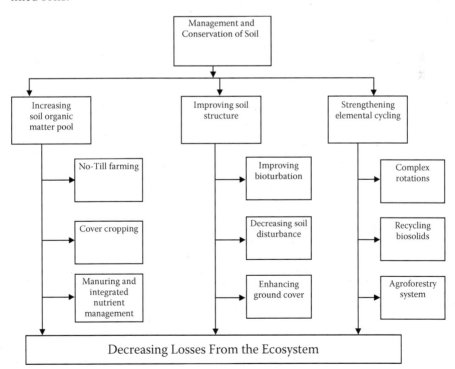

FIGURE 18.1   Processes and practices of enhancing soil quality.

ENHANCING THE SOIL ORGANIC MATTER POOL

Depletion of the SOM pool is an important process that sets in motion other degradation processes. In corollary, therefore, restoring the SOM pool is essential to improving soil structure, increasing the plant-available water capacity, reducing erodibility and decreasing risks of soil erosion and non-point source pollution, improving elemental cycling, and enhancing soil biodiversity (Figure 18.2). At a global scale, increasing the SOM pool has synergistic effects on achieving/advancing food security (Lal 2006), mitigating climate change (Lal 2004), and improving the water resources.

The SOM pool is an integral component of the terrestrial carbon pool (TCP) in the biosphere (Batjes 1996). Different components of the TCP include the carbon pool in the world's soils, forests, and wetlands (Figure 18.3). The soil carbon pool includes SOC and soil inorganic carbon (SIC) components. It is the transformation of the SOC pool from labile and dissolved fractions into stable humic substances and of precipitation of dissolved bicarbonates and carbonic acids into secondary carbonates that lead to stabilization/protection of the soil carbon pool. The forest/biotic carbon pool is comprised of these items and below-ground biomass and the detritus material. The depth distribution of the below-ground biomass and the relative abundance of recalcitrant substances (e.g., lignin, seubrin, phenols) are important to stabiliza-

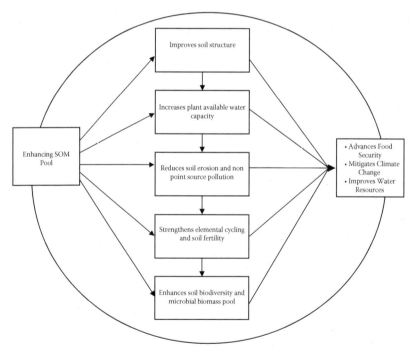

FIGURE 18.2 Interactive effects of enhancing soil organic matter (SOM) pool on global food security and mitigating climate change.

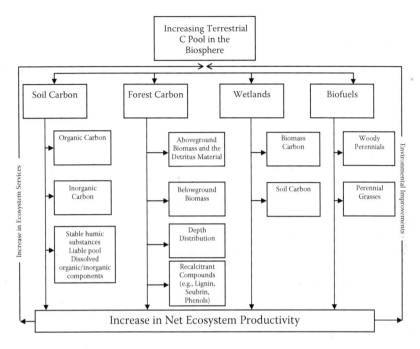

FIGURE 18.3 Strategies of enhancing carbon pool in the biosphere. NPP, net primary productivity.

tion of the TCP. Wetlands constitute a minor but an important component of the TCP, in terms of both the biomass and the soils' carbon.

Biofuels and their use are important to stabilization of the atmospheric abundance of CO_2 and are discussed in another section. Sources of biofuel feedstock, crop residues versus energy plantations, can have a significant impact on the SOC pool and TCP. Removal of crop residues as a source of lignocellulosic biomass for bioethanol production can have an adverse impact on soil quality and exacerbate soil erosion and physical degradation.

Several mechanisms of soil carbon sequestration (e.g., physical, chemical, biochemical) protect the SOM pool against microbial attack. The SOM pool is also protected from climatic and anthropogenic perturbations by translocation (illuviation, or vertical movement) into the subsoil and deposition/burial (lateral movement) into depressional sites (Figure 18.4). These protective mechanisms, while increasing the residence time of carbon in soil, have numerous positive impacts on soil and the environment. Improvements in soil quality lead to an increase in the net primary productivity, decrease in losses of water and nutrient, reduction in non-point source pollution, and increase in biodiversity, especially that of the soil (Figure 18.4). Increase in biomass input into the soil has numerous ancillary benefits and increases ecosystem services.

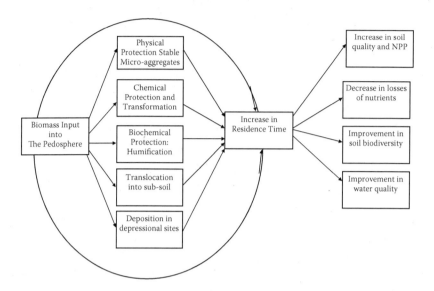

FIGURE 18.4 Mechanisms of soil carbon sequestration and protection against microbial processes.

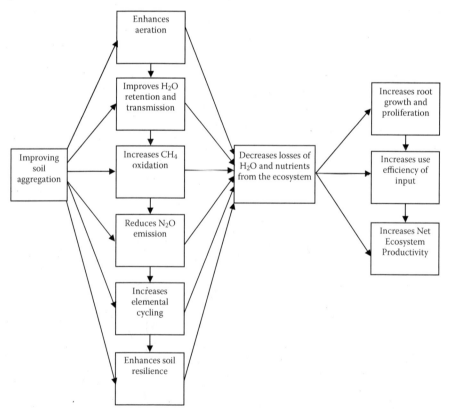

FIGURE 18.5 Positive feedback and synergistic impacts of improving soil aggregation.

MANAGING SOIL STRUCTURE AND ENHANCING AGGREGATION

There exists a strong and positive correlation between SOM pool and soil aggregation. Formation of stable micro- and macroaggregates is one of the several positive impacts of enhancing and managing the SOM in soils of agricultural and other managed ecosystems. Improvement in soil aggregation is essential to improving edaphic environments critical to plant growth and productivity. Increase in amount and stability of aggregates enhances aeration, improves water retention and transmission, increases CH_4 oxidation and soil sink capacity for absorbing CH_4, reduces emission of N_2O, strengthens the mechanism of elemental cycling, and enhances overall soil resilience (Figure 18.5). The attendant improvement in soil tilth enhances root growth and proliferation, increases use efficiency of input (water, fertilizer, energy), and accentuates net ecosystem productivity (Figure 18.5).

Soil structure and tilth are intimately linked to the choice of tillage methods, residue management, use of cover crops in the rotation cycle, frequency and weight of vehicular traffic, and application of biosolids (e.g., manure, crop residue mulch, sewage sludge). Conversion of plow tillage to no-till farming, use of crop residue mulch, and application of manure improve soil aggregation, total and macroporosity, and water transmission (infiltration) and retention (available water) capacity. Soils with favorable structure are less prone to crusting, surface sealing, compaction, and hard setting. Removal of crop residues for other uses (e.g., fodder, fuel, and construction/industrial material), excessive tillage, use of heavy machinery, and excessive and uncontrolled grazing cause a decline in aggregation and susceptibility to physical degradation. Maintenance of soil structure is important to sustainable management of agricultural soils.

NUTRIENT MANAGEMENT

Soil fertility depletion and chemical degradation are associated with the use of extractive farming practices. Soils of sub-Saharan Africa and those of South Asia under dry farming are severely depleted of their inherent fertility because plant nutrients harvested in crops and animal products are not replaced through application of chemical fertilizers or organic amendments. It is widely recognized that soils of sub-Saharan Africa have been subjected to negative nutrient balance at the rate of 30 to 40 kg of nitrogen/phosphorus/potassium hectare per year for decades because of the extractive farming practices. The data in Table 18.11 show that fertilizer consumption over the 40-year period (1961–2000) increased by a factor of 4.4 in the world (from 31.2 million tons in 1960–1961 to 136.4 million tons in 2000–2001), 43.2 in South Asia (from 0.5 to 21.6 million tons), 14.7 in East Asia (from 3.1 to 45.5 million tons), and 13.2 in Latin America (from 1.0 to 13.2 million tons). In contrast, however, fertilizer consumption in sub-Saharan Africa increased from 0.16 million tons in 1960–1961 to 1.2 million tons in 2000–2001 (by a factor of 7). In addition to the low rate of fertilizer use, the nutrient use efficiency is also low because of poor soil structure, susceptibility to crusting and compaction, and high losses by runoff, erosion, volatilization, and leaching.

TABLE 18.11
Regional and Global Fertilizer Consumptions (N+P+K)

Year	World	South Asia	East Asia	Sub-Saharan Africa	Lath America
	-----------------------------10^6 Nutrient Tons -----------------------------				
1961	31.2	0.5	3.1	0.16	1.0
1965	47.0	1.0	5.4	0.26	1.5
1970	69.3	2.8	8.3	0.44	2.9
1975	91.4	4.4	11.4	0.71	4.5
1980	116.7	7.4	21.1	0.96	7.5
1985	129.5	11.2	27.0	1.07	7.4
1990	137.8	15.2	36.5	0.25	7.9
1995	129.7	18.1	45.3	1.07	8.5
2000	136.4	21.6	45.5	1.23	13.2
2015	165.1	24.1	56.9	1.8	13.1
2030	188.0	28.9	63.0	2.6	16.3

Source: IFDC, 2002; Bruinsma, 2003.

In addition to a positive nutrient flux into the agricultural soils, it is also important to maintain a favorable elemental balance. The ratio of macroelements (nitrogen/phosphorus/potassium/calcium/magnesium) and micronutrients (zinc, copper, molybdenum) must be maintained at a favorable level, and the concentration of some toxic elements (aluminum, iron, manganese) must be kept low through use of balanced fertilizer, application of manure, and use of lime. Government policies of selling nitrogenous fertilizers at subsidized prices have led to mining of phosphorus and potassium in soils of South Asia. Such an elemental imbalance causes yield decline even when the supply of nitrogen in the root zone is adequate.

Similar to the increase in crop production in South Asia, improving nutrient supply through a judicious use of fertilizer input is also essential to enhancing crop yields in sub-Saharan Africa. The data on kilogram per hectare of fertilizer use is the lowest in sub-Saharan Africa and is projected to remain low (7 to 9 kg/ha) even by 2015 and 2030 (Table 18.12). Increase in fertilizer use in sub-Saharan Africa may necessitate an increase in area under supplemental irrigation and improvement in water availability through soil-water conservation and water harvesting and recycling.

WATER MANAGEMENT

Water conservation, recycling, and management aimed at enhancing the water use efficiency are crucial to increasing and sustaining productivity in dry farming or the rain-fed agroecosystems. Water management is especially important in the West

TABLE 18.12

Per Hectare Use of Fertilizer in Different Regions of the World

Region	1961/62	1979/81	1997/99	2015	2030
	----------------------- kg/ha of cropland-----------------------				
Sub-Saharan Africa	1	7	5	7	9
Latin America and the Caribbean	11	50	56	59	67
Near East & North Africa	6	38	71	84	99
South Asia	6	36	103	115	134
East Asia	10	100	194	244	266
World	25	80	92	—	—

Source: Adapted from Bruinsma, 2003.

African Sahel and in semiarid and arid regions of South Asia, Central Asia, and North Africa. The strategy is to conserve every drop of rain where it falls through minimization of losses by runoff and evaporation. Important RMPs for conserving soil water include mulching (crop residues, plastic) to reduce losses by evaporation, land forming and contour planting to reduce runoff, storing water in reservoirs and recycling, having a mixed/relay cropping system to utilize water from the subsoil or growing a short-duration crop on residual soil moisture, and providing the essential nutrients to enhance biomass production. The focus of soil and crop management is to decrease losses by runoff (red) and evaporation (blue) and increase green (transpiration) components of the hydrologic cycle.

Irrigated agriculture is a major factor responsible for a high and stable yield in India, China, Pakistan, and Egypt. However, only 5 to 10% of the irrigable cropland area in sub-Saharan Africa is presently irrigated (Table 18.13). Even the future

TABLE 18.13

Global and Regional Irrigated Cropland Area

Regional	1963	1980	1998	2015	2030
	------------------------------------10^6 ha -------------------------------				
Sub-Saharan Africa	3	4	5	6	7
Near East, North Africa	15	18	26	29	33
Latin America & the Caribbean	8	14	18	20	22
South Asia	37	56	81	87	95
India	25	39	58	63	70
East Asia	40	59	71	78	85
China	30	45	52	56	60
World	142	210	271	—	—

Source: FAO, 2001, 2005; Bruinsma, 2003.

projections of irrigation in sub-Saharan Africa are low and need to be aggressively explored and expanded. Future expansion of irrigation in sub-Saharan Africa can be achieved by developing small-scale projects rather than grandiose or large-scale irrigation schemes.

The most prevalent method of applying water to soils in Asia, Africa, and elsewhere is though flood irrigation. This is the most traditional method and is rather wasteful because of high losses in conveyance, evaporation, and seepage. With increase in water scarcity and competition from industry and urbanization (Johnson et al. 2001; Kondratyev et al. 2003; Gleick 2003), there is a strong need to improve water use efficiency. Among RMPs for improving water use efficiency, from the current range of merely 5 to 30%, are subirrigation and drip irrigation combined with fertigation. Losses by evaporation can be decreased through mulch farming techniques. Use of plastic mulch, while suppressing weed growth, can drastically reduce evaporation. Plastic mulch is widely used in East Asia (e.g., China, Korea, Japan) but rarely in South Asia and sub-Saharan Africa.

BIOFUEL

There is a strong, and rightfully justified, enthusiasm to use biofuel for power generation and transport (Brown 1998; EPA 2005; Weisz 2004; Pacala and Socolow, 2004). It is in this connection that the use of crop residues for producing lingocellulosic ethanol is widely being considered (Somerville 2006). It is important, however, to realize that traditional biofuels (crop residues and dung for cooking) are widely used in Africa and Asia and have severe adverse environmental, health, and agronomic implications. Venkatraman et al. (2005) reported severe adverse climatic impacts at the regional level because of the smoke and soot generated by traditional biofuels.

Crop residue, animal dung, and other biosolids (municipal waste, sludge) must be used as soil amendments rather than for short-term economic gains as traditional or modern biofuels (Wilhelm et al. 2004). Therefore, lignocellulosic materials for bioethanol must be grown by establishing energy plantations. Such plantations must be established on specifically identified land so that plantations do not compete with cropland and by specifically chosen and dedicated species so that crop residues can be returned to soil as amendment. There is a wide range of species that can be used for establishing energy plantations. These include short-rotation woody perennials (e.g., poplar, willow, mesquite, eucalyptus, leucaena, gliricidia, acacia); grasses (e.g., switch grass, elephant grass, guinea grass, napier grass, karnal grass); algae; and halophytes, which can be grown by irrigation with brackish water. Even the low-input and high-density prairie grasses can be used for producing biofuels (Tilman et al. 2006).

Soils of sub-Saharan Africa and South Asia are degraded and nonresponsive to input because their SOC/SOM pools have been severely depleted and are below the critical levels needed to maintain processes essential for plant growth and ecosystem services. Continuous removal of crop residues, for traditional or modern biofuels and other competing uses, can exacerbate soil and environmental degradation. These trends must be reversed for the urgency to break the agrarian stagnation and meet food demand of the 830 million food-insecure people and of 3.4 billion suffering from hidden hunger caused by poor nutrition.

CONCLUSIONS

The projected future increase in global population, from 6.5 billion in 2006 to 9.4 billion by 2050 and 10 billion by 2100, will entirely occur in developing countries of sub-Saharan Africa, South Asia, North Africa, Central America, and the Caribbean. These are also the regions where the soil resources are severely degraded, crop yields are low and stagnant, and 830 million food insecure and 3.4 billion prone to hidden hunger and malnutrition live. Furthermore, political instability and social/ethnic unrest make it difficult to implement meaningful and effective soil restorative programs. Yet, soil resources must be restored, improved, and used for generations to come. As the population increases and available per capita cropland and freshwater supply decrease, the urgency of adoption of recommended management practices increases. Such practices include conservation tillage, mulch farming, growing cover crops, adopting soil-specific management or precision farming, enhancing soil fertility though integrated nutrient management, water conservation and use of sub-irrigation or drip irrigation in conjunction with fertigation, and use of genetically engineered improved varieties. Crop residues, animal dung, and other biosolids must be used as soil amendments rather than as traditional or modern biofuels. The lignocellulosic biomass must be produced on energy plantations established on specifically identified land for growing dedicated crops. There is a strong need to restore degraded soils and expand the use of fertilizers and irrigation in sub-Saharan Africa. As has been the case in the past, those holding neo-Malthusian views will again be proven wrong. The needed increase in food/agronomic production will be achieved through restoration of degraded soils and adoption of recommended practices that enhance production per unit area, time and input of fertilizer, water, and energy. Soil resources must be improved, restored, and conserved for generations to come.

REFERENCES

Batjes, N.H. 1996. The total C and N in soils of the world. *Eur. J. Soil Sci.* 47:151–163.

Brown K.S. 1999. Bright future—or brief flare—for renewable energy. *Science* 285:678–680

Bruinsma, J., Ed. 2003. *World Agriculture: Towards 2015/2030. An FAO Perspective.* EARTHSCAN.

Cohen, J.E. 2003. Human population: the next half century. *Science* 302:1172–1175.

Department of Energy (DOE). 2006. *U.S. Climate Change Technology Program.* Washington, DC: Department of Energy.

Energy Information Administration (EIA). 2004a. *Annual Energy Outlook with Projections in 2025.* Washington, DC: Energy Information Administration.

Energy Information Administration (EIA). 2004b. *International Energy Outlook. Energy Information Administration.* EIA/DOE-OU84. Washington, DC: Energy Information Administration.

Energy Policy Act. 2005. *Title XV. Ethanol and Motor Fuels. Subtitle A, General Provision, Section 1501.* Washington, DC: U.S. Congress.

Fisher, G. and G.K. Heilig. 1997. Population momentum and the demand on land and water resources. *Phil. Trans. Roy. Soc. (B)* 352:869–889.

Food and Agricultural Organization (FAO). 2001, November. *FAO Stat.* Rome.

Food and Agricultural Organization (FAO). 2005. *Production Yearbook.* Rome: Food and Agriculture Organization for the United Nations

Gleick, P.H. 2003. Global fresh water resources. *Science* 302:1524–1528.

IFDC. 2002. *Global and Regional Data on Fertilizer Production and Consumption: 1961–2001.* Muscle Shoals, AL: IFDC.

Intergovernmental Panel on Climate Change (IPCC). 2001. *Climate Change 2001: The Scientific Basis.* Cambridge, U.K.: Cambridge University Press.

Johnson, N., C. Revenga, and J. Echeverria. 2001. Managing water for people and nature. *Science* 292:1071–1074.

Kondratyev, K.Y., V.F. Krapivin, and C.A. Varotsos. 2003. *Global Carbon Cycle and Climate Change.* Berlin: Springer-Verlag.

Lal, R. 1999. Soil management and restoration for carbon sequestration to mitigate the greenhouse effect. *Prog. Env. Sci.* 1:307–326.

Lal, R. 2004. Soil carbon sequestration impacts on global climate change and food security. *Science* 304:1623–1627.

Lal, R. 2005a. Soil erosion and global carbon budget. *Environ. Int.* 29:437–450.

Lal, R. 2005b. World crop residues production and implications of its use as a biofuel. *Environ. Int.* 31:575–584.

Lal, R. 2006. Enhancing crop yields in developing countries through restoration of soil organic carbon pool in agricultural lands. *Land Degrad. Dev.* 17:197–206.

Oldeman, L.R. 1994. The global extent of soil degradation. In D.J. Greenland and I. Szabolcs, Eds., *Soil Resilience and Sustainable Land Use.* Wallingford, U.K.: CAB International, pp. 99–118.

Pacala, S. and R. Socolow. 2004. Stabilization wedges: solving the climate problem for the next 50 years with current technologies. *Science* 305:968–972.

Pritchard, S.G. and J.S. Amthor. 2005. *Crops and Environmental Change.* New York: Haworth Press.

Rosegrant, M.W. and S.A. Cline. 2003. Global food security: challenges and policies. *Science* 302:1917–1919.

Somerville, C. 2006. The billion ton biofuel vision. *Science* 312:1277.

Tilman, D., J. Hill, and C. Lehman. 2006. Carbon-negative biofuels from low-input high-diversity grassland biomass. *Science* 314:1598–1600.

Venkatraman, C., G. Habib, A. Eiguren-Fernandez, A.H. Miguel, and K. Friedlander. 2005. Residential biofuels in South Asia: carbonaceous aerosol emissions and climate impacts. *Science* 307:1454–1456.

Weisz, P.B. 2004. Bad choices and constraints on long-term energy supplies. *Physics Today,* July:47–52.

Wilhelm, W.V., Johnson, J.M.F., Hatfield, J.L., Vorhees, W.B., and Linden, D.R. 2004. Crop and soil productivity response to crop residue management: a literature review. *Agron. J.* 96:1–17.

World Meterological Organization (WMO). 2006. *Greenhouse GOS Bulletin. The State of Greenhouse Gases in the Atmosphere Using Gl.* Geneva.

19 Advances in Water Science, Management, and Conservation

B. A. Stewart

CONTENTS

ABSTRACT

Water will become increasingly limited as the world's population increases to a projected 8 billion by 2025. Agriculture is by far the biggest user of water, but as population increases and industries grow, the competition among users will intensify. Although about 70% of the world's food is produced from rainfall, irrigation expansion during the past few decades is the primary reason that food production has increased at a faster rate than population. Worldwide, about 17% of the cropland is irrigated and accounts for 40% of the food and fiber production. Irrigation, however, uses much of the water abstracted from aquifers and surface water supplies, such as lakes and rivers. Irrigation accounts for nearly 70% of world water abstraction, and it is more than 90% in some agricultural economies in the arid and semiarid tropics. Agriculture must become more efficient in the use of both rainfall and irrigation water. There are essentially five ways that improvements can progress. The first, but perhaps the most challenging, is to increase photosynthetic efficiency. The second is to increase the amount of evapotranspiration, which is the sum of the water transpired through plant leaves and the water evaporated from the soil surface while the crop is growing. The third is to reduce evaporation of water during the period when a crop is not growing so that more water will be retained in the soil for use by a subsequent crop. Fourth, the harvestable portion of the crop can be increased. The fifth is to increase the portion of the field water supply that is used for evapotranspiration by reducing percolation, runoff, and evaporation.

INTRODUCTION

Water is often called the blood of life, but for an increasing portion of the world's population, it is life itself. Prema Ram, head of a village council in the Rajastan Desert of India, said "Ask for blood and we'll readily give it to you, but don't ask for water because our lives depend on every drop that falls from the sky" (Caryl, 2007). Almost 30% of the world's population lives in arid and semiarid environments where water is a severe constraint for producing food and fiber (Stewart et al., 2006). More than 1 billion people across the globe lack enough safe water to meet minimum levels of health and income (Rosengrant et al., 2002). Water will become increasingly limited as population increases to a projected 8 billion by 2025.

Agriculture is by far the biggest user of water. Although the amount of water required to produce cereal grain depends somewhat on the species and the climate in which it is grown, an average of 1000 tons of water to produce 1 ton of cereal is reasonable. Cereal production is an important benchmark because cereals directly supplied 57% of the calories in the global human diet in 2000 (Food and Agriculture Organization [FAO], 2003). Postel (1999) stated that crops get about 70% of their water directly from rainfall and 30% from irrigation. However, irrigation development is the primary reason that world food production has increased faster than world population during the past several decades. Between 1961 and 1997, world population increased by 89%, food production per person increased by 24%, while food prices fell by 40% in real terms (Wood et al., 2000). Worldwide, about 17% of the cropland is irrigated and accounts for 40% of the food and fiber production. Irrigation, however, uses much of the water abstracted from aquifers and surface water supplies such as lakes and rivers. On average, irrigation accounts for nearly 70% of world water abstraction—more than 90% in agricultural economies in the arid and semiarid tropics, but less than 40% in industrial economies in the humid temperate regions (FAO, 1996).

Irrigated area expanded rapidly during the twentieth century, and for much of the time it increased at a rate faster than the population (Table 19.1). Per capita area of irrigated land increased from 0.027 ha person^{-1} in 1900 to 0.38 in 1950 and reached a peak of 0.047 in 1980. Since 1980, population has increased at a faster rate than irrigated area, as reflected by the per capita area dropping back to 0.044 ha person^{-1}. This downward trend is of concern because there are a number of reasons that irrigation area will become increasingly difficult to increase. Postel (1999) stated that groundwater is being pumped faster than nature is recharging it in many of the world's most important food-producing regions, including parts of India, Pakistan, the north China plain, and the western United States. Postel further stated that there is simply little undeveloped water to tap, and that one of every 5 ha of irrigated land is losing productivity because of spreading soil salinization. There is general

TABLE 19.1
World Population and Irrigated Land

Year	Population (billions)	Irrigated Area (million ha)	Per Capita Irrigated Area (ha person⁻¹)
1900	1.5	40	0.027
1950	2.5	94	0.038
1970	3.9	169	0.043
1980	4.5	211	0.047
1990	5.2	239	0.046
2003	6.3	277	0.044

agreement that water scarcity is the biggest single threat to global food and fiber production. In her book, Postel (1999) made the case that water productivity (output from each unit of water abstracted from rivers, lakes, and underground aquifers) will have to double to have any hope of fulfilling the water requirements of 8 billion people while protecting the natural ecosystems on which economies and life itself depend. Indeed, this is a challenge that will not only require significant changes in how much of the world's water is managed and used, but in new scientific findings and technologies that will increase the efficient use of water.

DEFINITION OF TERMS

Although it is universally accepted that agriculture is the biggest user of water both from precipitation and that abstracted from rivers, lakes, and underground aquifers, the efficiency of water use by agriculture is not always expressed in the same manner and can be misleading. Therefore, it is very important that writers clearly define various terms, and perhaps even more important, readers must make sure they understand specifically what is meant by terms they see in an article. For example, water use efficiency is a term commonly used in the popular press and scientific literature, but it is a term that is often used in different ways. The following is a list of terms and their definitions as used in this chapter:

Transpiration: Transpiration is the evaporation of water from parts of plants, especially leaves but also stems, flowers, and fruits. Transpiration is a result of the need of the plant to open its stomata to obtain carbon dioxide from the air for photosynthesis. Transpiration also cools plants and enables mass flow of mineral nutrients from roots to shoots.

Transpiration ratio: The ratio of the units of aboveground biomass produced divided by the units of water transpired. For example, a plant with a transpiration ratio of 0.002 requires 500 g of water to be transpired to produce 1 g of aboveground biomass. Transpiration ratios are seldom determined under field conditions because it is difficult to separate water used by transpiration from water evaporated from the soil surface.

Water requirement: The inverse of the transpiration ratio. A plant with a transpiration ratio of 0.002 requires 500 g of water to be transpired to produce 1 g of aboveground biomass.

Evaporation: Evaporation accounts for the movement of water to the air from sources such as the soil or intercepted precipitation or irrigation.

Evapotranspiration: The sum of evaporation and transpiration from a defined area for a specified time divided by that area. The specified time is generally from date of planting a crop until harvesting the crop.

Potential evapotranspiration: A representation of the climatic demand for evapotranspiration; it represents the evapotranspiration rate of a short green crop, completely shading the ground, of uniform height and with adequate water status in the soil profile. It is a reflection of the energy available to evaporate water and of the wind available to transport the water vapor from the ground up into the lower atmosphere. Evapotranspiration is said to equal potential evapotranspiration when there is ample water available to the crop.

Field water supply: The sum of water that will become available to the crop during the growing season, the gross amount of seasonal irrigation, and precipitation received during the growing season. The gross amount of seasonal irrigation is the amount of water conveyed from the source to the field.

Irrigation efficiency: The net amount of water added to the root zone divided by the amount of water applied from the source.

Water use efficiency: Water use efficiency is the amount of harvestable crop product produced per volume of water used by evapotranspiration. For example, grain would be the harvestable product of crops like wheat, maize, and sorghum, while harvestable aboveground biomass would be the harvestable product for a forage crop. The water content of the harvestable product must also be considered.

PLANT GROWTH–WATER RELATIONSHIPS

It is important to understand some basic relationships between water use and plant growth before looking at some of the important advances made during the past few decades in utilizing water more efficiently for crop production. Climate is the most important factor for the efficient use of water for crop production, and this is true whether the crop depends entirely on precipitation, entirely on irrigation, or a combination of the two. There are four climatic characteristics that govern plant water use; simply stated, these are (1) how hot it is; (2) how sunny it is; (3) how windy it is; and (4) how dry the air is. These characteristics define the demand of the climate for water, and when water is not limited, the atmosphere will evaporate sufficient water to meet this demand. When a crop is growing and supplied with ample water, the amount of water transpired and evaporated from the soil surface is the amount of evapotranspiration. If the crop does not have an ample supply of water, it cannot meet the demands of the climate, referred to as the *potential evapotranspiration*, and the amount of actual evapotranspiration will be less because water is limiting. As

TABLE 19.2

Water Requirement (Grams of Water Required to Produce Grams of Biomass) of Alfalfa at Different Locations in the Great Plains

Location	Growth Period (1912)	Water Requirement	Pan Average[a] (mm d⁻¹)	Water Requirement/ Pan
Williston, ND	July 29 to September 24	518 ± 12	4.04	128
Newell, SD	August 9 to September 6	630 ± 8	4.75	133
Akron, CO	July 26 to September 6	853 ± 13	5.74	149
Dalhart, TX	July 26 to August 31	1005 ± 8	7.77	129

[a] Average daily amount of water evaporated from an open pan of water.

Source: L.J. Briggs and H.L. Shantz, 1913, *USDA Bureau Plant Industry Bulletin* 284.

water becomes increasingly limited, the actual evapotranspiration will also become increasingly more limited, and crop growth will decrease because of water stress. Some of the early work that showed the effect of climate on the amount of water required for plant growth was done by Briggs and Shantz (1917) and is summarized in Table 19.2. The locations reported in Table 19.2 move from north to south down the Great Plains, and the climate becomes increasingly hotter, as reflected by the higher amounts of water evaporated from an open pan. The water requirement for producing a gram of aboveground biomass of the same crop species was roughly two times greater at Dalhart, Texas, compared to Williston, North Dakota, and the pan evaporation amounts were in approximately the same ratios. These were some of the first data that could be used to show that evaporation and saturation deficit could be used to normalize the transpiration component.

The generalized relation between transpiration and yield of aboveground biomass and between evapotranspiration and agronomic yield is shown in Figure 19.1. The yield of biomass increases as a straight-line function of the amount of transpiration and the line intersects the origin. The slope of the line will vary with the crop and with the climate, but the relationship indicates that for every additional amount of water transpired by the crop, there will be a corresponding amount of biomass produced. The other line in Figure 19.1 representing the amount of agronomic yield as a function of evapotranspiration does not pass through the origin, but the relation is still a straight line after the threshold value of evapotranspiration has been met. Agronomic yield will depend on the crop and what the harvestable product is for that crop. For example, for crops like wheat, maize, and sorghum, the agronomic yield would be the amount of grain produced, while for a crop like alfalfa, it might be the amount of forage produced. The reason that the line does not pass through the origin is because there is an evaporation component of evapotranspiration. The relation shown indicates that for any agronomic crop there is a threshold amount of evapotranspiration that must be met before any agronomic yield is produced. However, once this threshold amount is met, agronomic yield is a straight-line function of evapotranspiration. Again, the slope of the line will vary depending on the crop and

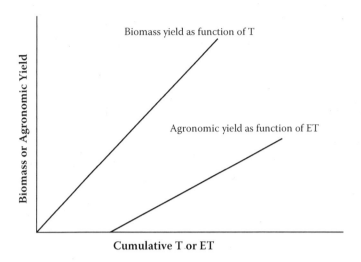

FIGURE 19.1 Generalized relation between production of biomass and transpiration and between agronomic yield and evapotranspiration. (Adapted from R.S. Loomis, 1983, in: H.M. Taylor, W.R. Jordan, and T.R. Sinclair, Eds., *Limitations to Efficient Water Use in Crop Production*. Madison, WI: American Society of Agronomy, Crop Science Society of America, and Soil Science Society of America, pp. 345–380.

the climate, but yield will increase proportionately with increasing evapotranspiration. Since water use efficiency was defined as the amount of harvestable crop product divided by the volume of water used by evapotranspiration, it is apparent from Figure 19.1 that the highest water use efficiency will always be at the highest yield. This is because the proportion of the evapotranspiration that makes up the threshold amount becomes smaller as the amount of evapotranspiration increases. Of course, there is a maximum amount of water that can be utilized as evapotranspiration, and this will ultimately determine the maximum yield for a given crop grown under a given environment.

The generalized relations shown in Figure 19.1 are useful in indicating ways to improve the efficient use of water in agriculture and serve as the foundation for discussions in this chapter. In theory, the efficient use of water can be enhanced several ways. First, the slope of the line showing the amount of biomass produced as a function of transpiration can be increased. This is essentially increasing the photosynthetic rate. Second, the amount of evapotranspiration can be increased, and this will increase the agronomic yield and water use efficiency. Third, management practices that decrease the evaporation portion of evapotranspiration will increase water use efficiency. Fourth, since the agronomic yield of a crop is the harvestable product that is usually only a portion of the aboveground biomass, an increase in the harvestable portion will increase water use efficiency. The harvestable product divided by the total aboveground biomass is defined as the harvest index. Fifth, evapotranspiration includes only evaporation and transpiration of water from a defined area during the growing season and is only a part of the field water supply. Management practices that increase the proportion of the total field water supply that is used for evapotranspiration can greatly increase the efficient use of water under both irrigated and rain-fed conditions. The discussions that follow address advances that have occurred for each of these approaches for improving water use by agriculture.

INCREASING PHOTOSYNTHETIC EFFICIENCY

Early work by Briggs and Shantz (1913) found striking differences between the water requirements of different plants growing in the same environment and for the same species growing in different environments. The millet, sorghum, and corn groups were found to produce significantly more dry matter per unit of water use than alfalfa and sweet clover. Also, the water requirement for the same species could be two times or greater depending on the climate. A promising finding of Briggs and Shantz was that there were measurable differences in the water requirements of different varieties of the same crop species, suggesting the possibility of developing through selection strains that were even more efficient in the use of water. However, the optimistic view of Briggs and Shantz has not materialized.

Tanner and Sinclair (1983) stated that while evapotranspiration efficiency had increased as a result of improved management, the increases had resulted mainly from increased transpiration as a fraction of the evapotranspiration, and transpiration efficiency had increased little, if at all. Richards et al. (1992) also concluded that plant breeding has indirectly increased water use efficiency because yield has increased with no additional water use. These increases have been substantial in all the major food crops. Richards et al. stated that most of the increases, however, resulted from improvements in harvest index values, and that there had been little increase in water use efficiency for biomass. They further stated that improvements in harvest index may be approaching theoretical limits in many crops and that if significant further gains are to be realized there must be increases in transpiration efficiency. There are many plant breeders and molecular geneticists working diligently to improve transpiration efficiencies, and major advances could come at any time.

INCREASING AMOUNT OF EVAPOTRANSPIRATION

As shown in Figure 19.1, the agronomic yield of a crop increases linearly with an increase in evapotranspiration. Of course, this relationship assumes that water is the most limiting factor. Evapotranspiration can be increased by a variety of management practices. In areas where precipitation is not adequate to meet the climatic demands of the crop and irrigation water is not available, the only way to increase evapotranspiration is to make better use of the precipitation. There are three components for successfully managing limited precipitation: (1) retaining the precipitation on the land, (2) reducing evaporation, and (3) utilizing crops that have drought tolerance and fit best with the rainfall pattern. Although these components have been known for centuries, progress in adapting them to specific areas and situations has been slow. However, technologies have emerged in recent years that have significantly increased the proportion of annual precipitation used for evapotranspiration, and these technologies, or the principles on which they are based, can be applied to other regions and countries.

There is considerable potential for increasing the proportion of annual precipitation used for evapotranspiration. A generalized overview of rainfall partitioning in semiarid regions is presented in Figure 19.2. Only 15 to 30% of the annual precipitation is used as transpiration, and this is the only water use that is directly related to crop yield. As much as 50% of the annual precipitation is lost as evaporation from the soil surface. The water that is evaporated from the soil surface during the growing of a crop is included along with transpiration as evapotranspiration and, as shown in Figure 19.1, results in an increase in agronomic yield. However, water evaporated from the soil surface during the period between crops is a total loss from the production system. In recent years, there has been a significant reduction in tillage. Even though most farmers have not moved entirely to no-till systems, there has been a marked reduction in both the number of tillage operations and their intensity. In many cases, herbicides are used to eliminate the need for tillage; the overall result has been an increase in the amount of crop residues remaining on the soil surface both during the period between crops and during the crop-growing season.

Unger and Baumhardt (1999) reported that grain yields of dryland grain sorghum, a major crop in the southern Great Plains, more than tripled in studies conducted at the U.S. Department of Agriculture (USDA) Conservation and Production Research Laboratory in Bushland, Texas, during the period from 1939 to 1997. They attributed about two thirds of the increase in grain yields to increases in plant-available soil water stored in the soil at the time of seeding the crop (Figure 19.3). There were approximately 75 mm more water stored in the soil profile at time of seeding grain sorghum after 1970 compared to prior years of the studies. There was a sharp increase in energy costs in the early 1970s, and this resulted in a rapid adoption, particularly in research studies, of reduced tillage and no-tillage systems. With less tillage, more crop residues remained on the soil surface, which reduced evaporation

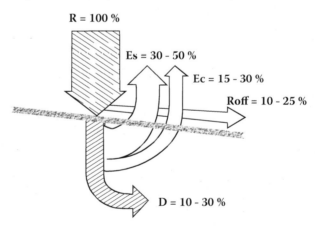

FIGURE 19.2 General overview of rainfall partitioning in semiarid regions. R, rainfall; Es, soil evaporation and interception; Ec, plant transpiration; Roff, runoff; D, drainage. (From P. Koohafkan and B.A. Stewart, 2008, *Water and Cereals in Drylands.* Rome: Food and Agriculture Organization of the United Nations.)

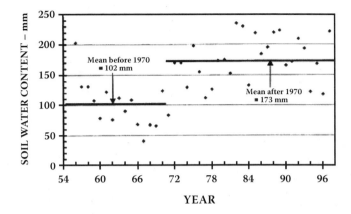

FIGURE 19.3 Average annual volumetric soil water content at planting time for dry-land grain sorghum in studies conducted from 1956 to 1997 at the U.S. Department of Agriculture's Agriculture Research Service (USDA-ARS) Conservation and Production Research Laboratory, Bushland, Texas. (From P.W. Unger and R.L. Baumhardt, 1999, *Agronomy Journal* 91:870–875.)

and increased soil water storage. The stored water supplemented the growing season precipitation and increased the amount of annual precipitation that could be used for evapotranspiration. As discussed regarding Figure 19.1, there is a direct relationship between the amount of water used by a crop for evapotranspiration and the agronomic yield.

Stone and Schlegel (2006) also showed that the amount of stored water present in the soil profile at time of seeding either grain sorghum or wheat under semiarid conditions in western Kansas directly affected grain yields. Less tillage and more plant residues remaining on the soil surface increased the percentage of precipitation occurring between crops stored in the soil profile that could be used by the subsequent crop. This in turn increased the amount of annual precipitation used for evapotranspiration and increased precipitation use efficiency.

DECREASING THE EVAPORATION PORTION OF EVAPOTRANSPIRATION

As defined, evapotranspiration is the sum of the water transpired by the crop and that evaporated from the soil between the time the crop is planted and is harvested. The amount of evapotranspiration can be determined relatively accurately by researchers, but it is much more difficult to measure evaporation and transpiration separately. However, it is well understood by the relationships shown in Figure 19.1 that the greater the amount of evapotranspiration that is used for transpiration, the greater the water use efficiency is because plant growth and yield are directly associated with transpiration.

Any management practice that increases the transpiration portion of evapotranspiration will increase the water use efficiency. Early in the growing season, when the amount of leaf area is small, the amount of water transpired by the plants is relatively small in proportion to the amount of water lost by evaporation from the soil surface. This is especially true if the soil surface becomes wet because of precipitation or added irrigation water. When the soil surface is wet, water is not limiting, so the amount of water that is evaporated depends on the climate—temperature, radiation, humidity, and wind. Bare soil that is wet will evaporate about the same amount of water as a well-watered grass crop will transpire because water is not limiting in either case. There is essentially no water lost by evaporation from the soil surface from a well-watered grass crop because the soil is completely covered by the crop.

These two situations are the extremes, and in most cases, transpiration by plants and evaporation from the soil surface are occurring simultaneously, with soil evaporation dominating when the plants are small and there is a lot of exposed soil and transpiration dominating later in the season after a full-crop canopy has developed. A full-crop canopy is generally considered one that has a leaf area index greater than three or four, which means the total area of all the leaves is three or four times greater than the area where the plants are growing. Leaf area indexes for agronomic crops can range from two or less for a crop like cotton or grain sorghum grown under dryland conditions to six or greater for corn or sugar cane grown with adequate water.

Although evaporation from the surface of a wet soil can be roughly equal to transpiration of a crop with a full canopy, the evaporation drops rapidly as soon as the surface of the soil dries because water becomes limiting. Evaporation from a soil with a dry surface depends on how fast the water from the wet soil beneath the surface can move to the surface where it can be evaporated. This will depend on the soil texture, which will control capillary movement of the water. After the soil dries to the point at which water no longer moves upward by capillarity, evaporation becomes much slower because then water in the soil must change from liquid to vapor before it can move to the surface where it will be evaporated.

Therefore, the amount of water lost from the soil by evaporation can vary greatly. As described, it occurs in three widely differing stages. Stage 1 occurs while the surface is wet; this is called the *steady-rate stage* because it is constant as long as the climate is constant because water is not limiting. Stage 2 occurs as the soil surface dries when water becomes limiting and the evaporation rate decreases as movement by capillary flow slows; this is called the *falling-rate stage*. Stage 3 is the vapor movement stage; the rate is slow because the water must change from liquid to vapor and move upward by diffusion to the soil surface before it can be evaporated. However, stage 3 drying can occur for long periods of time. During long periods of drought, soil profiles can become fairly dry, to depths of 1 to 2 m, requiring large amounts of water to replenish the profile once the rains return. This is why a common statement in drought prone areas is "one rain will not break a drought."

By understanding the principles of the evaporation and transpiration processes, scientists and producers have made significant advances in increasing the portion of evapotranspiration used by transpiration and decreasing the portion lost by evaporation. Although the benefit of leaving crop residues on the soil surface for increasing the amount of soil water between crops has been discussed, these residues also reduce evaporation from the soil during the crop-growing season, and this increases the portion of evapotranspiration that is used for transpiration. This can significantly increase agronomic yields in water-deficient areas and greatly increase water use

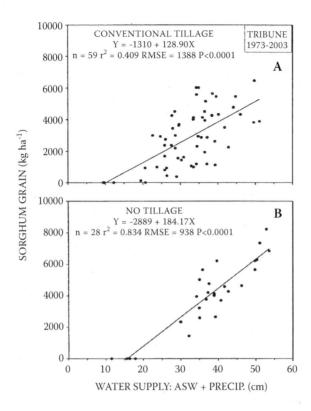

FIGURE 19.4   Grain sorghum yield at Tribune, Kansas, associated with water supply (plant-available soil water at emergence plus growing-season precipitation) for dryland conventional tillage (section A) and for no-till (section B) treatment groups. RMSE, root mean square error. (From L.R. Stone and A.J. Schlegel, 2006, *Agronomy Journal* 98:1359–1366.) ·

efficiency. Stone and Schlegel (2006) summarized almost 30 years of grain sorghum yields for western Kansas produced using conventional tillage compared to no tillage (Figure 19.4). There was a relatively straight-line relationship between the yield of grain and the water supply (available soil water plus growing season precipitation, and this was approximately equal to evapotranspiration), as would be expected in view of the generalized relationship presented in Figure 19.1. With no tillage, however, the slope of the line is greater, indicating that more grain was produced from a unit of additional water compared to that produced from an additional unit of water using conventional tillage.

INCREASING THE HARVESTABLE PORTION OF AGRONOMIC CROPS

The harvestable portion of agronomic crops can be increased by plant breeding and genetics and by management. Corn, rice, and wheat are the most widely grown grain crops in the world, and corn and wheat in particular are grown under a wide range of water conditions. Water stress is common in these crops and can have a major effect on the portion of the aboveground biomass that is grain. The ratio of the grain weight to the total aboveground biomass weight is the harvest index. The harvest index is employed as a criterion for screening cultivars and is a genetic characteristic. Prihar and Stewart (1990) suggested that the maximum harvest index value for corn was about 0.60 and about 0.50 for wheat. Evans (1980) stated that the yield increase of wheat through 1976 came from the increase in harvest index, which rose from 0.43 to 0.50. Pingali and Rajaram (1999) reported that wheat architecture had evolved over the past 45 years to sustain growth in genetic yield potential. Wheat grain yields increased by at least 15% in semidwarf wheats compared to yields of tall varieties largely due to an increase in harvest index.

Prihar and Stewart (1990) reviewed many experiments of wheat, corn, and grain sorghum and showed that the harvest index of these crops decreased with increasing stress, particularly water stress. Under severe drought conditions, the harvest index could approach zero because there would be little or no grain produced. Therefore, any management practice that reduces the stress of growing grain crops tends to increase the grain yield. Successful management practices under nonirrigated conditions include those involving plant densities, plant geometry, and row spacing. Under irrigated conditions, timing and amounts of water additions are critical. All of these practices will affect the harvest index values and therefore affect water use efficiency.

Passioura (1977) found a linear relationship between harvest index and the percentage of the seasonal water use after *anthesis*, the period during which a flower is fully open and functional. Harvest index values ranged from 0.2 when water use after anthesis was less than 5% of the season total to about 0.5 when water use after anthesis was 25 to 30% of the total. This shows that grain filling is a critical stage for producing grain. Bandaru et al. (2006) compared grain sorghum plants growing in clumps to those equally spaced in rows in a semiarid region under drought conditions and showed the plants in clumps used less water during the vegetative growth period, leaving more soil water for use during grain filling. Yields were increased by use of clumps, and harvest index values increased from less than 0.3 to more than 0.4. Although total aboveground biomass values were slightly less for the clumped plants, grain yields and water use efficiencies were higher because of the higher harvest index values.

INCREASING PORTION OF FIELD WATER SUPPLY USED FOR EVAPOTRANSPIRATION

When water is added to the soil by rainfall or irrigation, there are five things that can happen to the water. It can evaporate, percolate below the root zone, remain stored in the soil, be lost as runoff, or be used by transpiration by plants. Estimates of how precipitation in different climatic zones is distributed to runoff, evaporation, and evapotranspiration are presented in Figure 19.5. As the climate becomes drier, a much greater proportion of the total precipitation is lost to evaporation, and this reduces the water available for evapotranspiration. As discussed, the water evaporated from the soil during the growing season and the water transpired by the growing crop is evapotranspiration, and the agronomic yield of a crop increases linearly with increasing amounts of evapotranspiration as illustrated in Figure 19.1. The

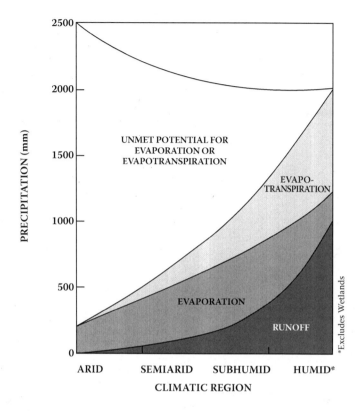

FIGURE 19.5 Estimated amounts of runoff, evaporation, evapotranspiration, and potential evapotranspiration for different climatic regions. (From V.M. Ponce, 1995, *Soil and Water Conservation Journal* 50:422–431.)

efficient use of water must maximize the portion of total water that is used as evapotranspiration, and this requires a reduction in evaporation during the year when a crop is not growing. If salinity is a problem, some water must percolate through the soil and move below the root zone to prevent salts accumulating to a detrimental concentration. This leaching requirement, however, should be kept as low as feasible to maximize the efficient use of water. If salinity is not a problem, efforts should be made to prevent movement of water below the root zone. Likewise, runoff should be prevented to the fullest extent feasible, particularly in areas where water is limiting.

A hypothetical water management example by Howell (1990) is presented in Figure 19.6 and is extremely useful in understanding water use efficiency under a variety of scenarios. In this example, the Q_o shows a case where the sum of water use from available soil water at time of seeding and rainfall during the growing season totaled 250 mm, and Q_m is the total field water supply of 1400 mm that includes the 250 already described plus 1150 mm of applied irrigation water. The yield of aboveground biomass produced and the yield of grain produced by a dryland crop are represented by P and Y, respectively. In this example, P was 6 Mg ha^{-1}, and Y was 2 Mg ha^{-1}. Therefore, the harvest index (grain yield/total aboveground biomass)

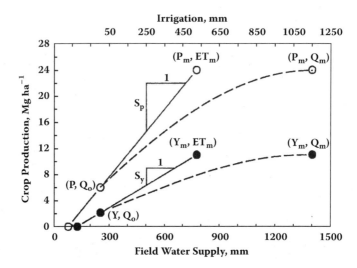

FIGURE 19.6 Hypothetical example illustrating a relationship between dry matter and grain production and field waters. (From T.A. Howell, 1990, in: B.A. Stewart and D.R. Nielsen, Eds., *Irrigation of Agricultural Crops.* Agronomy 30. Madison, WI: American Society of Agronomy, Crop Science Society of America, and Soil Science Society of America, pp. 391–434.)

was 0.33. The maximum P_m and Y_m values are 24 Mg^{-1} and 10 Mg^{-1}, respectively, and represent the yields obtained when water was not limiting. The WUE_{net} (yield of grain in kilograms)/(ET in cubic meters) for the dryland crop would be 2000 kg/2500 cubic meters or 1 kg m^{-3}. For the highest yield P_m, the WUE_{net} would be approximately 10,000 kg/7500 cubic meters or 1.33 kg m^{-3}. Theoretically, there is a straight-line relationship between water use and yield of either grain or above-ground biomass after the threshold amount has been supplied. Using the grain yield of this example, there was no grain produced until approximately 100 mm of water was utilized, but once the threshold value was reached, additional water utilized as ET water increased the yield about 1.5 kg m^{-3}. Therefore, the WUE_{net} is always the highest at the highest yield because the threshold portion of the water used becomes a smaller portion of the total water used for ET. The $WUE_{irrigation}$ may decrease with increasing yields. In the example, 1150 mm of irrigation water was required to produce the maximum grain yield of about 10 Mg ha^{-1}, but the addition of only 450 mm would have produced about 9 Mg ha^{-1}. Therefore, the $WUE_{irrigation}$ would be much lower for the maximum grain yield. The deviation of the dashed curves from the lines represents the combined effects of the irrigation hydrology (runoff, deep percolation, soil water recharge, spray evaporation, drift, etc.) with the effects of the irrigation water salinity, irrigation application uniformity, and the spatial variability of the soil physical parameters. In theory, with high-quality water and an efficient irrigation system, the deviation between the ET line and the dashed line representing field water supply can be close to each other. Howell (1990) stated, for the example depicted in Figure 19.6, that (1) the maximum water use efficiency (Y/ET) occurs at the point (Y_m, ET_m); (2) maximum irrigation use efficiency (Y/Q) occurs at a value of

Q of about 600 mm, which is considerably less than the 1150 mm necessary to produce maximum grain yield (this value can be graphically determined by the tangent on the curve to a line constructed through the origin); and (3) assuming a constant water cost, the maximum net profit will normally occur at a value of field water supply exceeding ET_m but less than Q_m (unless water is free) and will decrease as the water price increases for fixed land but increase with higher fixed production costs.

CONCLUSIONS/STUDY TOPICS

Agriculture is the biggest user of water in the world. Approximately 70% of all water abstracted from surface and groundwater supplies is used for irrigation, and this accounts for about 40% of the food and fiber produced. Increased irrigation during the past 50 years has been the primary reason that food and fiber production has increased faster than the increase of population. Irrigation expansion in the future, however, will be constrained because of limited water resources and increased development costs. Only 15 to 30% of rainfall is actually used by crops for transpiration. Runoff accounts for 10 to 25%, and the remainder is lost by evaporation from the soil surface. There are well-established relationships that can be used to develop management practices to increase the efficiency of water utilization from both irrigation and precipitation.

- Evapotranspiration is the sum of the water evaporated from the soil surface and transpired through the leaves of a growing crop between time of planting and harvesting.
- The agronomic yield of a crop increases linearly with an increase in amount of water used by evapotranspiration.
- The use of mulch can decrease the evaporation portion of evapotranspiration so that more water is available for transpiration because there is a direct relationship between plant growth and transpiration.

REFERENCES

Bandaru, V., B.A. Stewart, R.L. Baumhardt, S. Ambati, C.A. Robinson, and A. Schlegel. 2006. Growing dryland grain sorghum in clumps to reduce vegetative growth and increase yield. *Agronomy Journal* 98:1109–1120.

Briggs, L.J. and H.L. Shantz. 1913. The water requirement of plants: II. A review of the literature. *USDA Bureau Plant Industry Bulletin* 284.

Briggs, L.J. and H.L. Shantz. 1917. The water requirement of plants as influenced by environment. *Proceedings of the Second Pan-American Science Congress* 3:95–107.

Caryl, C. 2007. Cool, clear water. *Newsweek* October 1, p. 71.

Evans, L.T. 1980. The natural history of crop yields. *American Science* 68:388–397.

Food and Agriculture Organization (FAO). 1996. *Food Production: The Critical Role of Water*. Background paper for the World Food Summit. Rome: Food and Agriculture Organization of the United Nations.

Food and Agriculture Organization (FAO). 2003. *World Agriculture: Towards 2015/2030. An FAO Perspective*, J. Bruinsma, Ed. Rome: Food and Agriculture Organization of the United Nations.

Howell, T.A. 1990. Relationships between crop production and transpiration, evapotranspiration, and irrigation. In: B.A. Stewart and D.R. Nielsen, Eds., *Irrigation of Agricultural Crops.* Agronomy 30. Madison, WI: American Society of Agronomy, Crop Science Society of America, and Soil Science Society of America, pp. 391–434.

Koohafkan, P. and B.A. Stewart. 2008. *Water and Cereals in Drylands.* Rome: Food and Agriculture Organization of the United Nations.

Loomis, R.S. 1983. Crop manipulations for efficient use of water: an overview. In: H.M. Taylor, W.R. Jordan, and T.R. Sinclair, Eds., *Limitations to Efficient Water Use in Crop Production.* Madison, WI: American Society of Agronomy, Crop Science Society of America, and Soil Science Society of America, pp. 345–380.

Passioura, J.B. 1977. Grain yield, harvest index, and water use of wheat. *Journal Australian Institute Agricultural Science* 43:117–120.

Pingali, P.L. and S. Rajaram. 1999. Global wheat research in a changing world: options for sustaining growth in wheat productivity. In: P.L. Pingali, Ed., *Global Wheat Research in a Changing World: Challenges and Achievements.* Mexico, D.F.: CIMMYT, pp. 1–18.

Ponce, V.M. 1995. Management of droughts and floods in the semiarid Brazilian Northeast—the case for conservation. *Soil and Water Conservation Journal* 50:422–431.

Postel, S. 1999. *Pillar of Sand: Can the Irrigation Miracle Last?* New York: Norton.

Prihar, S.S. and B.A. Stewart. 1990. Using upper-bound slope through origin to estimate genetic harvest index. *Agronomy Journal* 82:1160–1165.

Richards, R.A., C. López-Castañeda, H. Gomez-Macpherson, and A.G. Condon. 1992. Improving the efficiency of water use by plant breeding and molecular biology. *Irrigation Science.* Available at: http://www.springerlink.com/content/m87270452192472t/ (accessed May 19, 2007).

Rosengrant, M.W., X. Cai, and S.A. Cline. 2002. *Global Water Outlook to 2025—Averting an Impending Crisis.* Washington, DC: International Food Policy Research Institute.

Stewart, B.A., P. Koohafkan, and K. Ramamoorthy. 2006. Dryland agriculture defined and its importance to the world. In: *Dryland Agriculture,* 2nd ed. Agronomy Monograph 23. Madison, WI: American Society of Agronomy, Crop Science Society of America, and Soil Science Society of America, pp. 1–26.

Stone, L.R. and A.J. Schlegel. 2006. Yield-water relationships of grain sorghum and winter wheat. *Agronomy Journal* 98:1359–1366.

Tanner, C.B. and T.R. Sinclair. 1983. Efficient water use in crop production: research or re-search. In: H.M. Taylor, W.R. Jordan, and T.R. Sinclair, Eds., *Limitations to Efficient Water Use in Crop Production.* Madison, WI: American Society of Agronomy, Crop Science Society of America, and Soil Science Society of America, pp. 1–27.

Unger, P.W. and R.L. Baumhardt. 1999. Factors related to dryland grain sorghum yield increases. *Agronomy Journal* 91:870–875.

Wood, S., K. Sebastian, and S.J. Scherr. 2000. *Agroecosystems: Pilot Analysis of Global Ecosystems.* Washington, DC: International Food Policy Research Institute and World Resources Institute.

20 Crop Science and Production Technology

Peter R. Hobbs and Larry W. Harrington

CONTENTS

ABSTRACT

In the last 40 years, world food production has risen dramatically, resulting in improved global food security. Production of the three main food grains (wheat, rice, and maize) increased threefold between 1961 and 2006 (from 642 million to nearly 2 billion metric tons). Increased food production was a result of increased acreage and higher yields (+250%) as a result of green revolution practices: stronger plants, more nutrients, better pest and disease resistance, and more irrigation. Production increases also occurred in other crops (e.g., vegetables, oil seeds, fodders, and fruits), although some dryland crops such as pulses and small grains lost ground. The purpose of this chapter is to identify new post-green revolution technology in crop science that will help achieve future global food needs. Future increased production will be required in many crops as well as cereals as consumers demand more nutritious food and more meat-based products. Important technologies discussed include major advances in molecular biology leading to more efficient development of improved varieties and improved crop management in irrigated, but also dry rainfed environments where many of the rural poor subsist (e.g., improved rainfall harvest). Conservation agriculture is described as one way to achieve the management efficiency. Precision agriculture, presently used in developed countries, combines remote sensing with modeling to better achieve efficient crop production. In developing countries, especially in dryland areas, microdosing techniques in water and nutrient management, improved innovation systems, and networks to accelerate the innovation process will be needed. To be successful, new innovations must be made available to all farmers through effective policy decisions, available credit to purchase the technology, and accessible markets for the extra production. Computers, the Internet, and cell phones to improve communications will be useful in future food production systems for all economic levels of farmer.

INTRODUCTION

During the last 40 years, world food production rose dramatically as a result of green revolution (GR) technologies, a combination of genetic, agronomic, and crop protection improvements. Production gains in maize were also seen as a result of hybrid seed development. Production of these three main cereal crops increased threefold between 1961 and 2006, from 642 to nearly 2000 million metric tons (Food and Agriculture Organization of the United Nations, n.d.). Population in the same period doubled from around 3 to 6 billion people. Increased food production was due to increased acreage (an extra 100 million ha counted as new land or where double and triple cropping became feasible) and higher yields (+250%). There have also been gains in yields and production levels for milk, fish, and other animal products, as discussed in other chapters in this book. Growth in cereal yields has been a major factor in maintaining global food security, although inequitable distribution and lack of income to buy food means that some of the very poor have remained hungry.

Continued growth in food production well into the future is essential to feed the world, especially in developing nations where population is still growing and in developed nations where land and water are being diverted for production of biofuels (discussed in Chapter 17). Food security will continue to be an issue. Food imports are unlikely to be the answer: Import costs will rise as fossil fuel prices rise and costs of shipping increase. Policies must continue to stimulate local and regional food production. Agriculture needs to increase production using new sources of technical change and more efficient use of natural resources. The rest of this chapter discusses some of the new post-GR production advances in crop science that will be needed to meet future food needs for cereals and other crops.

MOLECULAR BIOLOGY

Biotechnology is defined as any technique that uses living organisms to make or modify products, to improve plants or animals, or to develop microorganisms for specific purposes. Traditional biotechnology or plant breeding has been used for thousands of years by humans for selective breeding of better crop varieties. Farmers allowed natural crosses between related species to occur and then selected the seed of superior offspring. Today, in conventional breeding, plant breeders select the parents to cross from closely related species and varieties and after crossing, select the subsequent segregating plants over a number of generations to identify new lines with better yield, biotic, abiotic, and quality characters. There will be a continued future need to continue this selective plant breeding as humans strive to develop crops for the many different growing conditions or ecoregions that exist in the world.

In addition to the breeding techniques used in the past 50 years, future efforts will increasingly utilize the major advances in molecular biology or biotechnology that have occurred in the past 20 years. This molecular biology advance will give scientists better knowledge of the genetic background of various crops and interacting pests and diseases and therefore improve the genetic engineering efficiency and speed with which new lines are developed. Biotechnology involves the application of a wide range of genetic engineering techniques for the modification and improvement of plants, animals, and microorganisms that are of economic importance. This includes genomics, transgenic crops, tissue culture, and marker-assisted selection (MAS). This has given breeders the ability to move DNA (deoxyribonucleic acid) between more distantly related organisms and not just close relatives. This gene revolution has increased the efficiency of traditional plant breeding so that instead of many generations and years of selection of thousands of plants to develop a new variety, this can be done today and in the future in much less time and with much less space.

It was only in the nineteenth century that Mendel and Darwin demonstrated *hybridization*, the ability to mix species or varieties of closely related plants to create a hybrid that contained traits of both parents. In the twentieth century, this deliberate interbreeding of closely and in some cases distantly related species was used to produce new lines with desirable traits. The development of the modern wheat and rice varieties that led to the GR in the 1960s are examples of this hybridization. IR8 was one of the first modern rice varieties developed at the International Rice Research Institute in the Philippines. IR8 rice was developed by crossing Peta, a tall Indonesian variety with high vigor, seed dormancy, resistance to several insects and diseases, and widely grown in the Philippines, to Dee-geo-woo-gen, a high-yielding, heavy-tillering, short-stature variety from Taiwan. Norman Borlaug along with colleagues in Mexico developed similar short-stature, input-responsive, and biotic-stress-resistant varieties of wheat, which led to the 1970 Nobel Peace prize for Borlaug for this research. Plant breeders continue to develop even better rice and wheat varieties for biotic and abiotic stresses and better quality while maintaining yield potential using selected crosses and these new modern molecular tools. Some of these important biotechnology tools are discussed in the next section.

MARKER-ASSISTED BREEDING

Modern plant breeders use techniques of molecular biology to develop new, improved varieties. Use of molecular markers, MAS, or DNA fingerprinting allows plant breeders to screen large populations of plants and select for the trait of interest more efficiently. MAS allows the breeder to use a marker for indirect selection of important traits. The markers can be morphological, biochemical, or based on DNA variation. It has been used by breeders to develop varieties with increased productivity, disease resistance, abiotic stress tolerance, or quality. The process is based on using an easily identified marker linked to a trait of interest. For example, a drought gene would be closely linked to a marker allele that is easier to recognize. MAS tools will continue to be needed in the future to develop superior varieties.

GENOMICS

Watson and Crick received the Nobel Prize in physiology and medicine in 1962 for their discovery of the molecular structure of nucleic acids and their significance in transferring information in living organisms. Every living organism contains chromosomes that consist of strands of DNA. It is through the nucleotide subunits of the DNA that genetic information is translated into proteins and the building blocks of living organisms to produce new cells and individuals.

Since this discovery, great strides have been made to understand the genetic code and how it functions in determining new cells and individuals. *Genomics* is the study of an organism's entire genome or hereditary information encoded by its chromosomes. It results in the complete sequencing of one set of chromosomes of an organism. DNA was first sequenced in 1977, and the sequencing of genetic material took root in the 1980s and has gathered pace since then; today, many biological species, including viruses, bacteria, yeasts, insects, worms, fish, and even vertebrates have complete sequencing of their chromosomes.

The human genome was sequenced and completed in 2003 for all 24 distinct chromosomes. They consisted of about 3 billion base pairs and 20 to 25,000 genes. Various crops have had their genomes sequenced, including wild mustard (*Arabidopsis thaliana*, 2000), rice (*Oryza sativa*, 2002), poplar (*Populus trichocarpa*, 2006), and grapes (*Vitis vinifera*, 2007). Other crops being sequenced include wheat (*Triticum aestivum*), maize (*Zea mays*), tomato (*Solanum lycopersicum*), and potato (*Solanum tuberosum*).

Gene mapping provides a plant breeder with the location of specific genes of interest in relation to the various chromosomes. This has been accomplished by using microarray tools and bioinformatics. The microarrays use known specific sequences of DNA to scan the sequence of a genome to identify genetic variation at specific locations. *Bioinformatics* or *computational biology* uses mathematics, statistics, computer science, informatics, chemistry, and other disciplines to solve biological problems at the molecular level. It uses these tools to make sense of the large amount of information generated by chromosome sequencing. Bioinformatics also helps in gene sequencing to search for genes and regulatory sequences within a genome.

Functional genomics is a field of molecular biology that attempts to make use of the enormous quantity of data collected from genome projects to describe gene function and interactions. It focuses on the dynamic aspects of gene transcription as opposed to the static data available from gene sequencing. It also uses DNA microarrays and various electrophoresis and mass spectrometric tools to collect needed data. It provides the scientist with information concerning the location of specific genes needed for development of improved new varieties from a much wider spectrum of organisms than the closely related species that used to be the only source of new variation. This area of research will continue to be of immense value as humans better understand the variation, working, and location of genetic material needed to develop better crops for future food security.

TRANSGENIC CROPS

Classical plant breeders have been able to force hybridization between different plant species by making wide crosses and then using techniques such as embryo rescue, protoplast fusion, and mutagenesis to generate diversity that would not normally exist in nature. These plant-breeding techniques were not controversial and did not get wide publicity but did allow plant breeders to develop food crops with important traits. Wide crosses in wheat have been used for many decades to develop wheat varieties with resistance to various pathogens that can cause enormous crop losses. The first interspecific wheat crop developed by humans was named triticale and was a cereal hybrid developed by mixing chromosomes of wheat and rye.

Molecular biology allows breeders to develop transgenic crops similar to wide crosses, but with more control, through use of recombinant DNA. A piece of DNA containing one or more specific genes from nearly any organism, including plants, animals, bacteria, or viruses, is introduced into another crop species. This means that genes can be transferred between any species as opposed to only within the same species as normally occurs within plants during sexual replication. These resulting *transgenic* plants (the general public refer to these new plants as genetically modified organisms or GMOs) are developed through a technique called *genetic engineering.*

The development of genetically engineered crops was made possible by the discovery in the 1970s of restriction enzymes that can cut double-stranded DNA and ligases that enable pieces of DNA to be spliced together. The first transgenic organism developed was a bacterium. The first transgenic plant was developed in 1983, just 25 years ago. Today, the major transgenic crops grown commercially are either herbicide tolerant or insect resistant. Scientists are also working on other biotic and abiotic (mainly drought and salt tolerance) stress resistance and crops with improved nutrition (e.g., golden rice).

Transgenic plants are generated in the laboratory by adding one or more genes to a host plant's genome. This technique is called *transformation,* and it is usually achieved by direct bombardment of plant cells with a gene gun or by horizontal gene transfer with the aid of soil bacteria like *Agrobacterium tumefaciens.* The last bacteria acts as a vector and can carry the new genes attached to plasmids in their cells into the host plant cell, where they hybridize with the host chromosomes. In the gene gun system, DNA is impregnated onto gold particles that are then shot into plant pieces and cells. The DNA of the new gene then integrates into the chromosome of the host plant to produce transgenic plants with the new trait. In both systems, an additional antibiotic marker gene is often attached to the DNA strand in close proximity to the gene to be transferred and is used to enable easier selection of cells that have successfully received the gene of interest. Tissue culture is then used to grow the infected cells. The medium on the agar plate contains an antibiotic for the inserted marker gene, and only cells that have the antibiotic marker closely attached to the new gene survive for selection. These cells are then transformed into plantlets and grown out as crop plants containing the genetic trait somewhere in their chromosomes. This system facilitates genetic exchange in crops. It is far more precise because only a single (or just a few) gene is transferred with a specific trait. As a result, ancillary, unwanted traits that can accompany normal crosses are avoided.

This leads to more rapid development of new varieties. The entire span of genetic traits is available for use, expanding the range of useful traits available for development of new germplasm.

The two major transgenic traits that are found commercially today are *Bacillus thuringiensis* (Bt) and Roundup-ready (RR) (Roundup is a glyphosate herbicide) crops. Bt is a soil-dwelling bacterium that produces a *cry* protein toxin when it sporulates that specifically acts in the gut of certain insects and causes the gut of the insect to become leaky and thus kill the pest. There are several specific *cry* proteins that have specific activities against different orders of insects, including the Lepidoptera (moths and butterflies), Diptera (flies and mosquitoes), and Coleoptera (beetles). The main *cry* genes in commercial production are specific to the Lepidoptera and are used to control these insect pests in cotton (bollworm) and maize (earworm and stem borer). The gene that codes for the *cry* protein is introduced into the crop plant using the two techniques listed (gene gun or agrobacterium), where it gets incorporated into the crop genome. When the transgenic plant is eaten by a member of the caterpillar family, the gene enters the gut of the caterpillar, where it produces the insecticidal protein that causes the gut to become leaky and kill the caterpillar. This protein is very animal specific and does not react in the same manner in humans and thus poses no biosafety risk. Spores and crystalline insecticide proteins produced by the Bt organism are also marketed for pest control and used in organic farming for pest control. It is regarded as environmentally friendly with little or no effect on humans, wildlife, or beneficial insects like bees. Interestingly, Bt sprays are approved for use on organic farms, but transgenic crops are not. Since the gene for the *cry* protein is found in the chromosomes of the transgenic crop, all plant tissues produce the *cry* protein throughout crop growth and thus help protect the plant from insect attack. Molecular scientists can also engineer plants in which only specific plant tissue phases of the plants' life have the gene active or turned on, but this is not yet found in presently grown commercial transgenic crops.

A similar system is used in developing herbicide-resistant transgenic crops. In the case of RR crops, the gene was isolated from an *Agrobacterium* soil bacteria and incorporated into various crops, It imparted herbicide resistance to glyphosate, a broad-spectrum, translocated herbicide that kills most plants it enters. Of the various herbicides available for modern agriculture, it is considered to pose less health risk to humans and have fewer negative environmental effects. Glyphosate is definitely less toxic than other herbicides presently used to control weeds in agriculture.

In 2006, transgenic crops exceeded 100 million hectare (ha) globally, with soybean (60 million), maize (25 million), cotton (10 million), and canola (5 million) the major commercial crops using this technology (ISAAA, 2006). Herbicide tolerance has the largest hectarage (75 m), followed by Bt (15m) and the rest with both these traits. The major countries growing transgenic crops were the United States (55 m), Argentina (18 m), Brazil (11.5 m), and Canada (6 m). China and India also had significant acreages of these crops (3.5 m), mostly using Bt cotton. In 2006 in the United States, 89% of soybeans, 82% of cotton, and 61% of maize were transgenic. This means that many food items in supermarkets in the United States contain these materials.

An example of transgenic crop use in a developing country is found in India (Herring, 2007). Cotton is a major crop in that country for use domestically and for export. Prior to the introduction of Bt cotton, farmers in India had to apply large amounts of broad-spectrum insecticides to control the main pest, the cotton boll-worm. Surveys showed that 10 to 20 sprays were often used by farmers. In some years, these applications were not effective as insects developed resistance to the pesticide or farmers were sold spurious products. The chemicals were also toxic to the applicators, who in many cases were poor, landless labor. Bt cotton appeared on the scene in the late 1990s when Monsanto and a local seed company (Mayco) were given limited license to test Bt lines in secure fields by the government regula-tory organization. At the time, there was a moratorium on growing transgenic crops commercially in India. In 2001, there was a severe outbreak of the cotton bollworm that decimated much of the cotton and resulted in many farmers becoming bank-rupt, with several committing suicide. However, in Gujarat State, India, some fields remained green and productive (Herring, 2007). These fields had transgenic Bt cot-ton grown by some farmers who had obtained seed illegally. The government tried to burn all these plants and prevent them spreading to other farmers. But, the farm-ers refused to accept this, and eventually the government lifted its moratorium and approved Bt cotton in 2002. By 2006, Bt cotton in India had risen to 3.8 million ha. Farmers are finding they get more yield (less insect damage) at less cost (less pes-ticides purchased) and so more profit. So, instead of Bt cotton being the culprit for the suicides as suggested by some GMO opponents, it is actually the solution. This example is used to demonstrate that biotechnology and development of transgenic crops has great potential for resolving many world food problems in the future. The issue at the moment is a skeptical public, who probably need to be better informed about the advantages this technology can bring to farmers and better understand the issues of biosafety and ecological risks. Note that in the early 1900s the public were similarly against the use of pasteurization in milk. Today, it is accepted as a normal safe procedure.

There is still a lot of controversy over this technology. Issues include health, ecol-ogy, bioproperty, and politics. The first transgenic crops were planted commercially 10 years ago. In that time, many of the biosafety and ecological concerns of the public have not arisen. There were some scares about allergies and some reports of negative effects on rats fed transgenic crops, but none of these health issues have been any different from those with traditional crops. There are also coordinated regulatory frameworks in most countries that regulate transgenic organisms and require data on biosafety and environmental impacts before issuing a license to the company to continue development and release in agriculture. Transgenic crops are regulated much more severely than traditional crops, which highlights the public concern for this technology. Regulation also adds substantial cost to the development of these crops and the eventual cost to the end user. Ecological risks that include transfer of transgenes to local wild species and possible development of super species is another area of concern. Development of resistance to the transgene, especially if the transgenic variety resulted in farmers only growing that variety, also poses a problem. However, this is also the case for nontransgenic varieties and is addressed through resistance management. These concerns seem to be much less important than first thought 10 years ago, with little evidence it has occurred, although in the-ory it should happen. There are also issues of cost, ownership, and the domination of a few large corporations in the seed sector. However, the future of agriculture may require acceptance of this science since it may be the only way to find superior plants needed to produce the quantity of food needed for food security.

Tissue Culture

Tissue culture is a key biotechnology tool in developing transgenic crops. In tissue culture, a single cell is grown on a sterile medium (usually agar based) and is stimulated to multiply and produce *callus tissue*, a group of similar cells. Using hormonal and other additions to the media, the callus tissue can be tricked into forming embryos that then grow into whole plants. This results in clones of a plant that are all exact copies. When transgenic cells with a new trait plus an antibiotic marker are grown on agar containing an antibiotic, only the cells with the new gene survive and are easily selected. This technology is also very useful for developing virus-free plants, multiplication of plants in the absence of seeds, and the production of plants from seeds that otherwise have a low chance of germinating.

EFFICIENT CROP PRODUCTION

Improvements in crop science and production technology can contribute to sustainable food production in two different ways: by developing superior germplasm (as discussed) and by using agronomy to tap the full potential of this germplasm. *Agronomy* is the science of using plants to obtain food, fuel, fiber, and feed or, in other words, how to grow crops for maximum production and profit.

There are a number of well-defined steps involved in growing a crop: land preparation, seeding and plant establishment at proper spacing and population, fertilizing, managing moisture, crop protection (weeds, insects, diseases, and rodents) and harvest, threshing, cleaning, and storage. (Adding value to crop outputs through processing is not discussed in this chapter.) For the past hundred years or so, these farm operations have been carried out in ways that rely heavily on cheap energy from fossil fuels. Fossil fuels have also provided cheap nutrients. Urea, for example, is a major nitrogen-based fertilizer that is synthesized using large quantities of fossil fuel. In the present decade, however, the cost of fossil energy has risen dramatically and is likely to remain high. New production technologies are needed that use energy more efficiently.

There is general agreement that innovative production practices are needed that use energy, natural resources, and inputs more efficiently while reducing harmful impacts on the environment. This is becoming even more critical since agricultural land and water are increasingly in competition with other sectors of the economy. Rapidly growing urban centers are enveloping large areas of productive agricultural land and are requiring ever-larger quantities of water for direct consumption, hydropower, and industrial use. Agriculture, industry, and the environment all compete for the same scarce water resources (Comprehensive Assessment 2007).

CONSERVATION AGRICULTURE

Conservation agriculture (CA) is one approach to farming that can help meet food needs while using energy, inputs, and natural resources more efficiently. It is based on three pillars: minimal soil disturbance, permanent soil cover, and crop rotations (Hobbs et al., 2008).

Farming practices in the early 1900s often featured intensive plowing of soils. This led to minimal soil protection for lack of plant cover and excessive soil erosion by wind and water. Conservation tillage (minimum tillage and the use of crop residues for soil cover) was introduced in the United States in the 1930s to counter "dust bowl" land degradation. As energy costs increased and better no-till sowing equipment was developed, many U.S. farmers switched to CA, doing away with tillage altogether.

CA is adopted by farmers because it reduces production costs without reducing yields. Even under these conditions, however, there is often a long lag period between when the system is first introduced to farmers and when rapid adoption occurs. This is partly a problem of information transfer and availability of good equipment, but also overcoming the mindset of farmers who firmly believe that tillage is essential for good crop production. Farmers typically need to experiment, hear about, and see the technology and discuss it with fellow farmers before they are willing to risk investing in something as strange and unusual as no till (Harrington and Erenstein, 2005).

However, once this lag phase has been breached, widespread adoption can occur very rapidly. For example, no-till cropping was first introduced into Brazil to overcome erosion problems on sloping land in Parana and other southern states. No-till practices combined with residue retention reduced the soil erosion problem to acceptable limits. Farmers were able to visually see the benefits and reacted by purchasing improved no-till equipment. Information about CA spread very rapidly from one group of farmers to another. Ten years after its introduction, area under no till rose to 1 million ha. By 2005, more than 25 million ha of farmland in Brazil were under CA. This is a large proportion of the global area currently covered by CA, estimated at about 100 million ha (Derpsch, 2005). Apart from Brazil, CA is concentrated in the United States, Argentina, and Paraguay but is steadily increasing elsewhere, for example, South Asia.

The adoption of no till in South Asia has many parallels with the experience in Brazil. In northern India, the introduction of short-duration rice and wheat varieties allowed rice-wheat double cropping. Late planting of wheat after the rice harvest, however, caused by late harvest of the previous rice crop and poor and excessive land preparation, severely constrained wheat yields (Hobbs et al., 1997). In addition, *Phalaris minor*, a major weed in wheat, developed resistance to the herbicide (isoproturon) used by farmers for its control (Malik et al., 1998). Zero tillage was introduced in the mid-1990s to enable timely wheat planting and to reduce wheat land preparation costs. The idea was that the money saved by farmers on land preparation could be used to buy the new but expensive herbicides for *Phalaris* control.

Farmer experimentation showed that zero till did indeed allow earlier wheat sowing and higher yields, but it also resulted in reduced *Phalaris* germination because less soil disturbance meant fewer buried weed seeds were brought to the soil surface to germinate. At the same time, local equipment manufacturers developed cheap, locally adapted seed drills that performed well when planting wheat into unplowed rice paddies. The result has been a rapid increase in no-till wheat area. By 2005, between 3 and 4 million ha of wheat were sown using no till (Erenstein et al., 2007).

Scientists in Asia are now examining ways to grow both rice and wheat with no till. At present, rice is grown after puddling the soil (wet plowing), a practice that destroys much of the benefit obtained by using no till for wheat. A no-till rice-wheat rotation has the potential to substantially reduce water use in rice production, very important for water-scarce areas (Tang et al., 2004). Successful use of no till in rice, however, requires that weed control issues be resolved. Use of herbicide-tolerant transgenic rice may be the answer, although public concerns about transgenic plants and their effect on landless labor still need to be addressed.

Apart from reducing erosion, saving energy, reducing production costs, and (sometimes) increasing yields, CA often has other benefits. These include more efficient use of water, nutrients, and fossil fuel; lower greenhouse gas (GHG) emissions; less soil disturbance; improved soil biology (by promoting the growth of soil organisms); and improved soil physical conditions (improved soil structure, water infiltration, root penetration, and so on). Improved soil health facilitates nutrient cycling and fosters natural biological control of insect and disease pathogens. CA relies on soil organisms such as earthworms and fungal hyphae for natural tillage. These develop networks of channels within the soil for aeration and new root growth. There is evidence that a more diverse biological population helps restrict the activity of pathogenic fungi like *Fusarium*. Residue retention promotes a better habitat for beneficial insects that help control crop pests like rice stem borer (Hobbs 2007, Hobbs et al., 2008).

CA reduces GHG emissions by reducing the use of diesel fuel in land preparation and pumping water. Diesel savings can be as much as 50 to 60 L/ha in rice-wheat areas. By switching to nonpuddled, aerobic rice, methane emissions can also be significantly reduced. There is still a need to reduce nitrous oxide emissions through more efficient utilization of nitrogenous fertilizers. GHG emissions and climate change are further discussed in Chapter 16.

At times, concern is expressed that the use of zero tillage requires that herbicides be used instead of tillage for controlling weeds. In some instances, this may be true but in others not. In rice-wheat systems of South Asia, zero tillage led to a reduction in herbicide use (Singh et al., 2002). In these systems, wheat season and rice season weeds are very different. Such weeds as *Phalaris minor* only germinate in the cool temperatures of winter—and when seeds are brought to the surface by tillage. Data from South Asia indicate that fewer weeds grow under zero tillage than in plowed fields. Furthermore, after 3 to 5 years of using zero tillage (with the new herbicides), weed populations are reduced to the point at which farmers can grow wheat with no herbicides whatsoever. As noted, however, weed control in no-till rice is more of a problem. The extent to which herbicides are needed in no-till systems is location specific.

Widespread adoption of CA is heavily dependent on the local availability of suitable, reasonably priced implements capable of effective no-till sowing under local conditions. Very different kinds of implements are needed for systems powered by large tractors, small tractors, animal traction, and human labor. Note that no-till equipment developed for four-wheel tractors can be suitable for small farmers who are accustomed to renting-in tillage and establishment services. Such is the case in most rice-wheat areas of South Asia (Mehla et al., 2000).

ORGANIC FARMING

Some CA practices are used in producing "organic food," increasingly popular with consumers in developed countries aware of food safety issues. These consumers feel that organic food is safer, tastier, and better for their health. Typically, however, it is also more costly to produce and may be less affordable for the poor. Organic food is produced to meet specific standards; only farms meeting these standards are certified and allowed to label their produce as organic. Organic crops are grown without the use of conventional pesticides, artificial fertilizers, human waste, or sewage sludge; without processing with ionizing radiation or food additives; and without the use of GMOs or transgenic varieties. Many organic farmers use CA practices—no-till, soil cover, and rotations.

There is still debate regarding whether enough food can be produced organically to meet future food needs. As pointed out, use of zero till and soil mulch does improve soil health and promotes better nutrient recycling. But, is this enough to provide the nutrients needed to produce high yields without chemical fertilizers? Would more land area per unit of product be required to grow crops organically than farming with chemical inputs? Proponents of the organic movement say yes, assuming that sufficient amounts of organic nutrients can be made available to reliably sustain high crop yields over large areas. Experience in China suggests that successful reliance on organic inputs over large areas is difficult to maintain (Cheng et al., 1992). Interestingly, plants do not distinguish between nutrients such as nitrogen or phosphorus supplied by chemicals or those supplied organically. Nutrients are needed for plant growth and must be supplied either by the soil or as externally applied inputs, organic or inorganic.

The use of integrated pest management (IPM) has enabled farmers to produce crops with less reliance on chemical pesticides, the input that most affects food safety. In IPM, ecological processes (encouraging natural enemies of crop pests, reducing pest buildup through rotations, etc.) are used as much as possible to control pests, diseases, and weeds. IPM allows occasional judicious use of pesticides if all else fails (unlike organic farming, for which pesticides are banned completely).

A time may come when use of transgenic crops will be seen as a solution to the problem of crop protection and pest management. The example of the use of Bt cotton in India and China has shown that crops can be grown with little or no pesticide. Glyphosate-tolerant, transgenic rice would make growing aerobic rice much easier and would reduce the toxicity of herbicides applied, resulting in better soil health, less water use, and reduced GHG emissions. The use of new technology in food production needs to be assessed holistically and against the present systems of agriculture, rather than biopolitically.

PRECISION AGRICULTURE

Precision farming or precision agriculture is an agricultural concept that is becoming popular in developed countries as a way to handle in-field variability. It uses new technologies, such as global positioning systems (GPSs), sensors, satellites or aerial images, and information management tools such as Geographic Information Systems (GIS) to assess and understand field variability and precisely evaluate optimum sowing density, fertilizer, and other input needs. It also allows the farmer to more accurately predict crop yields earlier in the season and can help to better assess disease or lodging problems. Farmers are using this technology to maximize profits by spending money only in areas that need inputs, especially fertilizer. The farmer can vary the rate of fertilizer applied across the field according to the need identified by GPS-guided grid sampling, thereby optimizing its use. This will become an important farming practice in the future to help improve the efficiency of natural resource use and improve productivity.

DRYLAND AGRICULTURE

Sustainable food production for the future will continue to rely on favored agricultural areas where water is not a limiting factor. Practices (such as those described) that sustainably improve productivity in favored areas reduce the need to farm dryland, sloping, or other marginal areas. However, more than half of the world's farmland is still located in areas where rainfall is limiting either in quantity or in reliability. These dryland areas are home to many of the world's poorest farmers.

It is often said that the GR bypassed these dryland areas. This is not entirely accurate. GR-style genetic improvement has produced suitable materials with high-yield potential and some drought tolerance (Harrington et al., 2008), but because of lack of water, actual yields remain far below potential. Here, the key is higher water productivity—higher output per unit of water depleted. Once this is achieved, nutrient availability and improved biotic stress resistance become more important.

Water harvesting provides opportunities to improve productivity in dryland areas. Water may be harvested by using contour bunds, slowing water flows, channeling water to ponds and aquifers, and so on. Water harvesting is sometimes best introduced at the community level, in which case community-level management is needed to maintain water-harvesting structures and practices after development projects finish (Johnson et al., 2001). In India, there has been some success in improving farmer livelihoods through watershed projects that have increased water harvest, filled aquifers and soil profiles, and even made water available for partial irrigation. Crop yields have increased and yield variability across years has decreased, and through greater crop diversification and better markets, farmers have improved incomes.

CA will be as important in dryland as in favored production environments. No till and permanent ground cover have been shown to increase rainwater infiltration and improve water availability to plants. The challenge in dryland areas is to produce enough crop residue biomass to allow some to be left as mulch on the ground while still meeting animal feed requirements. Suitable seeding equipment is also needed to place the seed into moisture in soils that have had no tillage.

One example of how CA principles can be applied to dryland areas is found in southern Africa. Here, the traditional system is to plow the land using manual or animal power just after the onset of the rainy season. In part because of seasonal labor shortages, land preparation is often delayed, crops are planted late, plant stands are poor, fertilizer is not used, and weeding is performed either late or not at all (Waddington, 1999). In an improved CA system, farmers make planting basins (depressed areas in the field) during the dry season. When the rains begin, water accumulates in the basins. Farmers can immediately seed into the preprepared wet basins. Farmers are also encouraged to "microdose" fertilizers (use very precise application of very small amounts) and compost.

Timely planting, improved soil moisture in the basin, and better nutrient availability can lead to substantially higher yields while actually reducing peak season labor requirements. Rotating crops in the basins every year helps control pests and diseases. Weeds may still be a problem but can be partly controlled by mulching the land with crop residues. Farmers are encouraged not to burn crop residues but to use them for composting or as a surface residue cover. When open grazing of residues by cattle is allowed, however, residue retention may be difficult to impossible.

In animal-powered systems, the farmers use a ripper plow to open a furrow, where water is harvested. Burning is discouraged, and residues are left in the furrow. Planting is done by hand or behind the plow, with microdosing and concentration of nutrients near the seed. Both of these systems resulted in earlier planting, less labor, higher yields, and improved soil health over time (Bwalya, 1999). Interestingly, use of herbicide-resistant crops could make this system even more productive since weeds are still a major problem.

Introduction of improved two-wheel and small four-wheel tractors could also help accelerate the adoption of this more sustainable technology in the future. However, there are still many opponents to introduction of mechanized power into African farming. The one thing that is not contested is that present traditional systems have resulted in major soil degradation, nutrient loss, and declining yields, and that this must be reversed in the future.

EXTENSION OF TECHNOLOGY

New agricultural production technology is of little use unless large numbers of farmers adopt it and use it over large areas. Many models have been used in the past to address this issue. In traditional, linear extension systems, scientists develop technology, hand it over to extension agents, who in turn pass it on to farmers. This assumes that scientists fully understand farm systems and constraints faced by farmers. As this is not always the case, technology adoption using traditional extension systems is often disappointing.

To overcome low rates of adoption of agricultural technologies, the World Bank introduced in the 1980s a "training-and-visit" (T&V) system, intended to upgrade the technical capacity of extension workers and facilitate closer contact with farmer communities (Hussain et al., 1994). Farming systems research (FSR) was also introduced in the 1980s to make research more relevant to farmer's circumstances, but both still relied too much on a linear system of researcher-dominated research. In the 1980s and 1990s, more farmer participatory approaches were developed in which farmers are a key part of the research team, help define needs, plan research programs, and conduct trials (e.g., Ashby et al., 1995). The main limitation to this system was the intensive and expensive nature of the methodology and the difficulties of scaling up results to large areas.

Recent advances go beyond the notion of "extension" and focus on what have become known as "innovation systems." Any particular research entity is only one of many actors influencing technical change. Nevertheless, a research program can trigger or catalyze change through an approach that perceives innovation as a social process in which "agents" (researchers, stakeholders, end users) engage in "learning selection." In learning selection, agents try something new, for example, a new technology or a new institutional arrangement. They explore the innovation, try to make sense of it on their own terms, and adapt it more closely to their own needs. Ultimately, they decide whether to keep using it. The learning process enables them to select which (modified) innovations they will carry forward and which ones they will abandon. Learning selection, like natural selection, does not just happen once but rather iterates through numerous cycles. Research centers can stimulate and accelerate innovation systems and learning selection by using participatory methods, offering new technical prototypes to be subjected to learning selection, mapping and "weaving" networks, defining "impact pathways," and fostering interaction that leads innovation along these pathways (Douthwaite, 2002).

Impact pathways portray how research activities result in change by defining the causal chain of activities, outputs, and outcomes through which a project anticipates achieving its purpose and goal and by mapping the evolving relationships among project-implementing organizations, stakeholders, and ultimate beneficiaries.

In the future, new resource management technologies are likely to be more complicated, requiring more local adaptation. Facilitation of learning selection, fostering the development of suitable stakeholder networks, and anticipating impact pathways will become increasingly important. An example of this can be found in South Asia, where zero-till wheat was successful in the state of Haryana, and adoption was fast because farmers were part of the evaluation team, were given prototypes with which to experiment, and were offered opportunities to interact with and learn from each other—and because innovation systems included private sector implement manufacturers as well as farmers, extension workers, and scientists. In other states where this was not done, acceptance of zero till was much slower.

The last topic on extension relates to the availability of information to farmers. The potential for using electronic means of delivering information, even to remote villages, is gaining ground daily. In India, for example, there are public and private programs to make computers and the Internet available in each village. Local people are trained to use these modern facilities and provide information to farmers. The information could be about new varieties, better management, disease or insect issues, but also prices and markets for products. The need of the day is the development of relevant materials that can be posted on easily accessible Web pages that are useful for solving farmer problems. Similarly, many farmers or villages are also linked by cell phones, a situation that would have been hard to believe with land-line systems a few decades ago. This means that farmers now and in the future have the ability to communicate immediately with people who would have answers to their major production problems.

CONCLUSIONS AND STUDY TOPICS

Increased crop production will continue to be needed well into the present century to meet food security needs of a population that is still expanding. Added to this is the need to produce more nutritious and better quality food demanded by a population that is better educated and has better livelihoods and so can afford a greater diversity of crop and animal products. This will be a challenge to all involved in agriculture since many of the traditional ways of increasing crop production are already being used. Agricultural land area is also decreasing along with water and other natural resources as competition for land and resources for other human activities grow. The question is how to grow more food with fewer resources and still maintain low food costs for the growing urban population and provide a suitable livelihood for farmers.

Key topics needed to accomplish this task are as follows:

1. Optimal use of molecular tools by plant breeders to develop superior germplasm with increased productivity, biotic and abiotic stress tolerance, and quality.
2. Molecular tools include use of markers, genomics, bioinformatics, and functional genomics and tissue culture in developing transgenic crops with needed traits.
3. Improved crop management is needed that uses natural resources more efficiently and minimizes negative environmental impacts.
4. CA, organic farming, and precision agriculture are management systems that achieve these objectives better than existing agriculture. Equipment will be a key component of these systems.
5. Dryland areas need concerted efforts to make them more productive since they represent half the global farmland and are where many poor farmers live. Water harvesting and more efficient use of rainwater will be key in these areas.
6. Extending technology to farmers will be more complex in the future and will require facilitation of learning selection, fostering the development of suitable stakeholder networks, and anticipating impact pathways.
7. Modern communication tools such as the Internet and cell phones will become increasingly valuable for informing farmers of needed information.

REFERENCES

Ashby, J., T. Garcia, M. Guerrero, C.A. Quirós, J.I. Roa, and J.A. Beltrán. 1995. *Institutionalizing Farmer Participation in Adaptive Technology Testing With the "CIAL."* London: ODI. ODI Agricultural Research and Extension Network Paper 57.

Bwalya, M. 1999. Conservation farming with animal traction in smallholder farming systems: Palabana experiences. In *Conservation Tillage with Animal Traction*, P.G. Kaumbutho and T.E. Simalenga, Eds. Harare, Zimbabwe: Animal Traction Network for Eastern and Southern Africa.

Cheng, X., C. Han, and D.C. Taylor. 1992. Sustainable agricultural development in China. *World Development* 20(8):1127–1144.

Comprehensive Assessment of Water Management in Agriculture. 2007. *Water for Food, Water for Life: A Comprehensive Assessment of Water Management in Agriculture*. Molden, D. (Ed.). London: Earthscan and International Water Management Institute.

Derpsch, R. 2005. The extent of conservation agriculture adoption worldwide: implications and impact. Paper presented at Keynote for the Third World Congress of Conservation Agriculture, Nairobi, Kenya, October 3–7.

Douthwaite, B. 2002. *Enabling Innovation: A Practical Approach to Understanding and Fostering Technological Innovation*. London: Zed Books.

Erenstein, O., U. Farooq, R.K. Malik, and M. Sharif. 2007. *Adoption and Impacts of Zero Tillage as a Resource Conserving Technology in the Irrigated Plains of South Asia*. Colombo, Sri Lanka: International Water Management Institute. Comprehensive Assessment of Water Management in Agriculture Research Report 19.

Food and Agriculture Organization of the United Nations. FAOStat database. n.d. Available at: http://faostat.fao.org/site/567/default.aspx.

Harrington, L.W. and O. Erenstein. 2005. Conservation agriculture and resource conserving technologies—a global perspective. Paper presented at Conservation Agriculture—Status and Prospects, New Delhi.

Harrington, L.W., F. Gichuki, B. Bouman, N. Johnson, V. Sugunan, K. Geheb, and J. Woolley. 2008. *The CGIAR Challenge Program on Water and Food: A Synthesis of Activities and Achievements Through 2006*. Colombo, Sri Lanka: Systemwide Program on Collective Action and Property Rights (CGIAR) CPWF.

Herring, J.H. 2007. Stealth seeds: bioproperty, biosafety, biopolitics. *J. Dev. Studies* 43(1): 130–157.

Hobbs, P. R. 2007. Conservation agriculture: what is it and why is it important for future sustainable food production? *Journal of Agricultural Science, Cambridge* 145:127–138.

Hobbs, P.R., G.S. Giri, and P. Grace. 1997. *Reduced and Zero Tillage Options for the Establishment of Wheat after Rice in South Asia*. Mexico D.F.: RWC and CIMMYT. Rice Wheat Consortium Paper Series 2.

Hobbs, P.R., K.D. Sayre, and R.K. Gupta. 2008. The role of conservation agriculture in sustainable agriculture. *Philosophical Transactions of Royal Society B (UK)* 363:543–555.

Hussain, S., D. Byerlee, and P. Heisey. 1994. Impacts of the training and visit extension system on farmers' knowledge and adoption of technology: evidence from Pakistan. *Agricultural Economics* 10(1):39–47.

International Service for the Acquisition of Agri-Biotech Applications (ISAAA). Executive summary. Global status of commercialized biotech/GM crops: 2006. Brief 35-2006. Available at: http://www.isaaa.org/resources/publications/briefs/35/executivesummary/default.html.

Johnson, N., H.M. Ravnborg, O. Westermann, and K. Probst. 2001. *User Participation in Watershed Management and Research*. Washington, DC: Systemwide Program on Collective Action and Property Rights (CGIAR), International Food Policy Research Institute (IFPRI). Capri Working Paper 19.

Malik, R.K., G. Gill, and P.R. Hobbs. 1998. *Herbicide Resistance—A Major Issue for Sustaining Wheat Productivity in Rice-Wheat Cropping Systems in the Indo-Gangetic Plains*. New Delhi: Rice-Wheat Consortium for the Indo-Gangetic Plains. Rice-Wheat Consortium Paper Series 3.

Mehla, R.S., J.K. Verma, R.K. Gupta, and P.R. Hobbs. 2000. *Stagnation in the Productivity of Wheat in the Indo-Gangetic Plains: Zero-Till-Seed-Cum-Fertilizer Drill as an Integrated Solution.* New Delhi: Rice-Wheat Consortium for the Indo-Gangetic Plains. Rice-Wheat Consortium Paper Series 8.

Singh, S., A. Yadav, R.K. Malik, and H. Singh. 2002. Long-term effect of zero-tillage sowing technique on weed flora and productivity of wheat in rice-wheat cropping zones of Indo-Gangetic Plains. In Malik, R.K. et al. (Eds.). *Herbicide Resistence Management and Zero Tillage in Rice-Wheat Cropping System. Proceedings of the International Workshop, 4–6 March 2002.* Hisar, India. Chaundhhary Charan Singh Haryana Agricultural University, pp. 155–157.

Tang, Y., J.G. Zheng, G. Huang, and J. Du. 2004. Study on permanent bed planting with double zero tillage for rice and wheat in Sichuan Basin. Paper presented at the Fourth International Crop Science Congress, Brisbane, Australia, September 26–October 1.

Waddington, S. 1999. *Participatory Development of Tillage and Weed Control Technologies in Zimbabwe.* Harare, Zimbabwe: CIMMYT.

21 Animal Science and Production Technology

G. Eric Bradford, Wilson G. Pond,
and Kevin R. Pond

CONTENTS

ABSTRACT

Domestic animal production has been an integral part of human society for more than 10,000 years and remains so. This chapter addresses the importance of animals in society. Broad subject areas include (1) food animals, their products, and sustainable agroecology; (2) vertical integration of animal production systems; (3) plant-animal ecosystems; (4) animal well-being; (5) technologies and products that support globalization of food animal production, food safety, and animal health. We describe the interconnections of animal science with all of agricultural sciences in producing food in a sustainable environment. The food animal industry has major impacts on the global economy and on human well-being.

INTRODUCTION

Humans have evolved as omnivores. For most of human history, animal source foods were obtained by hunting. With the domestication of animals (see Chapter 1), hunting has been largely replaced by production of meat, milk, and eggs from domesticated animals; they represent an important contribution to human welfare. Animal source foods are highly palatable to most humans and are of higher nutrient density and bioavailability than many plant source foods. Through science-based production technologies, global consumption of meat, milk, and eggs has increased markedly in recent decades and is projected to continue. A straight-line increase in animal source food consumption has been observed for many decades in all developing countries as family incomes rise. This linear relationship between income and animal source food demand continues except in the economically advantaged countries, where animal source food demand plateaus as incomes rise.

Domesticated animals make other important contributions. They remain the major source of draft power for tillage and harvesting of crops and for transportation in many parts of the world. They provide wool, mohair, and fur, hides, and leather for clothing and other uses, and their products are used in industry, pharmaceuticals, and clinical medicine. Many advances in biological science and human medicine have been based on research with domestic animals. Animals have long been important to human culture as indicators of wealth and status and as religious symbols. Companion animals continue to play a major role in human societies. Guide dogs, dogs in rescue, drug and contraband detection, and cats in nursing homes and homes for the elderly are clear examples of the human-animal bond.

FOOD ANIMALS, THEIR PRODUCTS, AND SUSTAINABLE AGROECOLOGY

The trend toward increased per capita demand for animal source foods is occurring primarily in developing countries, which account for approximately 80% of the world population and, in many cases, have high population growth rates. Demand for animal source foods increases as income and urbanization rise. Large increases are projected in global demand for meat, milk, and eggs over the next two decades (Delgado et al., 1999; Delgado, 2003, Naylor et al., 2005, Steinfeld et al., 2006). In response to this demand, global meat production is projected to more than double from 229 million metric tons (mmt) in 1999–2001 to 465 mmt in 2050, and milk production is expected to rise from 580 to 1043 mmt (Steinfeld et al., 2006). Meeting this challenge will require global efforts to develop and use sustainable practices.

ANIMAL SOURCE FOODS IN THE TWENTY-FIRST CENTURY

The human population currently consumes about 29 g of animal protein per capita daily. Affluent societies far exceed this mean, while the poor in many parts of the world survive on almost no animal protein. Gilland (2006) and Cassman et al. (2003) projected that to maintain the current level of animal protein of 29 g per capita per day at the projected population growth rate, cereal production would have to rise from the current level of 2600 Mt to 3500 Mt by 2050. Gilland (2002, 2006) concluded that a modest improvement in the average diet in less-developed countries is possible; realization of this goal is a high priority for society.

Actions Needed to Maintain and Protect a Stable Global Environment

A report released by the Food and Agricultural Organization (FAO) and the Livestock, Environment and Development Initiative (LEAD) (Steinfeld et al., 2006) suggested sustainable practices to address the environmental challenge created by the projected increase in global food animal production in response to the expanding demand. These recommendations include

- Continue persistent and innovative soil conservation methods.
- Match appropriate controlled livestock-grazing measures to the specific ecosystems of vulnerable locations to prevent land degradation.
- Continue improvements in efficiency of livestock and crop production.
- Continue to improve efficiency of feed utilization in livestock and poultry.
- Adjust feed composition to reduce enteric methane emission in ruminants.
- Develop biogas plant initiatives to recycle manure.
- Improve efficiency of irrigation systems.
- Restrict large-scale integrated livestock systems concentrated near population centers. Site them only where effluent can be safely, efficiently, and completely cycled back to the crop or forage element of the production system.

These approaches and others are addressed in this chapter and elsewhere in the book.

Importance of Animal Disease Control

Viable systems are essential for effective control of the many infectious diseases that affect animals and that, in some cases, are also communicable to humans. The globalization of food animal production and animal source food distribution has created a greater need to improve such systems. Individual animal identification systems are evolving as a means of tracking intra- and intercountry movement of live animals and thereby disease agents. This enhances food safety and quality control in animal source foods. In the United States, the National Animal Identification System (NAIS) has been initiated in conjunction with the control or eradication of specific diseases in live animals. A country-of-origin labeling law (COOL) in the United States requires meat products sold at retail to be labeled to be able to trace the movement of the product back to the birthplace of the animal (Smith, 2007a, 2007b). Such monitoring systems are critical as globalization of animal agriculture proceeds.

Recent Improvements in Biological and Economic Efficiency of Food Animals

Advances in genetics, nutrition, and bioengineering have resulted in dramatic increases during the last two decades in biological efficiency (i.e., improvements in growth, reproduction, lactation, and efficiency of feed utilization of food animals and birds). These advances have resulted in large increases in the amounts and quality of animal source foods produced per unit of feed and other resources invested. Continued improvements in efficiency will be required in the coming decades to meet future needs. These improvements must be driven by robust funding for science and technology in plant and animal source foods and for training of new generations of scientists. These issues have been addressed by Ireland et al. (2008) and Reynolds et al. (2008).

Milk

Milk production in dairy cattle in the United States over a standard 305-day lactation period has increased from about 15,000 pounds in the mid-1980s to about 25,000 pounds in 2006. In many parts of the world, sheep, goats, water buffalo, reindeer, camels, and other animals are important sources of milk (Ullrey, 2000). Milk composition is qualitatively similar in all species, but all milk, regardless of species, has unique properties and constituents needed by all newborn mammals. Improved efficiency of milk production can be expected in all of these species through new technologies and more appropriate application of tested technologies.

Meat and Fish

Swine

In the past two decades, growth rate, feed utilization, and lean meat output in swine increased from 220 pounds body weight at 5 to 6 months of age and an average back fat depth of 1.5 inches in the mid-1980s to 270 pounds at 5 to 6 months of age and an average back fat depth of 1 inch or less.

Chickens

In the United States in the 1980s, it was the norm to market broiler chickens at 9 weeks of age. In 2005, broilers were marketed at 7 weeks and at a higher market weight.

Sheep

Similar improvements in production efficiency can be cited in lambs per ewe per year through accelerated lambing and the use of more prolific breeds and improved pasture management and nutrition.

Fish

Technological advances in commercial fish farming have increased pounds of fish per unit of feed through improved nutrition and management.

Beef Cattle

The average rate of weight gain during the finishing period is now approaching 4 pounds daily, compared to about 3 pounds daily two decades ago. This increase is due to a combination of improvements in nutrition, genetics, animal health, and environment.

Diverse climates create challenges for animal agriculture that are being addressed by research focused on adaptability of food animals to extremes in environmental temperature. Crossbreeding programs using heat-tolerant and disease-resistant *Bos indicus* cattle such as Brahman (Zebu) cattle from tropical zones mated with *Bos taurus* cattle from temperate zones results in improved reproduction and lactation of crossbred native tropical cattle raised in tropical environments (Herring, 2000).

OTHER DOMESTIC ANIMALS FOR FOOD, FIBER, AND HIDES

Many less-recognized domestic animals and birds are used for food, fiber, and hides around the globe. Uses of these lesser-known animal resources have been described (Ullrey and Bernard, 2000).

Small Animals and Birds

Small animals and birds include rabbits, turkeys, guinea pigs, quail, ducks, geese, and fish. These animals provide food for poor families in many parts of the world because of their small size, low feed and husbandry requirements, and ability to convert waste resources (seeds, insects, weeds, and aquatic plants) into meat and eggs.

Commercial duck farms are prominent in Asia and in Central and South America. Geese are found worldwide, but are particularly important in Asia and Central Europe. They are able to digest high-fiber diets so that they can be raised on less grain than other domestic fowl.

Aquatic Animals

Farming of fish, shrimp, and other aquatic animals (aquaculture) is one of the fastest-growing food production enterprises (Lovell, 2000; Engle and Stone, 2005). Channel catfish farming in the United States has grown from obscurity in 1970 to an annual production of more than 223,000 tons in 1996 and is still growing. Farming marine shrimp, primarily in South and Central America and Asia, is the fastest growing aquaculture industry in the world (43% of the global consumption). Ocean-netted pen cultures of salmon are produced in Norway. Other high-value marine species (e.g., sea basses, turbot, and yellow-tail tuna) are cultured commercially in Western Europe and Japan. Aquaculture is now recognized as a flourishing food production enterprise worldwide and is expected to supply a larger percentage of fishery products because of the continuing decline in the supply of wild-caught fish from the oceans. The percentage of edible lean tissue in fish is higher than in beef, pork, or poultry because fish contain less bone, fat, and connective tissue. A primary reason for the success of aquaculture is the research and technology base developed over the past few decades (Lovell, 2000).

Large Animals and Birds

Bison, Camels, and Llamas

The bison (American buffalo) was long the dominant ungulate (mammal with hoofs) in North America. It has been estimated that in 1860 there were as many as 60 million bison, having provided food and clothing to prehistoric human hunters for thousands of years. The popularity of bison meat has led to private herds totaling about 200,000 animals in North America.

Camels and llamas are known as *camelids*, The Old World camelids include the two-humped Bactrian camel (found principally in the cool desert regions of central Asia) and the one-humped dromedary camel (found in hot deserts of North Africa and western Asia). Both are used for transport, draft, meat, milk, fiber, and hides.

Llamas, prevalent in high altitudes of Latin America, are used predominantly for fiber.

Ratites

Ostriches and emus are large, flightless birds that have been domesticated for production of meat, leather, and plumes. They are able to digest high-fiber feeds to meet about one half of their maintenance energy requirements. Commercial ostrich breeding began in the United States in the 1980s; emu breeding began a decade later. Meat from these birds is marketed largely to restaurants.

VERTICAL INTEGRATION OF ANIMAL PRODUCTION SYSTEMS

The impact of vertical integration of animal agriculture on environmental sustainability and on the capacity for increased food production is often viewed negatively. Recent research and analysis has shown that well-planned and well-managed production systems improve rather than harm the environment and increase crop and animal source food production efficiency.

LAND USE AND GREENHOUSE GAS EMISSIONS FROM MILK AND MEAT PRODUCTION

Quantitative measurement of greenhouse gas (GHG) emissions by food animals is important for addressing their effects on global climate change (see Chapter 16). Information on dairy cattle and beef cattle is used here to illustrate the role of milk and meat production in contributing to GHG emissions. Mathematical modeling of the whole animal production system has been used to understand and manage GHG emissions.

Dairy Cattle

Casey and Holden (2005) used life-cycle assessment in dairy cattle to provide an objective estimate of GHG emissions and to evaluate emission management systems regarding carbon dioxide equivalents emitted per unit of milk produced. Of the total GHG emissions, 49% was animal enteric fermentation products, 21% crop fertilizer, 13% concentrate feed, 11% manure management, and 5% electricity and fuel consumption.

Expressing GHG emission on the basis of the quantity of GHG per unit of milk produced, it was determined that GHG emissions could be reduced 14 to 18% by selection for efficient, high-producing cows and an additional 14 to 26% by removal from the herd of inefficient and nonlactating animals. In dairy systems, about one half of the total carbon dioxide equivalent is from methane and one third from nitrous oxide. Use of mitigation systems such as intensive grazing in dairy herds reduces GHG emissions by 10% (Phetteplace et al., 2001).

Beef Cattle

In the case of beef, there are two major post-weaning production paradigms used in North America: (1) cattle in large feedlots fed a finishing diet of grain and forage and implanted subcutaneously or fed growth-promoting hormones and compounds that suppress rumen methane generation and (2) a traditional pasture-based beef production finishing system. Both systems have advantages and disadvantages, but each has different environmental impacts in terms of land used and emissions of GHGs per pound of beef produced.

Role of Growth Promoters in Beef Production Efficiency

Beef produced from cattle in feedlots and given growth-enhancing hormones requires less land during the finishing period than if cattle are kept on pasture. Acevedo et al. (2006) compared the average cost of production to achieve current market-grade standards for cattle in Iowa finished under five different feeding systems. Part of this study involved a comparison of the performance of feedlot cattle fed a conventional high-concentrate finishing diet with (W) or without (WO) growth-promoting supplements and methane generation suppressants. The results illustrated the improved growth performance, efficiency of feed utilization, and greater total beef production of W than WO cattle. Calf starting weight, postweaning average daily gain, final body weight, carcass weight, and total system beef production for the systems are shown in Table 21.1.

Greenhouse Gas Emissions Associated with Beef Production

Steinfeld et al. (2006) authored a report from the United Nations Food and Agriculture Organization (FAO) indicating that poultry and livestock are responsible for 18% of the carbon dioxide equivalent GHG emissions, of which 9% are carbon dioxide, 37%

TABLE 21.1

Comparison of Feedlot Cattle Performance Fed a Conventional High-Concentrate Diet[a]

	WO[b]	W[c]
Starting weight (lb)	475	475
Days on feed	329	303
Postwean daily gain (lb)	2.36	3.06
Feed:gain ratio	7.12	6.22
Final weight (lb)	1,251	1,401
Carcass weight (beef yield)	782	876
Total system beef production (lb)[a]	60,214	67,452

[a] Data calculated by Avery and Avery, 2007.

[b] Conventional high-concentrate diet without (WO) synthetic growth promoting hormones and methane generation supressants.

[c] Conventional high-concentrate diet with (W) synthetic growth promoting hormones and methane generation suppressants.

methane, and most of the remainder nitrous oxide. In beef cattle systems, the cow-calf herd emitted the most and feedlot cattle the least methane and nitrous oxide per unit product. Carbon dioxide emissions per unit product were the least for the cow-calf and greatest for feedlot cattle (Phetteplace et al., 2001).

The introduction of methanogenic suppressors for supplementation to the diet of finishing cattle has been a major benefit to the beef cattle industry and has reduced GHG emission into the environment. Most feedlot cattle are currently fed grain-based diets supplemented with methane-suppressing compounds (e.g., monensin and lasalocid).

Two Faces of Animal Agriculture

Animal agriculture now has two faces: (1) Concentrated animal feeding operations (CAFOs) use high-technology, high-volume, highly efficient *integrated animal production* in which a small minority produces the food for the vast majority of its populace in the area. Other smaller CAFOs are more *traditional animal production* units (some family owned, some owned solely or partially as corporations) that coexist with the larger CAFOs, which are often financed by external investors. (2) The other face of animal agriculture is represented by many smaller family-owned farms, many of which are located in developing countries. Economy of scale places these small farms at an economic disadvantage, which has resulted in consolidation of food animal production into fewer and larger farm operations.

Although average age of farm owners in North America continues to increase, younger farmers are more likely to be college trained and technically advanced in their farming and business expertise. Even with integration of agriculture, family farms represent more than half of current U.S. farm ownership, although many of these enterprises involve part-time farmers. These smaller farms continue to produce a significant proportion of the animal products sold in the United States. Organic farming enterprises (distinguished by use of manure rather than inorganic fertilizers and by adherence to the practice of no pesticide or herbicide use on crops) continue to increase in number and often sell their products at local farmer's markets, where customers are frequently willing to pay higher prices for foods advertised as organic.

Present and Future Animal Agriculture in Developing Countries

In developing countries, much of animal agriculture is based on local production on small farms, using meager resources and limited capital to provide for family or small market consumption. Worldwide, about 1.3 billion rural people in developing countries depend on livestock for a livelihood (Steinfeld et al., 2006). Livestock farming is critical for many of the poor in developing countries, offering pathways out of poverty and improving their nutrition and health (Randolph et al., 2007). Integrated animal production has become an economic force in many developing countries. Large-scale integrated confinement systems now account for 75% of the world's poultry supply, 40% of its pork, and 66% of all egg production (Bruinsma, 2003). The loss of linkage between livestock production and land, the natural resource base in animal source food production, has raised concerns about the environmental and resource costs of livestock production (Naylor et al., 2005). Environmental impacts of livestock in the developing countries have also been addressed by Nicholson et al. (2001). Naylor et al. asserted that "A re-coupling of crop and livestock systems is needed—if not physically, then through pricing and other policy mechanisms that reflect the social costs of resource use and ecological abuse," and that for the systems to work, there must be physical preservation of the environment. Naylor et al. stressed that policy measures should not compromise improving the diets of those in the developing countries or prohibit trade. Some of these issues are addressed in Chapter 23.

It has been predicted that increasing the supply of animal source foods in developing countries can be achieved by combining an increase in the number of animals with improvement in productivity and processing/marketing efficiency (Steinfeld, 2003). Limited land availability can curtail the expansion of extensive livestock production systems in some regions, but major improvements in the productivity of extensive systems without increasing land area devoted to it are possible in others. Most of the increase in livestock production will come from increased productivity by intensification and wider adoption of improved technologies in production and marketing. Steinfeld suggested that there is a danger that livestock production and processing may become dominated by integrated large-scale operations, which may displace small-scale farmers. On the other hand, well-managed and dynamic livestock enterprises may be a catalyst for improved rural economies and increased affordability of animal source food. The comparative advantage of grain-based beef production systems combined with the use of safe and highly effective growth-promoting substances in feedlots for finishing beef was described in the previous section. This new knowledge provides a strong incentive for development of supportive public policies and education of consumers concerning the compatibility of environmental stability with adequate and sustainable animal source food production.

RAMIFICATIONS OF INTEGRATED LIVESTOCK PRODUCTION SYSTEMS

The shift from traditional to large-scale, integrated animal production has significant ramifications. Advantages and disadvantages of integrated CAFOs include those discussed next.

Advantages

High-volume production. Large units can use volume buying of feed ingredients at lower cost per unit. They can afford to hire specialists such as veterinarians and nutritionists whose expertise is not generally affordable to small farm owners unless provided by governments with effective agricultural policies. Such expertise contributes to more efficient feed utilization, better herd health, superior genetics, and closer control of the environment in which animals are kept, whether provided by the private or public sector. Public university research/extension network has been responsible, to a large degree, for modernization, as in U.S. agriculture for more than a century. Large numbers of animals justify investment in mechanized feeding and waste removal to reduce labor costs. The opportunity to maximize throughput of animals per unit of living space in well-designed confinement facilities and the possibility for more precise control of marketing animals to coincide with high market price allow a higher net income to investors.

Less labor drudgery. Large units may permit a division of labor and a less confining 5-day work week, in contrast to the frequent pattern on small family-owned farms of a year-around 7-day work week.

Stronger market position. Through volume of production and often a more uniform product than that available from many small farms, large units generally have better marketing alternatives. The evolution of large food retail chains has contributed substantially to the move toward integrated systems.

Potentially better environmental stability. Integrated production units generally have greater capital resources and access to engineering expertise and can manage manure and waste more effectively than can a large number of small farm units collectively with the same total number of animals.

Disadvantages

Separation of feed production from the site where the animals are fed. Many integrated enterprises in developed and developing countries import feed from a distance rather than producing it on adjacent land, with associated costs of transport. This failure to return manure to the land on which the feed was produced may reduce soil organic matter and increase the need for chemical fertilizers on that land; this may be undesirable for long-term sustainability of crop production. The concentration of large quantities of nutrients, notably nitrogen and phosphorus, from the animal excreta produced in large feedlots requires adequate engineering systems for manure removal and processing to prevent environmental pollution. Furthermore, the potential for percolation of excess soil nutrients into groundwater and their accumulation must be addressed. If there is not adequate adjacent land to use the manure as fertilizer, economical systems must be developed to process it. For remedy, strategies such as (1) trapping methane produced during manure degradation for powering vehicles or home kitchen fuel (as practiced on livestock farms in many countries for decades) or (2) using manure as a biofuel source in agriculture may be useful approaches to animal waste utilization. However, such use must address, through innovative technologies, problems of nitrogen, phosphorus, and odor pollution of the environment.

Nuisance characteristics of large concentrations of animals. Some CAFOs, including large swine confinement units and large beef cattle feedlots and dairies, produce objectionable odors and reduce air quality for neighbors. Areas of the United States with low population density and reasonable proximity to feed resources and modern slaughter facilities (e.g., high plains of Oklahoma, Texas, Colorado) are attracting integrated livestock enterprises with reduced threats of lawsuits and fewer complaints related to objectionable odors and compromised environmental quality. Mitigation of the environmental and social impact of large concentrated livestock operations is an ongoing challenge for animal agriculture. Remedies that do not accelerate the depopulation of these areas and that might also protect the health and quality of life of the thousands of employees who live near these facilities should be sought.

Impact on communities. While CAFOs have freed those on traditional farms from very long hours of labor, they have had a major impact on rural communities. As family farms are replaced by large integrated units (often not in the same area), populations in rural areas decline; schools, libraries, banks, churches, and other essential services are lost, and communities shrink or disappear. Some individuals may shift from operating their own farms to working for the new enterprises (which may or may not improve their incomes). For many, it is a less-attractive lifestyle in spite of more regular hours. Others move to cities, again not necessarily considered a better life.

PLANT–ANIMAL ECOSYSTEMS

A balanced plant-animal ecosystem is fundamental to food production for, without plants, animals (including humans) could not survive. The many by-products of crops processed for human consumption provide a significant proportion of the total feed requirements to support a viable livestock industry. Awareness of the principle of the symbiosis of plants and animals in providing food for humans is needed for a full appreciation of the vital role of food animals in society.

FEED RESOURCES

Feed represents the major cost in animal production. Therefore, a nutritionally balanced diet fed at a level of intake sufficient for high efficiency of feed utilization by the animal is important for all of animal agriculture. More than 2000 feed products are available for animal feeding. National Academy of Science (NAS) Feed Composition Tables (2001) containing nutrient composition of each are updated frequently. New genetically modified varieties of forages and grains having increased disease and pest resistance, yields, and other economically valuable traits continue to become available each year, and other appropriate genetically modified organisms (GMOs) may be used. In 2008, the European Union and the U.S. Food and Drug Administration both approved the use of GMO crops as safe in animal feeds.

Although there is some competition, most plant products fed to animals either are not edible for humans or are produced in excess of demand for human consumption in a given location. Because of imperfect systems of distribution and differences in agricultural productivity, many crops, including corn (maize) and cereal grains (such as wheat, oats, or barley), oilseeds (such as soybeans), or other foods in surplus in the United States are in high demand for export for use by humans or in animal feeds. These feedstuffs include surpluses of cereal grains and other food crops; by-products of cereal grain and oilseed processing; substandard human food products; crop residues, and animal by-products.

Surpluses of Cereal Grains and Other Food Crops

A concern is the use of surplus grains for livestock, poultry, and aquatic animals. The remarkable increases in crop yields (the green revolution), for which the Nobel Peace Prize was received in 1970 by Dr. Norman Borlaug, has provided supplies of feed grains beyond amounts ever before produced. However, trends for increased corn diversion to ethanol production in the United States would modulate such surpluses. (Sources of ethanol from cellulosic plants are viewed by experts as a better choice of plant energy for ethanol production). The animal industry has been an important force in providing relief from crop surpluses and in providing highly nutritious animal source foods to improve human nutrition globally.

By-Products of Cereal Grain and Oilseed Processing

The milling of corn and cereal grains to produce flour, starch, gluten, and other products results in large amounts of by-products of value in animal feeding (Klopfenstein, 2005). Corn dry-milling produces meals and flour for human consumption and the by-product hominy (high in fiber) for cattle. Ethanol is also produced by a dry-milling procedure and by a more efficient wet-milling procedure. Large amounts of distiller's grain solubles, by-products of the brewing and ethanol production industries, are used in livestock feeding. Plant oils for human consumption are obtained from a variety of crops, such as soybeans, flax, cotton, canola, and olives, that are high in oil. The high-protein meal that remains after extraction of the oil from these seeds provides a valuable by-product for use by food animals.

Substandard Human Food Products

Turnips, beets, carrots, cassava, potatoes, and other root crops produced in surplus or not meeting human standards for marketing are available for animal production. Waste bakery products and other pastries, by-products of the potato- and citrus-processing industries, and surplus plant fats from the human food chain are marketable as constituents of animal feeds.

Crop Residues

Large quantities of low-quality straw and other residues are left in the field after harvest of cereal grains and corn. Straw is generally very low in digestible protein and very high in fiber and is of low feeding value by itself, but it may be used as the basal feed for beef cattle and other ruminants when properly supplemented with protein, minerals, and vitamins. Corn harvested in the field leaves behind the cob, stalks, and lost and broken seeds. These materials provide an otherwise wasted animal feed resource when animals are allowed to glean these harvested fields (animal manure is also directly applied to the soil for future crop production in this system). In cotton-growing areas, cottonseed hulls and cotton gin trash are used in substantial amounts for feeding ruminants. Other useful high-fiber by-products of the food-processing industry include pineapple juice press cake, rice mill feed, sugarcane bagasse, and citrus pulp.

Animal By-Products

Processing of meat, milk, and poultry for human consumption is associated with many value-added products used in animal feeds (Chiba, 2005). Meat and bone meal, whey, and feather meal are examples.

GRAZING PUBLIC RANGELANDS AND PASTURES

Land grazed by ruminant livestock represents 26% of the total global land area and nearly 70% of the world's land used for agriculture. Most grazed lands are too arid, steep, rocky, infertile, or cold for arable crop production but do produce a variety of grasses, shrubs, forbs, and other plants that are a potential source of feed for livestock. Most such plants are high in cellulose and lignin, which cannot be digested by humans or by nonruminant animals such as pigs and chickens, but can be utilized by ruminants, including cattle, sheep, goats, and buffalo, through microbial breakdown of the feed in their rumens. Therefore, grazing livestock permit the production of food for humans on land that would otherwise produce little or no human food. Because of the marginal character of most grazed lands, their productivity per unit area is much lower than that of cropped land, but they contribute more than 20% of global ruminant meat production and about 7% of global milk production. This represents a significant part of the total food supply and is critical to the survival and well-being of pastoral peoples.

Grazed lands have many functions apart from the production of animal source foods. They harbor a substantial proportion of the world's biological diversity, they are the site where much of the rainfall contributing to freshwater supply falls, and the open space they represent is highly valued by an increasingly urbanized human population. Grazing represents an important tool for vegetation management. However, improperly managed grazing can have negative effects on the environment and on biodiversity.

Grazing on Ranches in Arid Areas

Cattle, sheep, and goat ranching is widely practiced in North and South America, Australia, New Zealand, and some other countries. In general, ranching occurs in more arid areas and involves the maintenance of breeding herds for meat production (milk production more commonly occurs on better land and in higher rainfall areas). Offspring may be sent directly to slaughter at weaning, grazed in the same area following weaning until they reach harvest weight, or sent at weaning or at some later stage to feedlots for further feeding before harvest. Feedlots use forages, cereal grains, and a wide variety of crop-processing by-products to increase the amount and quality of meat, and their existence depends on a reliable source of "feeder" animals raised on less-expensive land.

Grazing on Improved Pastures in Higher-Rainfall Areas

In Europe, in higher rainfall areas in the United States, and in other temperate parts of the world, grazing more commonly occurs on improved pastures. The pastures are often on steeper or potentially more erosive land or included as part of a crop rotation system with food crops to benefit soil fertility and soil organic matter and thus contribute to sustainability of the agricultural system.

Transient Grazing Systems

In parts of Africa, and in some other parts of the world, transient grazing systems have long been and in some cases continue to be important. Such systems involve the systematic movement of animals during each year depending on plant growth and thus feed availability. A classical example is the grazing of livestock in the interior delta of the Niger River in West Africa during the dry season and their movement to desert areas during the rainy season when the interior delta floods. This system persisted for centuries but has been compromised by conversion of much of the dry season grazing land in the delta to crop production in response to human population growth. Also, some land has been lost to oil production and to reduced security due to wars and insurgence stemming from conflicts over oil.

Pastoral grazing systems on land that is too arid, cold, infertile, or steep for crop production occur on large areas of land and provide valuable animal products, often critical for a reliable food supply for pastoral peoples.

Grazing Systems Integrated with Cereal Production

In dry climates, including notably North Africa and much of the Middle East, live-stock grazing (largely sheep and goats) is closely integrated with cereal production by yearly alternating cereals with grazing. Grasses and forbs that grow during the rainy season of the fallow (noncereal) year are an important source of feed during that season. Following grain harvest, animals graze the stubble, including the straw, spilled grain, and weeds. Straw may be used to supplement diets late in the stubble-grazing season and at the start of the rainy season. In drought years when rain is inadequate to produce a grain crop, the cereal foliage may be grazed. Plant nutrients in crop residues and in plants other than cereals are largely returned to the land, contributing to sustainability of the system.

Adverse Effects of Some Grazing Systems

Problems associated with grazing systems include soil erosion due to loss of plant cover and overgrazing and impacts on wildlife habitat and biodiversity. Overgrazing can be avoided by adjusting stocking rate. Several factors make it difficult to achieve an optimum level of plant use for long-term sustainability. The wide seasonal and yearly variation in rainfall means that the optimum number of animals on average for a given area may result in underutilization of forage in a high-rainfall year and damage to the plant biota in a drought year. More detailed descriptions of man-agement of grazing lands are available (Bradford et al., 1999; Bradford, 2001). For pastoral peoples, wealth is often based on numbers of animals owned rather than on productivity of the herd or flock. The animals owned may be the primary buffer against crop failure or drought. Higher stocking rates than desirable ecologically have contributed to the tarnished reputation of livestock grazing as a cause of envi-ronmental degradation.

Complementary Role of Well-Managed Grazing Systems and Sustainable Food Production Systems

Grazing livestock in many areas of the world complement crop production and, when properly managed, contribute to sustainable food production systems. Additional science-based knowledge of optimum grazing management continues to accumulate, and its application significantly enhances the health and productivity of grazing lands and the multiple services provided by those lands.

ANIMAL WELL-BEING

Public interest in farm animal welfare was aroused in the mid-1960s by a committee of the British Parliament (Brambell, 1965), who issued a report on the welfare of farm animals kept under intensive husbandry systems. In the past four decades, farm animal welfare has continued as an issue and has received considerable attention in research related to animal behavior and physiological measures of stress in farm animals.

These issues include adaptability and social ethics for animals.

ADAPTABILITY OF DOMESTIC ANIMALS

For domestication of animals to succeed, it has been suggested (Curtis, 2000) that animals had to fit with human cultures, and that domestication was the evolutionary product of a mutual strategy for survival (Budiansky, 1992). It is widely accepted by livestock producers and the general public that animal comfort is important during all stages of the life cycle. The very essence of good husbandry requires respect for protection of animal health and well-being for the sake of both the animal and the producer. The economic reality is that healthy and content animals provide a higher return to the producer and a generally high-quality product.

Animal care guidelines are used in research laboratories and on farms to ensure that animals are treated humanely and afforded an environment conducive to their well-being. Innovative animal-handling facilities and devices to reduce stress have been designed (Grandin, 1993, 2005) and are widely used on ranches and in feedlots, animal slaughter facilities, and livestock operations during movement, transport, and marketing. Animal behaviorists are seriously engaged in discovering ways to promote stress reduction and ensure that dedicated farm animal care and husbandry remain an ongoing process.

SOCIAL ETHICS FOR ANIMALS

Rollins (2005) has enunciated the relationship between good animal husbandry and animal well-being, a portion of which is as follows:

> For most of human history, the anticruelty ethic and laws expressing it sufficed to encapsulate social concern for animal treatment for one fundamental reason: During that period, and today as well, the majority of animals used in society were agricultural, utilized for food, fiber, locomotion, and power.

He continued:

> Humans were in a contractual, symbiotic relationship with farm animals, with both
> parties living better than they would outside of the relationship. We put animals into
> optimal conditions dictated by their biological natures, and augmented their natural
> ability to survive and thrive by protecting them from predation, providing food and
> water during famine and drought, and giving them medical attention and help in birth-
> ing. The animals in turn provided us with their products (e.g., wool and milk), their
> labor, and sometimes their lives, but while they lived, their quality of life was good.
> Proper husbandry was sanctioned by the most powerful incentive there is—self inter-
> est! The producer did well if and only if the animals did well.

The same relationships and principles hold true in present animal agriculture,
including small farms and large integrated enterprises. Those engaged in food
animal agriculture are well aware of their obligation, as described by Rollins, to
continue the "contractual relationship" between humans and animals in ensuring that animal well-being is always protected based on applying the best current
knowledge.

TECHNOLOGIES AND PRODUCTS THAT SUPPORT GLOBALIZATION OF FOOD ANIMAL PRODUCTION, FOOD SAFETY, AND ANIMAL HEALTH

The Panel on Animal Health, Food Safety, and Public Health, National Academy
Press (National Research Council, 1999), examined the benefits and risks associated with drug use in food animal production and recommended, among other
issues, increased efforts related to drug resistance and drug residues in animals.
Technology and research by private industry, government and public and private
universities have resulted in a variety of products, including antibiotics, hormones,
metabolic modulators, and vaccines and other veterinary products that have been
tested and approved for use in food animal production to enhance growth, efficiency of feed utilization, and health. The use of these products in animal production has been instrumental in reducing costs of animal source food production and
improving animal health and well-being. Some of these products and their uses
are described here.

BOVINE SOMATOTROPIN FOR INCREASED MILK YIELD

Bovine somatotropin (bST) was approved for use in the United States about a decade
ago as a means of increasing milk yield and improving overall efficiency of production. Its safety and efficacy have been established through years of research and use
in the dairy industry. Consumer groups have challenged the use of bST in dairy
cattle, resulting in its curtailed use in some countries.

Hormone Implants and Feed Additives for Improved Growth and Beef Leanness

Synthetic (melengestrol, MGA; trenbolone acetate, TBA; zeranol) and natural steroid hormones (progesterone, estradiol, testosterone; all produced in significant amounts throughout the life of every human) are used in beef cattle, and sometimes in sheep, as a means of increasing growth rate, efficiency of feed utilization, and increased lean tissue mass and carcass leanness. MGA is administered daily in the feed, whereas each of the others is administered to an individual animal in a single dose as a pellet implanted early in the finishing period so that the hormone release is depleted long before animal slaughter, eliminating the possibility of tissue residues in the meat. More than 30 countries currently allow use of these hormones in beef production, and even European scientific groups have deemed hormones safe for use (Avery and Avery, 2007). Strict regulatory rules are enforced to ensure safety. This procedure has provided an effective means of improving efficiency of beef and lamb production.

Feed Additives for Improved Biological Efficiency and Animal Health

In addition to the nutrients contained in feedstuffs, many nonnutritive feed additives are used to improve performance. Industrialized countries have their individual programs for approval, regulation, and enforcement of use. In the United States, the Food and Drug Administration (FDA), Environmental Protection Administration (EPA), and Food Safety and Inspection Service (FSIS) are among the administrative agencies responsible for protecting the safety and efficacy of food and food constituents for humans and for approving new feed additives for use in animals. The European Union has similar programs, and parallel agencies in other countries have similar responsibilities. The main broad groups of feed additives approved for use in animal feeds are antibiotics, probiotics, enzymes, ionophores, and protein anabolic agents.

Antibiotics

For more than 50 years of extensive use worldwide, antibiotics (antimicrobial agents) have been added to livestock feeds at subtherapeutic levels to improve performance (Hays, 2005; Cromwell, 2005). Antibiotics are still as effective as at the beginning in increasing growth rate, efficiency of feed utilization, and decreasing mortality. Development of microbial resistance to antibiotics in animals and humans being treated with therapeutic levels (several times higher than the subtherapeutic levels used in animal feeding) of antibiotics is a serious problem in human patients, especially those confined in hospitals. This has created concerns about the possible relationship between subtherapeutic antibiotics in animal agriculture and development of antibiotic resistance to pathogens in humans. The existence of such a causal relationship has been difficult to establish given the complexity of the issue. A complete ban on the use of antibiotics in animal feeds would increase the cost of food animal production.

Probiotics

Probiotics are live microorganisms that may confer health benefits in the host animal (mainly cattle, swine, and poultry) and in humans when provided in appropriate amounts (Gilliland, 2005; Sanders et al., 2007). They have received attention as a potential alternative to antibiotics, but inconsistent results have been reported in animals. *Lactobacillus acidophilus, L. casei,* and *L. fermentum* have received the most attention in animals.

Enzymes

Some enzyme additives are useful in improving digestibility of nonstarch polysaccharides (carbohydrates). Supplementation of the feed with these enzymes may have use in improving utilization of these polysachharides in some cereal grains (Newbold and Hillman, 2005). The most promising application of added enzymes in the diet is that of phytases to improve the utilization of phosphorus in cereal grains. Reduction in excretion of unused phosphorus in the diet improves performance of pigs and poultry and may result in large decreases in phosphorus entering the environment and eutrification of streams and bodies of water This technology has been adopted by the feed industry (Lei and Porres, 2005).

Ionophores

A group of compounds, notably monensin and lasalocid, influence rumen fermentation in cattle when added to the feed and have had a major impact on animal agriculture by increasing milk yield in dairy cattle, reducing methane production, and improving feed utilization and growth in feedlot cattle. These dramatic responses have resulted in major saving of natural resources to animal agriculture (McGuffy, 2007). Ionophores are technically antibiotics but are usually classified separately because they act by very different mechanisms than other antibiotics and are not used much as antimicrobials for humans. There is also hope among manufacturers of these compounds and the companies that use them that, if broad-brush bans on antibiotics use were passed, ionophores might be spared.

Protein Anabolic Agents

A class of compounds known as β-adrenergic agonists (β-agonists, repartitioning agents) has been used for about two decades to enhance lean tissue and decrease fat in beef cattle and swine (Anderson et al., 2005). The β-agonists have been used safely for many years in human medicine as bronchodilators for asthma and as uterine relaxants in pregnancy to arrest premature labor. Approval by the U.S. FDA for use as feed additives resulted in their rapid acceptance to enhance carcass leanness, decrease fatness, and improve growth efficiency.

New Technologies in Food Safety and Tissue Preservation

Breakthroughs in Quick Freezing of Biological Materials for Long-Term Storage and Sustained Viability

The development of a new and unique freezing technology that freezes materials with little or no ice crystal formation has valuable application in the preservation of biological entities and meat products. Semen, embryos, and animal tissues have been frozen and thawed with excellent viability. This technology offers the promise for wide application in biology and agriculture, including genetic improvement, health, and food storage and distribution. Freezing of meat with later thawing as if "fresh" has implications for storage and distribution of animal source foods.

Breakthroughs in Eliminating Pathogenic Microorganisms from Meat and Other Food Products

New technologies for reduction or elimination of the pathogenic microorganisms often contaminating meat are being developed. Such innovations include specific wash procedures at the time of meat harvest and, more recently, the targeted use of selected strains of *Lactobacillus* bacteria that outcompete the pathogens, allowing for safer meat products. The bacteria are fed to cattle preharvest and are incorporated into ground meat products. The system is efficacious and is approved for use by the U.S. Food and Drug Administration. In addition, withdrawal of ruminants from high-grain diets and use of high-forage diets for a period before slaughter similarly reduces *Escherichia coli* O157 populations. This type of food safety intervention promises to have other important applications in the food industry.

Biotechnology

Broadly defined, *biotechnology* includes any technique that uses living organisms or processes to make or modify products, to improve plants or animals, or to develop microorganisms for specific purposes. The field has created new opportunities in agriculture and medical science. Much of this new knowledge is being applied as it emerges.

Artificial Insemination

One of the earliest uses of biotechnology was artificial insemination (AI) in livestock and poultry for genetic improvement (Flowers, 2005). This was especially effective in dairy cattle. By collecting semen from superior bulls, it was possible to make much faster progress in increasing milk production by selecting daughters with high milk production. AI is now used in beef cattle, swine, and other animals as a tool for rapid genetic progress in many heritable traits of economic importance.

Cloning Animals

Cloning is an asexual method of reproduction. It is the making of a biological copy of another organism with the identical genetic makeup. Animal cloning received worldwide attention with the birth of the first mammal cloned from an adult cell from the famous sheep, Dolly, in the United Kingdom (Wilmut et al., 1997). Ten years later, cattle, pigs, goats, cats, rabbits, and mice have been cloned (Lai and Prather, 2005).

Birth of identical twins in farm animals (and in humans) is an example of cloning of animals in nature. Nevertheless, cloning technology is of concern to the general public. However, meat from cloned animals has U.S. Department of Agriculture approval as safe for human consumption. Nutritional value of meat from cloned animals may be improved by the presence of superior genetic traits in the cloned animal for lean growth, efficiency of growth, and disease resistance. The high costs of cloning animals will limit this technology for the production of food animals for the foreseeable future.

Embryo Transfer

Nonsurgical embryo collection and transfer are now widely used in dairy and beef cattle for genetic improvement (Hasler, 2005). Embryos (usually more than one resulting from superovulation) from a genetically superior donor cow are transferred at a few days of age to a recipient cow whose genetic background may be of little concern but whose high milk supply is important to the foster calf. This technology has had an important positive impact on the livestock industry.

Transgenic Animals

All organisms (microbes, plants, animals) continually undergo genetic alteration by mutation and natural selection or by intentional genetic selection for phenotypic traits. The development of recombinant DNA technology has made it possible to isolate single genes, modify their nucleotide structures, make copies of these isolated genes, and insert copies of the genes into the genome of animals and plants (Pursel and Wall, 2005). Gene transfer has been accomplished in mice, pigs, sheep, and rabbits. Theoretically practical applications of transgenic technology in food animal production include improvements in milk production and composition, growth rate, disease resistance, carcass composition, reproductive performance, and prolificacy.

CONCLUSIONS AND STUDY TOPICS

1. Humans have evolved as omnivores.
2. Large increases are projected in global demand for meat, milk, and eggs, particularly in the developing countries. Global meat production is projected to more than double over the next two decades.
3. The challenge of producing enough food from plant and animal sources to meet the needs of all is perhaps the largest challenge facing society in the coming decades.
4. Food animal production on a crowded planet must be based on practices conducive to sustainable agroecology. Actions needed in food animal production to ensure sustainable animal agriculture include
   a. Develop improved livestock grazing measures in vulnerable locations to prevent land degradation
   b. Continue to improve biologic and economic efficiency in food animal production
   c. Continue to develop technologies in manure management and use in crop production and in biogas plant initiatives to recycle manure

    d. Adjust feed composition to reduce enteric methane emission in ruminants

    e. Pursue viable systems for control of infectious diseases that affect animals and, in some cases, humans

    f. Foster balanced plant-animal ecosystems

    g. Develop technologies and products that support globalization of sustainable food animal production, food safety, and animal health

ACKNOWLEDGMENT

Eric Bradford passed away in 2007, leaving a legacy of leadership in international animal agriculture as a scientist, teacher, and mentor to many students and young scientists (including chapter coauthor Kevin Pond). During the early planning stage of this chapter, Eric contributed many ideas and knowledge that helped determine its ultimate subject matter and structure. We are indebted to him for these efforts and will cherish his memory and our long friendship.

Wilson and Kevin Pond

REFERENCES

Acevedo, N., J.D. Lawrence, and M. Smith. 2006. *Organic, Natural, and Grass-Fed Beef: Profitability and Constraints to Production in the Midwestern U.S.* Ames: Leopold Center for Sustainable Agriculture, Iowa State University.

Anderson, D.B., D.E. Moody, and D.L. Hancock. 2005. Beta adrenergic agonists. In: W.G. Pond and A.W. Bell, Eds., *Encyclopedia of Animal Science.* New York: Dekker, pp. 104–107.

Avery, A. and D. Avery. 2007. The environmental safety and benefits of growth enhancing pharmaceutical technologies in beef production. Hudson Institute Center for Global Food Issues. Available at: http://www.cgfi.org (accessed April 21, 2008).

Bradford, G.E. 2001. Prospects for the western range livestock industry. *Proc. Western Section A.S.A.S.* 52:1–7.

Bradford, G.E., R.L. Baldwin, H. Blackburn, K. Cassman, and K. Crosson. 1999. *Animal Agriculture and the Global Food Supply.* Task Force Report 135. Ames, IA: Council for Agricultural Science and Technology.

Brambell, F.W.R. 1965. *Report of Technical Committee to Enquire into the Welfare of Animals Kept Under Intensive Husbandry Systems.* Command Paper 2836. London: H.M. Stationery Office.

Bruinsma, J. 2003. *World Agriculture: Towards 2015/2030: An FAO Perspective.* London: Earthscan.

Budiansky, S.R. 1992. *The Covenant of the Wild: Why Animals Chose Domestication.* New York: Morrow.

Casey, J.W. and N.M. Holden. 2005. Analysis of greenhouse gas emissions from the average Irish milk production system. *Agric. Sys.* 86:97–114.

Cassman, K.G., A Dobermann, D.T. Walters, and H. Yang. 2003. Meeting cereal demand while protecting natural resources and improving environmental quality. *Annu. Rev. Environ. Res.* 28:315–358.

Chiba, L.I. 2005. By-product feeds: animal origin. In: W.G. Pond and A.W. Bell, Eds., *Encyclopedia of Animal Science*. New York: Dekker, pp. 180–183.

Cromwell, G.L. 2005. Feed supplements: antibiotics. In: W.G. Pond and A.W. Bell, Eds., *Encyclopedia of Animal Science*. New York: Dekker, pp. 369–371.

Curtis, S.E. 2000. Animal states-of-being. In: W.G. Pond and K. R. Pond, Eds., New York: Wiley, pp. 232–250.

Delgado, C.L. 2003. Rising consumption of meat and milk in developing countries has created a new food revolution. *J. Nutr.* 133(11 Suppl 2). 3907S–3910S.

Delgado, C.L., M. Rosegrant, H. Steinfeld, S. Ehui, and C. Courbois. 1999. Livestock to 2020: the next food revolution. In: *Food, Agriculture, and the Environment*. Discussion Paper 28. Washington, DC: International Food Policy Research Institute. pp. 1–67.

Engle, C.R. and N.M. Stone. 2005. Aquaculture: production, processing, and marketing. In: W.G. Pond and A.W. Bell, Eds., *Encyclopedia of Animal Science*. New York: Dekker, pp. 48–51.

Flowers, W.L. 2005. Biotechnology: artificial insemination. In: W.G. Pond and A.W. Bell, Eds., *Encyclopedia of Animal Science*. New York: Dekker, pp. 130–132.

Gilland, B. 2002. World population and food supply. *Food Policy* 27:47–63.

Gilland, B. 2006. Population, nutrition and agriculture. *Population and Environment* 28:1–16.

Gilliland, S.E. 2005. Probiotics. In: W.G. Pond and A.W. Bell, Eds., *Encyclopedia of Animal Science*. New York: Dekker, pp. 754–756.

Grandin, T. 1993. *Livestock Handling and Transport*. Wallington, U.K.: CAB International.

Grandin, T. 2005. Animal handling-behavior. In: W.G. Pond and A.W. Bell, Eds., *Encyclopedia of Animal Science*. New York: Dekker, pp. 22–24.

Hasler, J.F. 2005. Embryo transfer in farm animals. In: W.G. Pond and A.W. Bell, Eds., *Encyclopedia of Animal Science*. New York: Dekker, pp. 329–331.

Hays, V.W. 2005. Antibiotics: subtherapeutic levels. In: W.G. Pond and A.W. Bell, Eds., *Encyclopedia of Animal Science*. New York: Dekker, pp. 42–44.

Herring, A.D. 2000. Beef cattle. In: W.G. Pond and K.R. Pond, Eds., *Encyclopedia of Animal Science*. New York: Dekker, pp. 335–372.

Ireland, J.J., R.M. Roberts, G.H. Palmer, D.E. Bauman, and F.W. Bazer. 2008. A commentary on domestic animals as dual-purpose models that benefit agricultural and biomedical research. *J. Animal Sci.* 86:2797–2805.

Klopfenstein, T.J. 2005. By-product feeds: plant origin. In: W.G. Pond and A.W. Bell, Eds., *Encyclopedia of Animal Science*. New York: Dekker, pp. 184–186.

Lai, L. and R.S. Prather. 2005. Biotechnology: cloning animals. In: W.G. Pond and A.W. Bell, Eds., *Encyclopedia of Animal Science*. New York: Dekker, pp. 133–135.

Lei, X.G. and J.M. Porres. 2005. Phytases. In: W.G. Pond and A.W. Bell, Eds., *Encyclopedia of Animal Science*. New York: Dekker, pp. 704–707.

Lovell, R.T. 2000. Aquatic animals. In: W.G. Pond and K.R. Pond, Eds., *Introduction to Animal Science*. New York: Wiley, pp. 510–525.

McGuffy, R.K. 2007. Ionophores in ruminant diets may spare energy. *Feedstuffs* 79:20–26.

National Academy of Sciences. 2001. *Feed Composition Tables*. Washington, DC: National Research Council.

National Research Council. 1999. *The Use of Drugs in Food Animals: Benefits and Risks*. Washington, DC: Panel on Animal Health, Food Safety, and Public Health, National Academy of Sciences, National Academy Press.

Naylor, R., H.H. Steinfeld, W. Falcon, J. Galloway, V. Smil, E. Bradford, J. Alder, and H. Mooney. 2005. Losing the links between livestock and land. *Science* 310:1621–1622.

Newbold, L.J. and K. Hillman. 2005. Feed supplements: enzymes, probiotics, yeasts. In: W.G. Pond and A.W. Bell, Eds., *Encyclopedia of Animal Science*. New York: Dekker, pp. 376–378.

Nicholson, C.F., R.W. Blake, R.S. Reid, and J. Schelhaus. 2001. Environmental impacts of livestock in the developing countries. *Environment* March, pp. 7–71.

Phetteplace, H.W., D.E. Johnson, and A.F. Seidl. 2001. Greenhouse gas emissions from simulated beef and dairy livestock systems in the United States. *Nutr. Cycling Agroecosyst.* 60:99–102.

Pursel, V.G. and R.J. Wall. 2005. Biotechnology: transgenic animals. In: W.G. Pond and A.W. Bell, Eds., *Encyclopedia of Animal Science*. New York: Dekker, pp.149–151.

Randolph, T.F., E. Schelling, D. Grace, C.F. Nicholson, J.L. Leroy, G.C. Cole, M.W. Demment, A. Omore, and J. Zinsstag. 2007. Role of livestock in human nutrition and health for poverty reduction in developing countries. *J. Animal Sci.* 85:2788–2800.

Reynolds, L.P., J.J. Ireland, and G.E. Seidel, Jr. 2008. Editorial: "Brain drain" and loss of resources jeopardize the continued use of domestic animals for agricultural and biomedical research. *J. Animal Sci.* 86:2445–2446.

Rollins, B.E. 2005. Animal agriculture and social ethics for animals. In: W.G. Pond and A.W. Bell, Eds., *Encyclopedia of Animal Science*. New York: Dekker, pp. 16–18.

Sanders, M.E., G. Gibson, H.S. Gill, and F. Guarner. 2007. *Probiotics: Their Potential to Impact Human Health*. CAST Issue Paper 36. October, pp. 1–20.

Smith, R. 2007a. COOL burdens all. *Feedstuffs* 79(26):1, 5. Miller Publishing Co.: Minnetonka, MN.

Smith, R. 2007b. NAIS pilot projects show merit. *Feedstuffs* 79(35):9. Full report available at: NAIS Program Updates, http://www.animalid.aphis.usda.gov/nais/newsroom/spotlight.shtml.

Steinfeld, H. 2003. Economic constraints on production and consumption of animal source foods for nutrition in developing countries. *J. Nutr.* 133(11S-II):4054S–4061S.

Steinfeld, H., P. Gerber, T. Wassenaar, V. Castel, M. Rosales, and C. de Haan. 2006. *Livestock's Long Shadow: Environmental Issues and Options*. Rome: Food and Agriculture Organization (FAO) and Washington, DC: Livestock, Environment Development Initiative (LEAD).

Ullrey, D.E. and J. Bernard. 2000. Other animals, other uses, other opportunities. In: W.G. Pond and K.R. Pond, Eds., *Introduction to Animal Science*. New York: Wiley, pp. 553–583.

Wilmut, I., A.E. Schnieke, J. McWhir, A.J. Kind, and K.H. S. Campbell. 1997. Viable offspring derived for fetal and adult mammalian cell. *Nature* 385:810–813.

Section VI

Global Food Security

22 World Population and Food Availability

Shahla Shapouri and Stacey Rosen

CONTENTS

ABSTRACT

Through the twentieth century, population grew at an unprecedented rate. During this time, developing countries transitioned to low death rates, thereby fueling global population growth. In developed countries, birth rates and death rates generally declined at the same rate, leading to nearly stagnating population growth rates. As a result of this difference in growth paths, 80% of the population growth since 1900 has taken place in developing countries, particularly in the world's poorer countries. During the last century, growth in global food production surpassed population growth, leading to an overall improvement in per capita food consumption at the aggregate level. In developing countries, the key factors that influenced diets were income growth and urbanization; these factors led to increased demand for higher-value farm products, such as animal proteins, fruits, and vegetables. During 1970 to 2005, daily per capita calorie availability in developing countries increased at three times the rate of developed countries. However, the available food has not been distributed evenly. At one extreme, there is the United States, where per capita food availability exceeded 3500 calories per day in 2005. At the other extreme is sub-Saharan Africa (SSA), where per capita food availability averaged only 2300 calories. The paradox of the increase in global food availability and persistent hunger stems from income inequality both among and within countries. One of the key variables determining the outlook for food availability, the growth rate of world population, is on a slowing growth path. More than 90% of the population growth will be in developing countries. Improvements in global per capita food availability are expected to continue, but gains will be slower than historical rates. While there will be adequate food to keep up with demand at the global level, millions will continue to go hungry because of national unequal distribution of income that limits food purchasing power.

TREND IN POPULATION GROWTH

Through the twentieth century, population grew at an unprecedented rate. Between 1800 and 1900, population increased from 600 million to 1.6 billion. However, in the following century, the pace of growth accelerated dramatically such that by the year 2000, global population increased more than 3.5 times, reaching the 6 billion mark. In fact, it took only 12 years for the population to increase from 5 to 6 billion (United Nations, 2007). This jump took place despite several important developments: there was a decline in the global rate of population growth; fertility rates in 88 countries fell below the level required for long-term replacement of their populations; the AIDS disease that spread across East and Southern Africa resulted in a significant cut in population growth rates in several countries in these regions. The global population growth rate has averaged about 1% per year since 1990, about half the rate experienced during the 1950s and 1960s. Despite this slowdown, the annual net addition to global population is about 57 million. In other words, in a period of 6 years roughly 340 million people—larger than the population of the United States in 2007—are added to the total.

During the twentieth century, population growth in developed and developing countries followed a distinct path. Developing countries, which had been characterized by high birth and death rates, transitioned to low death rates, thereby fueling global population growth. In developed countries, birth rates and death rates generally declined at the same rate; in some countries, mainly in Europe and Japan, however, birth rates declined even faster than death rates, leading to nearly stagnating population growth rates. This difference in growth paths led to a lasting change in the dynamic of populations during the last century; 80% of the population growth since 1900 has taken place in developing countries, particularly in the world's poorer countries.

In addition to differences in population growth paths by income class, there have been variations in population growth among regions, particularly in the developing countries. According to the United Nations' population statistics, the population growth rate has declined since the early 1960s, but not at the same rate in all regions. This decline was much faster in Asia, followed by Latin America, and then Africa. At the subregion level, the slowdown in population growth in North Africa was similar to Asia. The smallest change has occurred in SSA, where historical annual growth of 3% is projected to decline to 2.2% per year through the next decade, most of which can be attributed to the AIDS pandemic. According to a U.N. report, HIV/AIDS prevalence levels in SSA countries have risen to catastrophic levels. According to the report:

> Previously, demographers routinely assumed gradually rising life expectancy and improved health conditions in all countries, an assumption generally backed by the actual trends. AIDS has changed all that and, for the first time in the history of modern population projections, rising mortality levels caused by an intractable disease has become a key element of projections (United Nations, 2007).
>
> In Botswana, for example, life expectancy for men in 2005 was 33.7 years, while in the absence of AIDS it would have been about 40 years higher. In countries such as South Africa, Lesotho, and Zimbabwe, life expectancy for men and women declined by about 25 years.

Despite the HIV/AIDS pandemic, SSA's population nearly tripled to an estimated 650 million between 1960 and 2000; in 2008, the region's population was roughly 760 million. The prolonged rapid population growth combined with inadequate investment in productive capacity left the region behind all other developing regions in terms of income growth and other social indicators such as education and health. The problem is that low education levels, scarce employment opportunities, and high population growth are part of a vicious cycle that is difficult to break. Factors such as agrarian structure, poverty, stagnant rural incomes, and religious and cultural beliefs are believed to be important determinants of a family's demand for children. In SSA countries, where production systems tend to be labor intensive rather than capital intensive, there is a strong incentive for large families because the ability to increase cultivated area is positively correlated with family size. The importance of family size is highlighted by the fact that most of the food production activities are handled by women and children. In this system, a man can significantly increase his income by having several wives and many children. He also depends on a large family for old age security since he cannot mortgage or sell land to which he has only user rights. Until the time that an additional child is more expensive than the income and labor that it contributes, households will have few incentives to restrict family size. Therefore, to reduce population growth, policies to increase agricultural productivity and investments in market infrastructure to facilitate the function of markets are essential. If the adoption of new agricultural technologies is not supported, labor will remain the main input in production, and large families will be a key to survival. The outcome of this trend, as we have seen in low-income, food-deficit countries, will be slow or no growth in per capita food supplies, stagnant or deteriorating caloric intake, and declining nutritional status.

TREND IN GLOBAL FOOD AVAILABILITY

During the last century, growth in global food production surpassed population growth, leading to an overall improvement in per capita food consumption at the aggregate level. The main food production surge occurred during the 1950s and 1960s as new high-yielding crop varieties were adopted. Productivity growth allowed for the release of resources, labor in particular, to the rest of the economy and was responsible for a decline in the prices of major agricultural commodities. These factors led to a growth in urbanization, which resulted in an increase in availability of inexpensive labor, which in turn became the cornerstone of industrial development. The acceleration of industrial growth led to higher rates of urban migration and the decline in agriculture's share of the global economy. Industrial-led economic growth fueled food demand and improvement in diets.

The combination of income growth and declining real food prices not only resulted in increased food availability but also changed the composition of diets. Globally, daily per capita calorie availability increased by 17% between 1970 and 2005—from 2432 to 2852. In 1970, cereals accounted for more than half of calories consumed (Figure 22.1). Sugar was the next largest commodity group with a 9% share. Vegetable oils held a near 6% share of the diet. Meat accounted for 5.4% of the total. While cereals and sugar continued to account for nearly 60% of the global diet, there were some notable changes in composition that took place between 1970 and 2005. The most significant change was for vegetables, whose share nearly doubled but remains quite small at under 3%. The growth was supported by the expansion and improvement of the global transportation system, which facilitated trade in perishable products. The second highest growth was for meat, 80%, whose share exceeded 8% in 2005. The vegetable oil share of the global diet increased by 67%, reaching nearly 11% by 2005. In contrast to these increases, consumption of animal

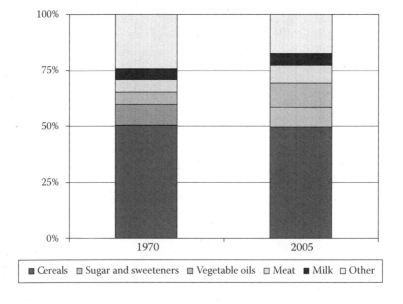

FIGURE 22.1 *A color version of this figure follows page 198.* Global diet composition.

fats, pulses, and root crops declined during this time period. Overall, globalization opened markets for products, and many farmers were able to capitalize on these changes by supplying a wide variety of products in growing and evolving markets.

The global increase in calorie availability has led to excess food consumption in many developed countries. Consumers in these countries, principally the United States, Canada, and the European Union, are struggling with the health consequences of this high level of food consumption. Calorie availability for people in these areas increased 9% between 1970 and 2005, reaching an average of more than 3400 calories per day (Figure 22.2). This level of availability far exceeds the U.S. Department of Agriculture (USDA) recommended range of 2000 to 2800 calories for a moderately active adult (USDA, 2005). The cereals' share of the developed country diet has changed little over time—remaining just below 40% (Figure 22.3). Meat now holds the second largest share of the diet at more than 12%, marking a more than 30% increase in share since 1970. Fruits increased at an even greater rate, but this category remains small with less than a 4% share of the developed country diet. Two of the bigger developments with respect to shares of these diets were actually decreases. The largest decrease was for animal fats, whose share fell more than 80% to roughly 1% in 2005. This decline clearly reflects the influence of research on adverse health effects, such as cardiovascular disease and obesity, associated with consumption of these fats. The other large decline, more than 20%, occurred in the share of sugar, which fell from 13.4% in 1970 to 10.4% in 2005. This decline is likely due to the increased use of sugar substitutes such as high-fructose corn syrup and artificial sweeteners such as saccharin and aspartame.

In developing countries, the key factors that influenced diets were income growth and urbanization. The rates of urbanization in developing countries were two to three times higher than their population growth. Urbanization usually coincided with increased incomes, which in turn influenced diets, often leading to increased

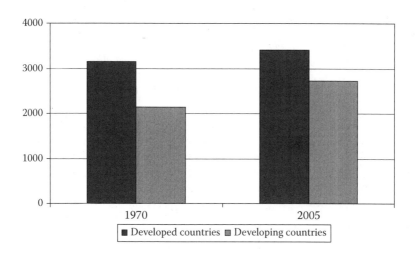

FIGURE 22.2 Calorie availability: developed versus developing countries.

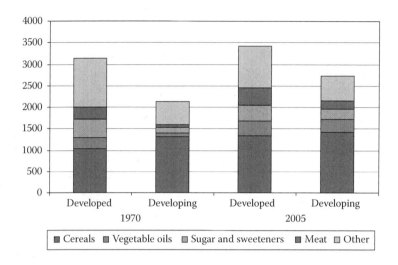

FIGURE 22.3 *A color version of this figure follows page 198.* Diet composition: developed versus developing countries, 1970 and 2005.

demand for higher-value farm products such as animal proteins, fruits, and vegetables. During 1970 to 2005, daily per capita calorie availability in developing countries increased by 27%, three times the rate of developed countries. Although cereals continued to dominate the diet of developing countries, the 8% rate of growth in cereal consumption was obviously much lower than the overall increase in calorie consumption of these countries.

In 2005, the cereals' share of the developing country diet was 52%, on average, compared to 63% in 1970. Within the cereal group, wheat consumption grew rapidly, 57%, reflecting the increased role of traded products in the diet; consumption of rice and millet declined. From 1970 to 2005, developing countries' per capita consumption of vegetable oils more than doubled (11% share of this diet in 2005), while meat, milk, and eggs increased by almost 2.6-fold (12% diet share). Sugar and sugar product consumption increased by about 60%.

It is important to note that the trends in food availability in developing countries are highly driven by the economic growth of Asian countries. Rapid income growth, especially in China and India, boosted food availability of developing countries as well as global statistics (measured by average calorie intake) to record levels in recent years. The growth in availability was for all basic food items: cereals, meat, vegetable oils, and so on. Average daily calorie availability by the developing countries of Asia climbed 33% since 1970–1972, including China, with growth of nearly 50%, and India at more than 20%.

The benefits of this global growth, however, were not distributed equally among all countries and regions. In Latin America, for example, a large increase in food availability in the 1970s had slowed by the 1980s, when calorie intake held steady at about the world average. SSA has by far the world's lowest intake levels, with per capita availability in the region averaging less than 2300 calories in 2005. Within this region, there are large variations. Eritrea, Burundi, and the Democratic Republic of Congo have an average per capita availability of less than 1700 calories, indicating severe food insecurity. Conversely, countries such as Nigeria and Ghana have surpassed 2800 calories.

ADEQUACY OF FOOD

As discussed, at the aggregate level, world food production grew faster than population over the past several decades. However, the available food has not been distributed evenly. At one extreme, there is the United States, where per capita food availability exceeded 3500 calories per day in 2005. At the other extreme is SSA, where per capita food availability averaged only 2300 calories. The paradox of the increase in global food availability and persistent hunger stems from income inequality both among and within countries. The gap in per capita income across countries has grown over time, and now there are concerns over widening income gaps in the most populous emerging market economies (i.e., China and India).

Hunger persists even in well-off countries, but in general, hunger is deep and most severe in countries where food supplies are low at the national level. According to FAO, in 2005 about 800 million people, or 20% of the population of developing countries, were undernourished (FAO, 2006a). In SSA, this share equaled 37%. Even in countries where per capita income is relatively high, poverty and hunger continue to be major problems, largely due to skewed income distribution and the resulting implications for food security for the lower-income groups. Latin American countries have the most skewed income distribution. On average, the highest income quintile (20% of the population) in this region holds nearly 57% of the national income, while the lowest-income group holds less than 5%. In Colombia and Guatemala, these numbers are even more distorted as the highest-income group holds more than 60% of total income, while the lowest-income group holds less than 1%. These resulting low-income levels severely constrain purchasing power of the lower-income groups and thereby hinder these populations from purchasing a nutritionally adequate food basket.

PERFORMANCE OF COMPONENTS OF FOOD CONSUMPTION: PRODUCTION, TRADE, AND FOOD AID

In almost all countries, domestic food production accounts for the bulk of food consumption, despite the growing role of food trade. This means that performance of domestic production is a key to improving food consumption. In SSA, for example, domestic production accounts for 90% of food availability on average. For lower-income developing countries, performance of the food and agriculture sectors is critical because, in addition to nutritional needs, a large share of their populations depends on agriculture for their livelihoods.

TRENDS IN FOOD PRODUCTION

The steep growth in food production during the last century was historically unprecedented. This in large part was achieved during 1950 through 1970 as a result of the introduction of high-yielding crop varieties and energy-intensive agriculture. This period is called the green revolution, a time when world grain production expanded by about 3.5% per year. This rate far surpassed the global population growth of 2.3% per year. The success of the green revolution lay primarily in its increased use of fertilizers, pesticides, irrigation, and improved seeds, all of which raised crop yields. Growth in food production has slowed since 1970, averaging about 2.3% per year in 1970 to 2006 (FAOSTAT, online database), but nevertheless remained positive on a per capita basis, 0.7% per year. The production performance varies by country and region: Food production growth was much higher in developing countries than developed countries at 3.6% versus 0.74% per year, respectively, during 1970 to 2006. The high population growth in developing countries limited the per capita gains to 1.63% per year; per capita growth in developed countries averaged 0.12% per year. Food production performance among the lowest-income group (50 countries according to the U.N. classification) was impressive, about 2.3% per year, but again, because of their high population growth rates, output on a per capita basis actually declined by about 0.3% per year. Since domestic food production accounts for the bulk of consumption for this group of countries as financial constraints limit import capacity, the hunger situation improved little in most of these countries.

Growth in grain production stems from three sources: expansion of arable land, increase in cropping intensity (i.e., multiple cropping), and growth in yields. During the last three decades, the growth in global grain yields has exceeded that of total grain production: 1.6% per year versus 1.3%. This has translated into a decline in the production area allocated to grains. This pattern was more pronounced in developing countries, where about 90% of the growth in grain production was due to growth in yield (2.3% and 2% growth per year, respectively). Of the lowest-income countries in the developing country group, the picture was reversed in that nearly 90% of the growth in grain production was achieved through area expansion, and only 10% was due to increases in yields. In SSA and the lower-income Latin American countries, expansion of arable land has been the force behind production growth. Grain yields per hectare in SSA are the lowest in the world, measuring about one third of world averages. Yields for corn—a staple crop for many countries in the region—have increased negligibly since the early 1970s and currently equal about a third of world levels (Figure 22.4). SSA's reliance on area expansion for growth in output is unsustainable because much of the land being brought into production at this point is of poor quality and limits the performance of yields. The low quality of land influences farmers' investment decisions, and the risks associated with these investments are magnified because of their limited financial position. Overall, low-income farmers encounter a vicious cycle in that they are faced with limited financial capacity to purchase inputs and new technologies that can improve increased yields, so they expand area by moving to marginal lands and thereby intensify the problem of land degradation.

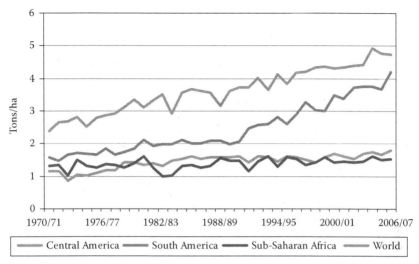

Source: USDA, FAS.

FIGURE 22.4   Corn yields.

A clear example of inadequate use of inputs is the case of fertilizer in SSA; the region accounts for only 1% of global consumption. In many low-income countries, particularly in SSA and Latin America, almost all fertilizer is imported, and insufficient foreign exchange constrains availability. In addition, fertilizer use is most productive on irrigated area or areas with sufficient moisture. Therefore, in regions such as these that suffer from or are vulnerable to dry periods, fertilizer use would not achieve the results that would be experienced in areas without similar adversities. Consequently, increased fertilizer use in these regions might be limited to irrigated areas or regions where rainfall is more predictable.

FAO reports that grain yields in developing countries are more than twice as high in irrigated areas as in rain-fed areas (FAO, 2004). Agriculture accounts for 70% of freshwater use worldwide and over 90% of withdrawals in low-income developing countries. Globally, irrigated area increased nearly 1.4% per year between 1980 and 2002, although the growth rate has declined over time. Growth in developing countries exceeded this rate, and currently more than a quarter of arable land area in developing countries is irrigated. It is estimated that about half of the grain production in developing countries is grown on irrigated land area.

The highest growth in irrigated area in the developing world has occurred in Asia, particularly Bangladesh, Nepal, and Vietnam. In East and Southeast Asia, more than 28% of arable land is irrigated. In the Latin American and Caribbean countries, nearly 13% of arable land is irrigated. Irrigation is severely limited in the most nutritionally vulnerable region, SSA. The region accounts for less than 2% of the world's irrigated area. Less than 4% of its arable land is irrigated. Irrigation requires investment in equipment and maintenance—all factors that are elusive in most of SSA. The World Bank and other international institutions have reduced their investments in major irrigation projects and are concentrating more on improving water management at the local level. Improved water management not only has a direct effect on crop growth but can also increase the efficiency of other inputs. As mentioned, the principal factor limiting yield response to fertilizer use is the inadequate supply of water during the growing season.

ROLE OF TRADE IN FOOD AVAILABILITY

Food imports have become an increasingly important component of diets in both developed and developing countries as food self-sufficiency has declined during the last few decades. The changes in self-sufficiency varied by country grouping; the higher-income developed countries became more dependent on imports of fruits and vegetables, while developing countries became more dependent on imports of staple commodities such as cereals and vegetable oils. From 1970 to 2003, import dependency grew the most among the least-developed countries (LDCs) as compared to other income groups. In 2003, LDCs relied on cereal imports for 17% of their consumption (compared to 8% in 1970), 45% for sugar and sweeteners (compared to 18% in 1970), and 55% for vegetable oils (compared to 9% in 1970).

The growing import dependency of these countries implies rapid growth both in quantity and in variety of traded food. Between 1970 and 2003, the highest growth occurred in the poultry market—imports grew 18-fold at the global level. This growth was fueled by developing countries, where imports were negligible in 1970 but soared to nearly 400,000 tons by 2003. Growth in developed countries was quite strong as well, jumping 13-fold. Strong growth also took place in the vegetable oils market, where imports increased more than eightfold. Imports of meat, fruits, and vegetables grew in the range of four- to fivefold.

The high-income developed countries continue to be the largest importers of food, although their market share has declined through time. In 1970, the share of the import market held by developed countries ranged from nearly two thirds for grains to more than 90% for fruits and meat. While the developed countries' share of imports of fruits and vegetables remained quite high by 2003, their share of grain and vegetable oil imports declined to about 40%. Income and consumption growth in developing countries, in particular in populous countries such as China and India, are in part the factors behind the growing import share held by the developing countries. The growth in food trade also was stimulated by trade liberalization policies adopted by many of these countries. And, while these policies have not been fully implemented, they have led to a significant decline in import tariffs and other trade barriers both in developed and developing countries.

ROLE OF FOOD AID

While global trade has increased rapidly during the last few decades, not all countries were able to import adequate food to meet their needs. Some of the lowest-income countries with low levels of nutritional intake have low import dependency simply because they cannot afford to import. Food aid has been used as an international instrument to reduce world hunger. It was first provided to developing countries in the 1950s when the United States came under pressure to dispose of grain surpluses. For producers and exporters, food aid became a desirable policy choice when stockpiled commodities lost most of their market value. Compared to the decades of the 1960s and 1970s, however, world food supplies have recently tightened, and therefore most commodities are no longer treated as if they are free goods.

The global quantity of food aid has fluctuated during the last two decades, and its share has declined relative to both total exports of the food aid suppliers and total food imports of low-income countries. Despite this trend, it has been a common practice for the world's major producers to supply food aid. All donors cite humanitarian relief as their basic food aid distribution criterion, but economic and political considerations have also played an important role in allocation decisions. The commodity mix of food aid usually reflects the export profile of the donor country and tends to vary with yearly fluctuations in availability. The United States, the European Union, Canada, Japan, and Australia are the major food aid donors. The United States is the principal contributor, providing more than half of global food aid. The notable success of food aid has been in dealing with food emergencies. For example, during the first stages of Somalia's civil strife in the early 1990s, food aid contributed to about 70% of their consumption. During Rwanda's civil strife of the mid-1990s, food aid contributed to roughly a third of food consumption.

FOOD AVAILABILITY OULOOK IN THE CONTEXT OF EMERGING ISSUES

One of the key variables determining the outlook for food availability, the growth rate of world population, is on a slowing growth path. The U.N. population assessment report projects a slowdown in population growth, from 1.7% per year during 1970–2000 to about 1% through 2030 and to 0.5% per year by the middle of the century (UNFPA, 2007). Despite these changes, the absolute increase in world population is not small as it is projected to rise about 50 million people per year during the next three decades, followed by a 30- to 40-million increase by 2050. The concern lies in the fact that more than 90% of the population growth will be in developing countries. Among regions, population growth in SSA is expected to be the highest, accounting for about half of the world population growth by 2050. Conversely, population in East Asia is projected to stagnate and then decline about 0.2% per year during 2045 to 2050. During the same period, SSA will account for 18 million of the 26 million people added annually to global population.

By 2050, population in the Democratic Republic of Congo is expected to increase fivefold and exceed 200 million. Population growth could nearly match that in Niger, which may increase its population fourfold to more than 50 million. In countries such as Burkina Faso, Mali, Chad, Malawi, Somalia, and Zambia, populations are projected to double by 2050. In 1950, populations in these countries were in the range of 2 to 5 million compared to 2050 projections that are in the range of 30 to 100 million people. Ethiopia, one the most vulnerable countries to food insecurity, had a population of 18 million in 1950, but it is projected to swell to 186 million by 2050, a near tenfold increase. The concern is that most of the countries with high population growth rates also have chronic food insecurity issues and are among the LDCs in terms of income. In sum, during the next century, global population growth will persist in countries with limited resources to accommodate the growth, thereby possibly leading to an increase in food-insecure people. Historically, high population growth was closely associated with higher demand for food.

NUTRITION OUTLOOK

According to most studies (Rosegrant et al. 2001, FAO 2006a,b; USDA 2007), improvements in global per capita food availability are expected to continue, but gains will be slower than historical rates. According to FAO, per capita food availability at the global level is projected to average 3130 calories per person per day by 2050. Intake in developed countries is projected to increase only slightly, while in developing countries availability is projected to increase from 2700 calories per person per day in 2005 to more than 3000 calories by 2050 (Table 22.1). However, there is considerable variation among regions. In Latin America, the Caribbean, and East Asia, calorie consumption is projected at around 3200 calories in 2050. The percentage of population that is undernourished in those regions is projected to fall under 3%. In contrast, calorie consumption in SSA is projected to remain the lowest in the world at 2830 calories despite rising at the fastest rate—17% between 2015 and 2050. As a result, nearly 6% of the region's population will remain undernourished.

Although there will be adequate food to keep up with demand at the global level, millions will continue to go hungry because of national unequal distribution of income that limits food purchasing power. According to World Bank projections, with the exception of East Asia, per capita income growth in all regions will be much higher during 2005 to 2015 than during the 1990s. East Asia, despite projections of slower income growth, will have the highest annual per capita income growth of all regions, 5.2% per year (World Bank, 2000). South Asia, with deep poverty levels, is expected to make a major leap and is projected to have the second-highest annual per capita income growth in the upcoming decade, 4.7% per year. Growth in SSA, while expected to improve from its historical declining trend, is projected to be the lowest

TABLE 22.1
Developing Countries: Projections of Per Capita Food Consumption and Undernourished Population, 2050

Region	Per Capita Consumption (kcal/person/day)	Population Undernourished (%)
Developing countries	3070	3.9
Sub-Saharan Africa	2830	5.8
Latin America and Caribbean	3200	2.6
South Asia	2980	4.1
East Asia	3230	2.9

Source: United Nations FAO World Agriculture: Towards 2030/2050 (Interim Report, June 2006).

in the world, about 1.7% per year until 2015. The number of people in the region living on less than a $1 per day rose from 227 million to 303 million between 1990 and 2002, the largest increase in the world (World Bank Development data, 2007). This trend is projected to continue but at a slower rate. Moreover, the region's absolute poverty number is projected to surpass that of South Asia, the region that has always had the largest number of poor people. In SSA, nearly 40% of the region's population is projected to fall below the poverty line by 2015. The region's poverty, in addition to low level of per capita income, is largely due to the region's highly skewed income distribution. Regionally, after Latin America, SSA has the highest income inequality in the world. The poorest 20% of the region's population hold less than 6% of total income, while the wealthiest 20% hold over 50%.

USDA Economic Research Service (ERS) medium-term (2016) projections of the number of hungry people and food gaps show a similar path to the World Bank poverty projections, which indicate improvements in Asia but some deterioration in SSA (USDA, ERS, 2007). These results indicate that the food security situation is projected to remain precarious or even worsen in 17 of the 37 SSA countries. In more than half of these countries, population growth is projected to be among the highest in the world, ranging from a low of 2.9% per year in the Democratic Republic of the Congo and Mauritania to a high of 3.8% in Somalia. These countries simply do not have the resources to raise food supplies at rates adequate to compensate for the large increases in population. In addition, the harsh climate in many countries leads to frequent drought, which is a chronic problem with respect to domestic production in many countries. Production shortfalls in SSA are problematic because most countries have limited financial capacity to import food. Exacerbating the situation is civil strife and political instability, which disrupt agricultural activities and discourage investment. Civil unrest and border conflicts in countries such as the Democratic Republic of Congo, Sudan, Somalia, and Ethiopia for the last several decades have no clear end in sight.

An important cross-country issue is the growing obesity epidemic stemming from increasing consumption of high energy-dense foods such as oils and sugar (Popkin, 2001). Historically, this issue was limited to developed countries, but it has now expanded to developing countries, including the LDCs. This trend is disturbing because obesity-related health issues such as heart disease and diabetes will put additional financial pressure on already-taxed health care systems in these countries. In developing countries, annual per capita health expenditures are only 5% of those in developed countries, and in the LDCs it amounts to less than 1% (World Bank Development data, 2007).

FOOD PRODUCTION OUTLOOK

Growth in food demand will be slow due to lower population growth rates as well as the fact that diets in several highly populated emerging markets such as China and Brazil have transitioned to the point at which their food consumption is now similar to that of developed countries. While per capita income is projected to grow, higher food prices may offset that potential impact. By all accounts, the growing interest in biofuels is expected to result in higher staple food prices, representing a departure from the historical trend (Shapouri and Rosen, 2007). It is too early to project the long-term influence of the biofuel push on food prices, but the impact of higher food prices on consumption in the higher-income, mature markets will be limited. The food share of total expenditures in these households is very low; for example, in the United States this share is less than 10%. However, for the LDCs, where more than half of household income is spent on food, growing food insecurity due to higher food prices is an increasing concern. Higher food prices would benefit countries or farmers with adequate agricultural resources where production and exports could be expanded, but would hurt those who import food and benefit from low food prices. According to the USDA-ERS medium-term projections (2006 to 2016), the rapid expansion in global production of biofuels will change the price relationships among agricultural commodities (USDA, 2007). Increased demand for corn (ethanol) and soybeans (biodiesel) will influence prices of other grain and vegetable oil crops because of crop area substitution or their feed value. According to USDA projections, the 2007 increase in grain prices will be followed by a declining trend such that less than one third of the price spike of 2006–2007 is retained by 2016. As for LDCs, high food prices could spur domestic production and thereby reduce hunger because domestic production accounts for most of the food consumed in these countries. However, the net result depends on the magnitude of the supply response to the price increase and the supporting economic policies in the areas of technology adoption and other services to improve the functioning of markets.

World food production is projected to grow during 2030 to 2050, but at a lower rate than the last three decades. Most of the historical gains in global production of grains (the key component of the global diet) were due to yield growth, and this is expected to continue in the future (FAO, 2006b). Higher yield growth is necessary to maintain production growth in the future due to constraints on expanding cropland. The expansion in production of oil crops, which was much higher than grains in the past, also will face constraints on land availability. These constraints exist despite the potential of large unused land area in SSA and Latin America. The reason for this is that most of the land in these regions has low soil fertility, has high soil toxicity, or is in hilly areas. Using information on soils, climate, and land cover, researchers at USDA's ERS compared the quality of cropland by country and region. They found that the quality of cropland is lowest in SSA. In 12 of 38 SSA countries studied, less than 1% of cropland was classified in the top three land-quality classes, and the median share of cropland that was classified in the top three land-quality classes in SSA countries was about 6%. This is quite low compared with a median of 16% in Asia (where 7 of 17 countries studied have more than a quarter of their land in the top three classes) and 27% in Latin America (where 12 of 19 countries studied have more than a quarter of their land in the top three classes). In contrast, the median share of high-quality cropland was 29% in the high-income countries (where 13 of 22 countries studied had more than a quarter of their land in the top three classes). Expanding crop production to where the quality of land is low leads to soil mining and declining yields. The disparity in land quality is reflected in the differences in yields—in absolute level as well as growth—both historically and those likely in the future. In 2004, grain yields in developed countries were six times higher than those of the LDCs and more than two times higher than those of the developing countries.

Grain yields doubled in both developed and developing countries during 1970 to 2004, but in the LDCs, many in SSA, the growth was small, only 16%. The principal constraint to improving crop yields in the LDCs is the limited access to essential production inputs, such as water. Agriculture is by far the biggest user of water, accounting for 70% of global water withdrawals. Growing demand for water for irrigation as well as industrial and domestic uses is placing increasing pressure on prices. In fact, prices are expected to exceed the level that is profitable for staple food production in most countries. SSA's irrigated area as a share of total agricultural area stagnated during the 1990s and measures only about 3% of total crop production land. In Latin America, this share exceeded 11%, and in Asia it was 20%.

To increase yields, farmers must switch from traditional seeds to high-yielding varieties. However, most of the new high-yielding varieties are grown in the irrigated land areas and require agrochemicals such as fertilizer and disease-preventing pesticides. Currently, most of the producers in Africa are too poor to purchase agrochemicals. Africa's share of global pesticide consumption is very low, about 1 kg/ha per year compared to high-end users such as Japan and Holland at 21 kg. Many countries with high levels of fertilizer use are experiencing the environmental problems associated with intensive use of fertilizers. On the opposite end of the spectrum is SSA, where the crops are mining the soil of its nutrients because they are not being replaced with plant residues, manures, or fertilizers. This means potential to increase yields for staple crops consumed by the poor is significant. The recent increase in food prices could provide an opportunity for the countries to increase their food production. Unfortunately, in most of these countries, agricultural supply response to higher prices is low because of inadequate access to information, high costs of inefficient marketing systems, and limited capital for investment.

Food Trade Outlook

Food imports are expected to grow with globalization and expansion of trade agreements among countries. During the last three decades, trade in staple food such as grains, vegetable oils, and meat increased by three- to fivefold. The growth in imports was not limited to staple food as it expanded to a variety of commodities, including semiprocessed and processed foods. Structural factors, such as income growth and urbanization, were behind these trends, and they are expected to continue during the next century. Increased global access to the communication infrastructure of the twentieth century, along with global trade liberalization, were responsible for the rapid transition of developing country diets. The changes in diets that evolved over more than century in Western countries have taken place in only a few decades in developing countries. The increase in imports of processed foods at the global level— 4.5-fold—during the last three decades is a good indicator of such an evolution. In the LDCs, imports of commodities such as concentrated fruit juice, frozen vegetables, and canned meat that were either not imported or imported in negligible quantities through the 1990s increased by more than 10- to 20-fold. Among the imported processed foods, a continuation of the high import growth of hydrogenated fats (animal and vegetable) is of concern because it has the potential of raising blood cholesterol, which is a risk factor in the development of heart disease (Popkin, 2001).

Developing countries, on average, can afford the growing costs of food imports, but the future food import capacity of the LDCs is less certain due to their financial limitations. To assist many of these countries, food aid is expected to continue its historical role of augmenting food supplies. However, the global quantity of food aid has fluctuated during the last two decades, and its share has declined relative to total food imports of low-income countries. The share of food aid in total grain imports was around 20% in the early 1990s but declined to about 6% in 2005. This means the fundamental forces that could influence the future of global hunger will be domestic food production and trade.

In sum, trade is expected to continue to play an increasingly important role in global food availability. One important factor influencing the future role of trade is trade negotiations and the subsequent potential for further reductions in trade barriers. Reduced market intervention has reduced food import costs and expanded the role of trade in terms of both quantity and variety. The other factor that could lead to an increase in global food trade is the continued growth in urbanization, which increased by about 35% during 1970 to 2005. This growth was most pronounced, a 63% increase, in developing countries. If this high rate of growth continues, more than two thirds of the developing world's population will live in urban areas by 2030. Urban populations tend to consume a greater variety of foods than those in rural areas because of their higher incomes and the fact that they are exposed to more information and advertising.

CONCLUSIONS

The slowdown in global population growth will dampen growth in food demand, thereby reducing pressures on resource needs, particularly land use. Per capita food consumption at the global level is projected to increase moderately until the middle of the twenty-first century. Consumption of higher-valued commodities such as meat, dairy products, and vegetable oils will increase, while consumption of grains will likely stagnate or even decline. Globalization and the increase in access to information will lead to increased diet variation and the expansion of the role of trade in the global food system. Trade in processed foods in particular is expected to remain strong, but the deepening of capital market liberalization and higher transportation costs could result in increased investment in food processing in populated countries such as India.

Improvements in diets will not be uniform across all countries or income groups. Although the food situation in developing countries (on average) is projected to improve, for the least-developing countries, the improvement will be significantly slower. The problems in these countries are multidimensional: high population growth, deep poverty, and poor agricultural resources, which constrain expansion of food production. The implication for many of these countries is that they may not be able to break the path of food insecurity for a significant segment of their population for some time to come.

For developed and developing countries alike, the issue of obesity will remain serious and costly. The reason is that, contrary to expectations, increased food availability and variety have not resulted in improvements in the quality of the diet.

Increased consumption of foods that are energy dense, high in fat and refined carbohydrates, and low in fiber is altering the traditional diet in many countries. While the sedentary urban lifestyle requires much less energy-dense foods, the global increase in calorie availability has led to excess food consumption, the adverse health effects of which have only recently become apparent in developed countries. The obesity problem is expected to continue, and its associated health costs could be overwhelming for developing countries. However, public nutritional education efforts, particularly in schools, can put countries on a path to slow the obesity epidemic.

The worldwide interest in crop biomass for biofuels will affect countries differently. Those countries with abundant agricultural resources and capital will certainly benefit. For example, many countries in Latin America will benefit from the higher demand for biofuels and higher agricultural prices; these factors could work to reduce food insecurity and poverty in some of these countries, which are among the poorest in the world. There is also concern over the deepening of food insecurity because of higher food prices and transportation costs in areas with low incomes with poor agricultural resources. It is not clear in what way or for how long this new industry will influence the future food market, but given the growing interest level, the issue surely requires further monitoring and analysis.

STUDY TOPICS

Despite slowing considerably through time, global population growth remains around 1% per year; this means that nearly 57 million people are added to the total population each year.

During the last century, growth in global food production surpassed population growth, leading to gains in per capita consumption.

Daily per capita calorie availability, at the global level, increased 17% between 1970 and 2005.

While cereals and sugar continue to account for the largest share of the global diet, there have been some marked shifts over time; the meat share of the diet rose 80% from 1970 to 2005, while the vegetable oil share rose 67% during the same time period.

The global increase in calorie availability has led to excess food consumption in many developed countries, which has had adverse health consequences.

Urbanization and higher incomes resulted in an increase in daily per capita calorie intake in developing countries between 1970 and 2005 that was three times the rate of developed countries.

Between 1970 and 2005, the cereals' share of the developing country diet fell; at the same time, consumption of vegetable oils in these countries doubled, while that of meat, milk, and eggs jumped almost 2.6-fold.

Domestic production accounts for the bulk of food consumption, meaning performance of production is key to improving food consumption; growth in production stems from expansion of arable land, increase in cropping intensity, and growth in yields.

While domestic production is important, imports have played an increasingly important role in the diet in both developed and developing countries.

Between 1970 and 2003, the highest import growth occurred in the poultry market, for which imports grew 18-fold at the global level.

Improvements in global per capita food availability are projected to continue, although at a slower rate than the historical period; growth in intake in developing countries is expected to exceed that of developed countries.

REFERENCES

Food and Agriculture Organization of the United Nations (FAO). FAOSTAT Statistical database, Rome.

Food and Agriculture Organization of the United Nations (FAO). *World Agriculture: Towards 2015/2030*. Rome: FAO, 2004.

Food and Agriculture Organization of the United Nations (FAO). *State of Food Insecurity*. Rome: FAO, 2006a.

Food and Agriculture Organization of the United Nations (FAO). *World Agriculture: Towards 2030/2050 (Interim Report)*. Rome: FAO, June 2006b.

Food Security Assessment Report. GFA-18. USDA, Economic Research Service, Washington, DC. June 2007.

Popkin, B. The nutritional transition and obesity in the developing world. *Journal of Nutrition*, 131:871S–873S, 2001.

Rosegrant, M., M. Paisner, and S. Witcover. *2020 Global Food Outlook: Trends, Alternatives, and Choices*. Washington, DC: International Food Policy Research Institute, August 2001.

Shapouri, S. and S. Rosen. *Energy Price Implications for Food Security in Developing Countries*. GFA-18. USDA, Economic Research Service, June 2007.

World Bank. *Global Poverty Report*, World Bank, Washington DC, 2000.

World Bank Development Data. Washington, DC: World Bank, December 2007.

United Nations *World Population Prospects: The 2006 Revision*. New York, 2007. *Population Statistics*. New York: UNFPA, 2007.

U.S. Department of Agriculture. *USDA Dietary Guidelines for Americans*. Washington, DC: USDA, 2005.

U.S. Department of Agriculture. *USDA Agricultural Projections to 2016*. Washington, DC: USDA, February 2007.

U.S. Department of Agriculture. Production, Supply and Distribution database, Washington DC.

23 Changing Food Supply, Demand, and Marketing Issues:
What Affects Price and Affordability?

Miguel I. Gómez, Charles F. Nicholson, and Paul E. McNamara

CONTENTS

ABSTRACT

Increasing food production and declining prices are a necessary, but not a sufficient condition to ensure food availability and affordability for all. The primary reason is that complex interactions between farmers, food distribution systems, and consumers influence availability and affordability of food. A systems perspective is useful to understand and address factors affecting global food supply and demand. Using this perspective, we examine the main trends and issues in food production, distribution, and consumption. Although food production has increased faster than population in the past, there is no consensus among researchers regarding future trends. Recent developments such as global warming, HIV/AIDS, wars, and biofuels may have substantial impacts on food availability and affordability in the future. The ability of the food-marketing system to make food available and affordable depends on the level of value added desired by consumers and by the costs of carrying out multiple marketing activities. Recent food marketing trends that are contributing to better performance of the food-marketing system include specialization in marketing activities, adoption of information technologies, economies of scale in food distribution, improved infrastructure, and better segmentation of the end consumer. In addition, factors shaping food demand that will influence food adequacy in the future include the pace of population growth, the distribution of the population across age groups, income level, income distribution, and geographic location (e.g., urban, rural). Three cases illustrate the complexity of relationships involved in the global food supply chain. The first case discusses the demand-driven livestock revolution and its links to changing diets, the second case explores critical links between the rising biofuels sector and food prices, and the third case describes the impact of modern food retailers on trade local supply chains in developing countries. We conclude with a discussion of policy alternatives to ensure adequate food for all.

INTRODUCTION

The last 30 years have seen both increased production of calories per capita for the world as a whole (Gilland, 2002) and declining real prices (that is, adjusted for inflation) for grains such as rice, wheat, and corn (Pinstrup-Andersen et al., 1999). These observations have been taken as an indicator that food production at a global level is adequate for human needs. Despite these positive trends, many millions of people continue to suffer from undernourishment of "macronutrients" such as energy or protein or micronutrients like iron, vitamin A, or iodine. This suggests that neither total production of calories nor declining prices are sufficient to eliminate undernutrition. In essence, it is both the *availability* of food and its *affordability* by all households that are required to achieve adequate food for all. Moreover, past trends may not accurately indicate our planet's future food production. Concerns about the limits to food production and predictions of widespread nutritional crises have been expressed at least since the time of Thomas Malthus two centuries ago and continue to be commonly heard in the early twenty-first century. New challenges continue to emerge with the potential to influence the availability and affordability of food, making the future of global food supplies and their prices uncertain.

The objective of this chapter is to review the basic factors that influence the availability and affordability of food at a global level. A central idea is that food availability and affordability result from complex interactions between the supply chain for food (food production, processing, transporting, and selling) and the demand for food (which is influenced by a variety of factors, including income, urbanization, and policies). We introduce the concept of a dynamic food system, discuss recent trends in food production, marketing, and consumption and indicate likely future challenges. Case studies on the rapid growth in demand for livestock products, the impacts of growing biofuels demand and the modernization of food retailing in the developing countries illustrate the integration of production, marketing, and consumption in specific settings. Because government policies are often employed to improve food- and nutrition-related outcomes in both developed and developing countries, we also review basic policy approaches as well as their likely impacts.

FOOD SYSTEM CONCEPTUAL FRAMEWORK

Food systems are complex, comprising many actors and activities in multiple locations. In addition, food systems are dynamic, in constant evolution through the decisions of consumers in their local supermarket, farmers (or governments representing them) thousands of miles away, and (usually) multiple food processors, wholesalers, and retailers in between. This complexity and dynamism make it challenging to understand how food systems function and to intervene effectively when this is considered necessary to achieve desired outcomes (such as adequate food for all). The purpose of this section is to introduce overall concepts involved with food supply and demand in both developed and developing countries and to provide an organizing framework for the rest of the chapter.

To address the complexity of food systems, an explicit systems perspective can be useful. A *system* can be defined as "any set of interrelated elements" (Meadows and Robinson, 1985). A system consists of two essential components: *elements* (visible or measurable objects of flows) and *relationships* (connections postulated to exist among the elements). For a food system, the essential elements include the different actors, activities, operating environment, and information flows. The relationships among these elements are complex but can be represented in a useful manner by a diagram depicting the elements and the hypothesized relationships[1] among them (Figure 23.1). In the diagrammatic representation of the system, text indicates elements, and the connecting arrows indicate postulated relationships. The arrows indicate hypothesized causality, and the plus or minus signs indicate the likely direction of influence. For example, the arrow beginning at "Household Income" and pointing to "At Home Food Consumption" with a plus sign indicates that an increase in household income would result in (cause) an increase in at-home food consumption, all other things being equal.[2]

Although the diagram is visually complex, it is important to remember that the real-world system is even more complex. The diagram illustrates a number of important concepts. First, the multiple actors directly involved in the food system include crop and livestock producers, food processors, food marketers, and consumers. Each of these actors responds to a variety of factors, particularly prices (and income for

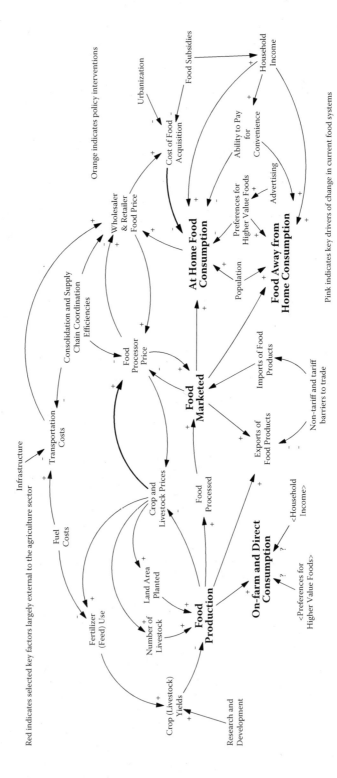

FIGURE 23.1 *A color version of this figure follows page 198.* Conceptual diagram of a generic food system, indicating selected drivers, external factors, and interventions.

consumers). Each actor provides a service or function that must be compensated. The food system involves physical (i.e., physical movement and processing of food products), economic (e.g., prices), and information (e.g., consumer preferences) elements that interact to determine overall outcomes. The actors are linked through the outcomes of their decisions, so that developments or interventions designed to affect one (or more) actors often have effects throughout the system.[3]

Figure 23.1 also indicates the need to distinguish between different food-marketing channels and types of food consumption. These include direct from farmer, retail food outlets, and food consumed away from home. The importance of these market channels and types of food consumption varies between developed and developing countries. Retail food and food away-from-home consumption are more important in the higher-income countries and on-farm and direct consumption relatively more important in the developing world. In addition, factors often considered "external" to the food sector, such as fuel costs or infrastructure investments, can have significant impacts on food consumption and the food supply chain. In what follows, we distinguish between the food supply chain (food production and food marketing) and consumer demand. It is important to recognize that although this separation facilitates the discussion of the factors influencing their ongoing evolution, each of these components influences the others.

PRIMARY AGRICULTURAL PRODUCTION: TRENDS AND ISSUES

Food production has been used as a central indicator of global food adequacy for many years. Many international organizations concerned with food and nutrition use a calculation such as "global calories produced divided by population" to indicate whether global food production is adequate (e.g., Pinstrup-Andersen et al., 1999), although this number does not indicate much at all about the number of people with adequate food access. However, food production is a necessary if not sufficient condition for food adequacy. Although food production systems are both diverse and complex, a simple algebraic equation can be used to help focus on key production issues:

Food Production per Year = Land Area × Yield per Crop × Crops per Year   (23.1)

This equation indicates that for food production to increase, either the land area planted (and harvested) or the intensity of land use through increased yields or more crops per year must increase.[4] A similar equation can be specified for livestock production based on animal numbers and the productivity per animal (in producing either other animals or products like milk). Global land areas planted in grain have actually declined during the past two decades due to the breakup of the former Soviet Union (but have been roughly constant in the rest of the world). Crops per year are often limited by climate and thus are more difficult to modify. The largest impacts on food production in the past have come from increased yields. Yields for grains[5] have increased continuously for the past 40 years, increasing at nearly 3% per year during the time of the green revolution in the 1960s and 1970s, but more slowly (1 to 2% per year) since then. Although this slowing of yield increases has been a cause for concern, the Food and Agriculture Organization of the United Nations (FAO, 2002) and Gilland (2002) attributed these more to low prices (and thus incentives for using yield-increasing technologies like fertilizer) than inherent technological limits. These yield improvements have resulted from the increased use of fertilizers, expanded irrigation, biotech crops, and in some cases, retirement of less-productive land.

Although the recent past suggests that world food production has more than kept pace with population growth in recent times, the key question is whether that can continue in the future. This has been the topic of numerous studies in recent years; viewpoints range from very pessimistic (Meadows et al., 2004), to uncertain (Döös, 2002), to cautiously optimistic (FAO, 2002; von Braun et al., 2005). Meadows et al. used a systems simulation model to suggest that the world economy as a whole will undergo a major collapse during this century, and the food production will also be markedly affected. Döös reached the conclusion that practically none of the factors having an important influence on food production can be specified with a high degree of accuracy due to insufficient knowledge of processes (e.g., soil degradation) and insufficient data. The last two studies also used simulation modeling with different assumptions to assess future world food supplies. In contrast, they suggested that—globally, at least—land areas will grow during the next couple of decades, that land areas will be adequate, that yields will continue to increase as fertilizer use increases (and will continue to be the dominant factor in increasing production), and that water will be adequate. As a result, food prices are not expected to increase rapidly over the next two decades. Both studies noted, however, that much of their projected growth in food production is dependent on a variety of factors, including further development of environmentally friendly production technologies and appropriate policies (especially for trade in agricultural products), and that regions will be affected differently.

In addition, a number of other recent developments have been identified as having the potential to markedly affect global food production (Table 23.1). These include diverse effects, most of which arise outside the agricultural sector itself. These issues and others yet unforeseen may have significant impacts on future food production. As a result, the most certain conclusion that can be reached is that the food production situation needs to be continuously monitored, and efforts at research and development for new food production technologies—particularly those that combine with environmental protection or restoration—need to continue. As Gilland (2002) put it, there is a need not only to continue to increase yields but also to slow population growth, but at present "technological optimism and ecological pessimism are both misguided."

FOOD MARKETING AND DISTRIBUTION: TRENDS AND ISSUES

A food supply chain consists of all parties and their supplied activities that help firms create and deliver food products to the end consumer. The primary task of marketers and distributors in this chain is to take food products from the farm gate and make them available and affordable to the end consumer. As Figure 23.1 illustrates, this is a complex process that involves both a large number of firms and multiple activities, including transportation, storage, processing, packing, selling, and financing. Several studies pointed out that although there may be enough food to feed the world's population, the food-marketing system may hinder food availability and

TABLE 23.1

Future Issues in Global Food Production

Issue	Discussion or Impact
Global warming	Changes in temperature and precipitation patterns as well as the CO_2 fertilization effect will probably increase food production in developed countries and decrease it in developing countries (FAO, 2002).
HIV/AIDS	Food production can be reduced due to loss of income to purchase inputs, the need to sell farm assets to pay for treatment, and loss of skills and knowledge when household members die (von Braun et al., 2005).
War and conflict	Conflicts result in reduced agricultural productivity; estimated losses during 1970–1997 totaled more than $121 billion (von Braun et al., 2005).
Biofuels and energy prices	Increases in demand for biofuels may divert land and crops from food uses to energy uses (see the section on bioenergy, agriculture, and global food prices).
Environmental degradation	Soil degradation and deforestation are concentrated in resource-poor areas where growth in yields has been slower than average (von Braun et al., 2005). "Agroecological" approaches to food production are under development (Pinstrup-Andersen et al., 1999).

affordability (FAO, 2004). For instance, poor infrastructure for food storage and distribution may make food extremely expensive to some consumers, or the level of value added provided by the food supply chain may be too low (or high) relative to the needs of consumers.

According to Kotler and Keller (2006), there are multiple benefits offered by the global food supply chain. First, it generates cost savings due to specialization in marketing activities and reduces exchange time. Consider what would happen if a grocery store received direct shipments from every manufacturer that sells products in the store. This delivery system would be chaotic as hundreds of trucks line up each day to make deliveries, many of which would consist of only a few boxes. Instead, a better distribution scheme would have the grocery store purchasing its supplies from a grocery wholesaler that has its own warehouse for handling simultaneous shipments from a large number of suppliers. The wholesaler will distribute to the store in the quantities the store needs and on a schedule that works for the store. Other benefits offered by members of food-marketing systems include ability to understand customer needs regarding assortment and convenience, to perform an active selling role using persuasive techniques, and to provide information to the primary sector regarding consumer preferences. Moreover, the food-marketing system plays an important role in bulk breaking. For example, it is common in "mom-and-pop" stores in less-developed countries that edible oil is sold by the spoon. This provides an important benefit to low-income consumers whose daily purchasing power does not allow them to buy larger quantities, even if they have to pay (per unit of volume) higher prices buying "by the spoon."

An ideal food-marketing system is one that efficiently meets the needs of the end food consumer. An executive working in a large multinational in New York has different demands than a floating sundries distributor in Vietnam's Ke Sat River. The global food-marketing system must meet the needs of both. Thus, food marketing is as important as food production to ensure food affordability. Similar to food production, a simple equation can help explain relevant food-marketing issues:

Profits of the Food Marketing System = (Quantity Sold × Unit Price) − Cost   (23.2)

This equation indicates that the performance of the food-marketing system depends on the price that consumers are willing to pay for food, the costs of bringing food from production to consumption sites, and the quantity demanded. This section focuses on trends and issues regarding the first two items, while the next section explores the last. There are two primary ways in which a private food-marketing system can improve its performance. The supply chain can add value to the product for consumers who are willing to pay higher prices, or it can reduce the costs incurred in the transport, storage, processing, and distribution of food. Global food supply chains continually review prices along the chain and the costs of bringing food from the farm gate to the consumer's table. We consider these two issues in turn.

Food Prices and Value-Added Activities

Some consumers are willing and are able to pay for higher food prices. Consider, for example the desire of time-starved consumers in the United States for such convenience foods as precut salads, ready-to-eat entrees, and meals away from home. The U.S. Department of Agriculture (USDA) tracks annually the costs of processing and distributing food or the "marketing bill," which is calculated as the difference between what consumers spend on food each year and what farmers were paid. Consider the 2002 marketing bill (Figure 23.2). In that year, total U.S. consumer spending on food produced, processed, and distributed domestically totaled roughly $700 billion. In that year, only 19 cents of every dollar spent went to the farmer, while the rest (81 cents) covered the cost of transforming products at the farm gate into food products and getting them to supermarkets and restaurants (Elitzak, 2004).

FIGURE 23.2   The 2002 marketing bill. (From H. Elitzak, 2004, *Amber Waves* 2:43.)

The figure indicates that labor and packaging are the largest component of the marketing bill. The size of the marketing bill is affected by changes in the amount and type of products consumers buy. For example, restaurant meals have more marketing costs associated with them and are therefore more expensive than foods at grocery stores. So, as consumers spend more at restaurants, the marketing bill increases in value. Similarly, as consumers purchase more highly processed food products, such as microwave-ready dinners, relative to less-processed fruits, vegetables, and meats, the value of the marketing bill increases. Over the last two decades, the marketing bill has increasingly taken a larger share of the consumer food dollar, growing from 73 to 81% of consumer food spending from 1982 to 2002 (Elitzak, 2004).

The production and distribution of highly value-added foods as illustrated in the marketing bill have important consequences for the economy as the food supply chain tends to be labor intensive in comparison to other industries.

COSTS IN THE FOOD SUPPLY CHAIN

The second important variable for global food supply chains regards the costs of bringing food from the farm gate to the end consumer. In particular, one needs to consider the costs incurred by all firms involved in the global food supply chain, including transporters, assemblers, processors, distributors, and resellers. In some cases, the emphasis of supply chains is on reducing costs to make food affordable to certain segments of consumers. For instance, food firms are finding great market opportunities in targeting low-income consumers in emerging economies (Prahalad and Hammond, 2002; Prahalad and Lieberthal, 2003). With sales growth harder to come by in a competitive world, food companies are seeking expansion among low- and middle-income classes in developing countries. The food supply chain is responding with the development of low value-added products that meet the needs of such large populations at affordable prices, directed primarily to emerging middle classes in urban areas that are constrained by income levels. Recent trends that are contributing to decrease the costs of food distribution include

- *Reduction in the price of inputs employed in food distribution.* Under competitive market conditions, reduction in input prices are transmitted to the end consumer in the form of lower food prices. International transportation costs, for example, have decreased substantially since the 1980s. Hummel (1999) argued that containerization in ocean transport is shown to have changed the composition of freight rates (lowering the cost of distant relative to proximate travel), particularly since the 1990s.
- *Improvements in information technology.* In the past few decades, there has been a revolution in information technology that has led to lower communication costs as a result of technological improvements and increased competition. Information technologies have allowed the functional integration of many interdependent activities associated with the flow of foods, resulting in substantial improvements in the design of global food supply chains and enhanced food distribution system performance (Byrd and Davidson, 2003; Lewis and Talayevsky, 2004).

- *Gains from economies of scale.* Economies of scale characterize production and distribution processes in which an increase in the size of the firm results in lower long-run average cost. Recent years have witnessed the rapid consolidation of the food distribution system worldwide. For instance, in most European countries, the market share of the five top supermarket companies was above 70% in 2004 (McLaughlin, 2006). Supermarket concentration in emerging countries is increasing rapidly, and the market share of the top five companies in 2003 reached 53.7, 45.3, and 40.5% in Chile, Argentina, and Brazil, respectively. Although increasing size of retail firms may decrease costs in food supply chains, it is possible that retailers use the size to their advantage by exercising market power with their suppliers and consumers.

THE CONSUMERS: TRENDS AND ISSUES

What factors shape global food demand, and what forces will determine whether there is adequate food for all in the future? Perhaps the most important demand factor is simply the size of the population. Along with this factor come other demographic dimensions, such as the age profile of the population. Income levels and the distribution of income also importantly shape demand and help determine whether people can purchase and obtain access to the food they require. Increasingly, particularly as consumer incomes increase, convenience and time, whether it is time spent in purchasing or in preparing food, become significant determinants of demand, hence the rise in food away from home and prepared foods. Also, as consumers achieve higher income levels and meet basic nutritional needs, they tend to place emphasis on food quality, freshness, and variety as well as on how the food is produced.

POPULATION AND DEMOGRAPHICS

Over the last century, the world grew from a population level of 1.6 billion to 6.1 billion people, an unprecedented increase in global population of 4.5 billion. At the present time, the world adds roughly 82 million more people to its overall population each year, with the bulk (80 million) added to the less-developed countries (Population Reference Bureau, 2007). The combination of public health (particularly in the areas of sanitation and nutrition) advances and medical improvements have led to longer life expectancies, and these, along with high fertility levels, explain the current population growth rates.

INCOME AND REAL PURCHASING POWER

Since the mid-1970s, food security analysts and researchers have emphasized the role of income, purchasing power, and access to food as principal reasons explaining hunger and food insecurity. The importance of poverty and a lack of income in explaining hunger were heightened by statistics that demonstrated that while globally sufficient food quantities may exist, poor people simply lack sufficient economic resources to purchase food. Moreover, since many of the poorest people live in rural areas and they have access to only small amounts of land or are landless, their incomes and ability to earn money play a large role in determining their food security.

Economists measure the impact of changes in income on the amount of food demanded with the concept of the income elasticity of demand. *Income elasticity of food demand* is defined as

$$E_i = \frac{\%\ \text{Change in quantity of food demanded}}{\%\ \text{Change in income}} \qquad (23.3)$$

Here, E_i denotes the income elasticity of demand. An income elasticity of 0.8 for food means that as income goes up by 10%, the quantity of food demanded by the consumer will increase by 8%. Of course, if the income elasticity is less than 1, then the proportion of the budget allocated to food will decrease as income rises. Figure 23.3 shows the income elasticities for the commodity groups food, housing, medical care, and education across countries by gross domestic product (GDP; a measure of income per person) per capita (plotted data from Seale et al., 2003). As can be seen, the income elasticity of food decreases from a measure of about 0.8 to a low of 0.15 as countries move from low levels of GDP per capita to higher levels. The income elasticities for food pertain to aggregate data at the country level. Nevertheless, a similar pattern holds at the individual or household level within a country. This variation in food-purchasing behavior, along with the fact that in poor settings people spend a high percentage of their overall income on food, helps explain the differential impact across the income spectrum of an income shock, such as a drought or a decline of a fishery, on household food insecurity.

Figure 23.4, Figure 23.5, and Figure 23.6 illustrate that as incomes increase, the proportion of the total household budget allocated to food declines, and the mix of food products purchased changes. The composition of the food basket changes from low-value food goods, such as rice and low-cost sources of carbohydrates (cassava, corn, etc.), to higher-value foods such as meat, fruit, vegetables, and prepared or fast foods. This general pattern of change has implications for both food producers and consumers, as well as for the overall food security situation in developing countries. With real incomes rising in countries such as China and India at a rate of between 5 and 9% per year in recent years, as well as in many other developing countries in Africa, Asia, and Latin America, rapid changes in the nature of food demanded occurs. In addition, as Seale et al. (2003) reported, lower-income countries have higher-income elasticities for high-value foods (meat, dairy, and fruit and vegetables) compared to higher-income countries. This implies that as income growth continues, poorer consumers will demand more of these goods.

This pattern of moving toward a diet that increasingly includes high-value foods is accentuated by the simultaneous demographic shift of population from rural to urban areas. As people move from rural to urban areas, we observe a shift in preferences toward Western-style foods (e.g., breads from rice, takeout food and fast food, packaged foods from global brands or in imitation of global brands) and high-value foods. These demand factors drive growth in agriculture in developing countries, where

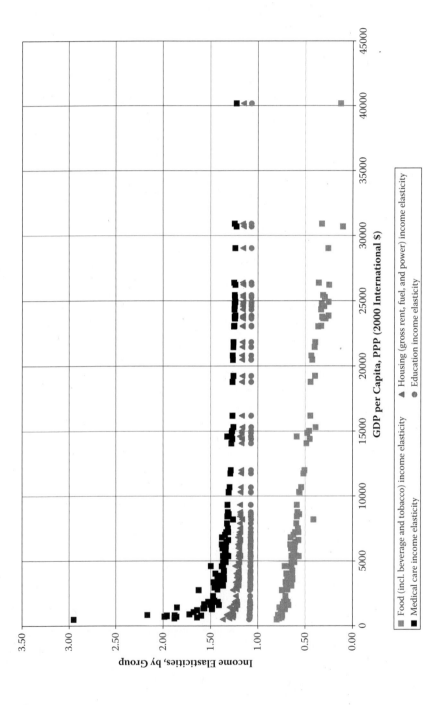

FIGURE 23.3   Income elasticities for major commodity groups across countries, by income. GDP, gross domestic product.

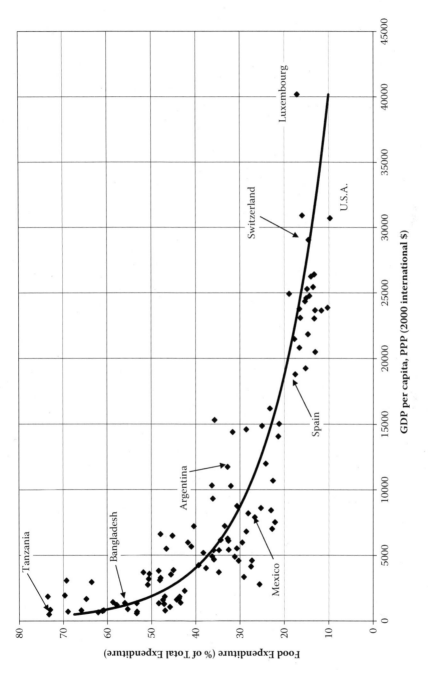

FIGURE 23.4   Gross domestic product (GDP) per capita and food expenditure.

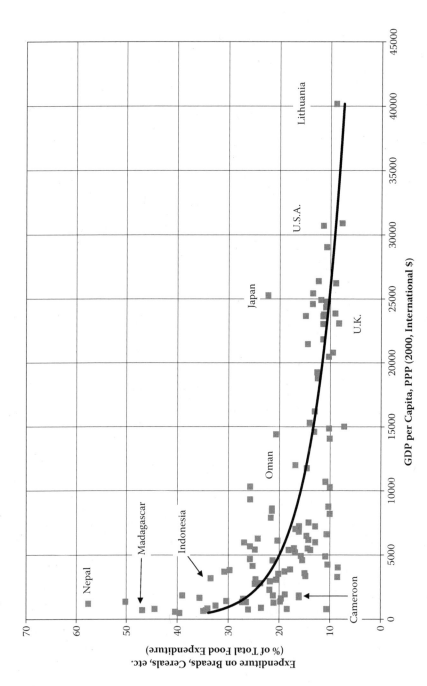

FIGURE 23.5 Gross domestic product (GDP) and expenditure on bread and cereals.

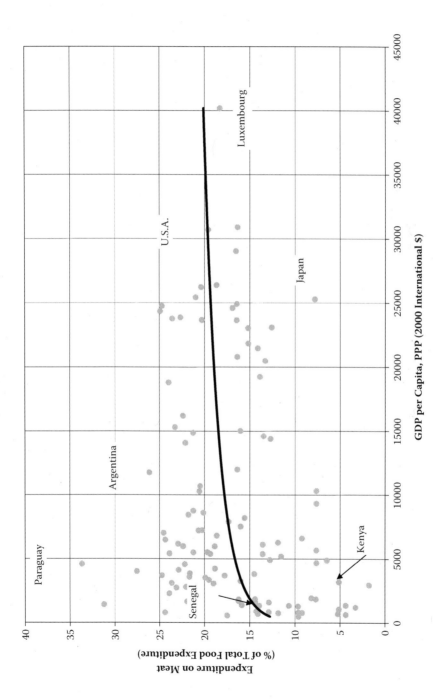

FIGURE 23.6 Gross domestic product (GDP) and expenditure on meat.

the annual growth in production of high-value foods outpaces the rate in developed countries (von Braun, 2007). For example, milk production increased by 4% from 2004 to 2006 in developing countries compared to 0.3% in developed countries. Similarly, meat production increased by 4.0% in developing countries compared to 0.6% in developed countries over the same time period.

Furthermore, this variation in price elasticities by income implies food price increases will have a greater impact on developing country consumers than consumers in higher-income countries. This will particularly hold true for urban consumers, who will not benefit on the income side from food price increases in developing countries. Although the distinct distributional impacts within a country will depend on its unique market structures and relationships, recent evidence from southern Africa shows that in a period of grain shortages large farmers unequivocally benefited from higher average prices, while urban consumers (especially poor consumers) and rural consumers (including small farmers who purchase a portion of their food in the market) suffered welfare losses (Jayne et al., 2006). Hence, the mental image of food-insecure rural people in developing countries increasingly needs to be amended to include urban dwellers living at the bottom of the income pyramid.

CASES ON EFFECTS OF SUPPLY CHAIN TRENDS ON FOOD PRICES AND AFFORDABILITY

THE LIVESTOCK REVOLUTION AND DEMAND TRANSFORMATION

The "livestock revolution" (LR) is "a complex series of interrelated process and outcomes in production, consumption and economic growth" (Delgado et al., 1999). Unlike its predecessor the green revolution, the LR is being driven by dramatic changes in demand for livestock products. Per capita demand for livestock products has been relatively stagnant in developed countries but is expected to grow rapidly in developing countries during the next 15 years. A case in point is milk and dairy products consumption. Delgado et al. predict that total consumption of milk products in the developed world will increase about 7% by 2020. In contrast, milk demand in the developing countries is expected to increase 132% by that year. This rapid growth implies two major changes in the structure of world food demand: The principal consumers of livestock products will be in the developing world, and the value of livestock production is expected to surpass that of grains by 2020. These changes are being driven by population growth, by growth in incomes, by urbanization in the developing world (which increases the range of food choices and results in greater dietary diversity), and by the declining real prices for both livestock products and cereals during much of the past three decades. Growth in incomes has allowed many (but certainly not all) consumers in the developing world to reach a satiation point of sorts with consumption of grains and tubers, and they now seek greater dietary diversity.

The growth in demand for livestock products will result in dramatic transformation of the production and marketing components of the supply chain. To meet the rapidly growing demand, there must be an accompanying transformation of livestock supply, including increases in land area devoted to forage production, increased animal numbers, and increased productivity of animals and land. Increasing productivity will require developing, adapting, and disseminating new technologies and production systems. In addition, because most livestock products are more perishable and not as easily stored as grains, developments in the processing and transportation will also be necessary. Technological, institutional, and regulatory structures throughout the marketing chain will be challenged to evolve rapidly.

The LR is expected to create opportunities for livestock producers in the developing world given that production growth in the developed world will not keep pace with growing global demand. However, whether small-scale producers can benefit from increases in demand depends in large measure on the nature of transformations in the postfarm supply chain. Concerns about the LR include potential adverse effects on the poor in the developing world due to increases in grain prices driven by increased demand for animal feed and allocation of cropland to fodder production, potential negative environmental impacts, and public health systems in developing countries. Research related to these concerns (e.g., Delgado et al., 1999; Nicholson et al., 2001; Randolph et al. 2007) has suggested that many of them can be addressed through a combination of policies of institutional strengthening and research that allows the livestock sector to contribute more to food security and poverty alleviation while addressing environmental and public health concerns.

BIOENERGY, AGRICULTURE, AND GLOBAL FOOD PRICES

Bioenergy is energy generated from the production of fuels of biological and renewable origin. These biofuels include traditional fuel sources such as fuel wood, charcoal, and manure, but more recently a great deal of interest has been shown in the production of liquid biofuels such as ethanol and biodiesel. Biofuels are touted as having the potential to reduce carbon emissions (although net carbon balance varies greatly; Hazell and Pachuri, 2006), to have production potential in a wide variety of countries, and to increase the demand for agricultural products that were believed to be in "chronic oversupply"[6] (Hazell and Pachuri, 2006).

Although this increased demand for agricultural production may be beneficial to some farmers, it has potentially important implications for global food supply, demand, and prices. Increasing demand for crops to be used as "feedstock" for biofuel production can increase the prices of those crops, and these effects may be further exacerbated by the allocation of land to the production of nonfood crops grown specifically for conversion into biofuels. Higher prices for food will tend to negatively affect the poor (who are net buyers of food), but this may be offset to some extent by cheaper energy or by employment opportunities generated in the production of biofuels.

The extent to which increases in biofuel demand will increase food prices was examined by Rosegrant et al. (2006). They found that, under assumptions of fairly aggressive demand growth for ethanol and biodiesel, prices for key food crops that can be used for biofuels production would increase between 10 and 135% (Table 23.2). The development of new technologies in the conversion of nonfood feedstocks (using what is called *cellulosic conversion technologies*) to biofuels and faster productivity increases in food crop production would result in lower—but still reasonably large—price increases.

TABLE 23.2

Predicted Percentage Price Increases in Feedstock Crops in 2020 Due to Aggressive Growth in Biofuel Demand with Alternative Technological Assumptions

Feedstock Crop	Biofuel Demand Growth with Baseline Productivity Growth[a]	Biofuel Demand Growth with Increased Productivity Growth[b]
Cassava	+135	+54
Corn	+41	+23
Oilseeds	+76	+43
Sugar beet	+25	+10
Sugarcane	+66	+43
Wheat	+30	+16

[a] Assumes that bioethanol and biodiesel replace 20% of gasoline by 2020 in most of the world except Brazil, the European Union, and the United States (for which other projections are used). Productivity of these crops is assumed to grow through 2020 but not be influenced by their demand as biofuels.

[b] Assumes the same demand growth but also increases in cellulosic conversion technologies and higher productivity growth for these crops.

Source: From M.W. Rosegrant, S. Msangi, T. Sulser, and R. Valmonte-Santos. December 2006, in: *Bioenergy and Agriculture: Promises and Challenges.* Washington, DC: International Food Policy Research Institute (IFPRI). Brief 3 of 12, 2020 Vision for Food Agriculture an the Environment, Focus 14.

Reducing the trade-offs between biofuel crops and food production will require research and development activities to increase the yield of energy per unit land, a greater focus on food crops that also generate by-products suitable for biofuel production, expanded production of biofuel crops that grow in less-favored areas, increased productivity of the food crops themselves, and removing barriers to trade in biofuels (Hazell, 2006). These actions will reduce the amount of land required for both food and fuel production, reducing the potential negative impacts of biofuels on food consumers.

MODERNIZATION OF FOOD RETAILING IN DEVELOPING COUNTRIES: IMPLICATIONS FOR THE GLOBAL SUPPLY CHAINS

According to Regmi and Gehlhar (2007), in recent years there has been an unprecedented growth in supermarkets in developing countries. Such growth is taking place mostly in many Latin American and Asian countries, in which consumer demand for higher-valued processed food products have increased as a result of rising incomes. The implications of such trends are closer geographical integration and increasing centralization of food supply chains. Only 15 years ago, the supermarket sector in most Latin American and Asian countries consisted of small firms financed by domestic capital operating primarily in the major cities' upscale neighborhoods.

In contrast, by 2005 supermarkets had penetrated the middle-class and the poorer urban neighborhoods and rural areas, particularly in Latin America. Whereas supermarkets accounted for 15 to 30% of the national food retail sales before 1990, they now account for about 50 to 70% of the retail sales in many Latin American countries. This corresponds to a level of growth experienced in the United States during a 50-year period (Table 23.3). Although at somewhat a slower pace, the growth of modern food retailing in Asia resembles that of Latin America. However, supermarkets in Asia are growing at a faster rate compared to supermarkets in Latin America (Reardon and Berdegué, 2002; Hu et al. 2004).

Regmi and Gehlhar (2005) also noted that the expanding supermarket sector in developing countries has been led by multinational companies that are often foreign owned. In Latin America, for instance, such large multinational firms as Carrefour, Wal-Mart, and Royal Ahold represent nearly 80% of the top five supermarket chains in most large countries. These companies have also made substantial investments to grab market share in the Asian markets in recent years. The success of multinational retailers in the global retail market is due primarily to their ability to offer a wide assortment of low-price products. Therefore, the presence of large multinational food retailers is having evident impacts on food prices and affordability, in particular for urban populations. These companies have large-scale operations with sophisticated supply chains reaching multiple countries, allowing them to handle large volumes of

TABLE 23.3
Supermarkets' Share of Retail Sales in Selected Countries

| Country | Previous | | 2001 |
	Year	Percent	Percent
United States	1930	5–10	80[a]
Latin America			
Argentina	1985	17	57
Brazil	1990	30	75
Chile	NA	NA	50
Guatemala	1994	15	35
Mexico	NA	NA	45
Asia			
China (urban)	1999	30	48
Indonesia	1999	20	25
Korea	1999	61	65
Malaysia	1999	27	31
Thailand	1999	35	43

[a] 2000.

Sources: T. Reardon and J.A. Berdegué, 2002, *Development Policy Review* 20(4):317–334; D. Hu, T. Reardon, S. Rozelle, P. Timmer, and H. Wang, 2004, *Development Policy Review* 22(5):557–586; A. Regmi and M. Gehlhar, 2005, *Amber Waves* 1:12–19.

products efficiently, resulting in cost savings for the whole supply chain. These savings result in lower food retail prices, assuming that multinational retailers do not exercise market power in developing countries. The changes in food retailing are producing a growing centralization of procurement. Therefore, large distribution centers are displacing a large number of smaller centers and intermediaries in developing countries. Regmi and Gehlhar (2007) indicated that the Carrefour distribution center in Brazil's São Paulo may serve nearly 50 million consumers, and an Ahold wholesaler in Costa Rica may control the entire distribution for Central America for Ahold.

The rapid globalization of food retailing can lead to more food trade and to more prominent global food supply chains. The need for year-round supply of food products has encouraged joint venture partnerships and strategic alliances among firms in the Northern and the Southern Hemispheres, increasing food trade. Likewise, alliances with multinational retailers provide untapped export opportunities for both small and large food producers and manufacturers in both developed and developing countries. Nonetheless, the presence of large multinational retailers may also discourage trade. The reason is that food manufacturers have incentives to expand their manufacturing capacity into the region where multinational retailers make investments. Consequently, the presence of a global retail firm may promote more local processing from domestically produced raw products, thus having substantial impacts on local food supply chains.

FOOD AFFORDABILITY AND POLICY: ACCESS TO FOOD FOR ALL

As political leaders and governments, nongovernmental organizations (NGOs), academics, and others interested in the issues of equitable development, food security, and nutrition security consider access to food, increasingly the view is through the rubric of human rights. The human rights view of food security and nutrition security is rooted in the notion of freedom from want, which was emphasized by President Franklin D. Roosevelt in 1941 and, subsequently, came to form the basis of the Universal Declaration of Human Rights, which was adopted by the United Nations in 1948. The four freedoms that Roosevelt detailed were freedom from want, freedom of worship, freedom of speech and expression, and freedom of fear of physical aggression.

The understanding of freedom from want in this human rights view is a broad one, and it emphasizes the nature of the relationship between claim holders and duty bearers. Claimants and bearers can include governments, international actors (such as foreign governments and international agencies), NGOs, households, communities, families, parents, and other individuals. The obligation of the duty bearers is to "respect, protect, facilitate and fulfill the rights of claim holders" (Haddad and Orshaug, 1998). Furthermore, the human rights view sees the development agenda of a process of increasing the recognition of human rights and the strengthening of the civil processes that allow the full exercise of these rights in a given country. This broad view does not place all the emphasis on the state or governmental responsibility to fulfill the specific right that people have to food. Instead, the emphasis, particularly in the context of development, is on the duty of bearers to recognize and protect the rights of people with respect to productive resources such as land holdings and access to education. Furthermore, by protecting people from discrimination, the state can be ensuring access to the income necessary to have better food security.

POLICIES AND PROGRAMS TO ENSURE FOOD FOR ALL

Governments interested in ensuring access to food for the food insecure have a range of policy options at their disposal from direct transfers of cash or food to indirect measures, such as subsidized food shops or agricultural policies that favor consumers. In the long run, the best option to improve food security and ensure adequate food for all is to pursue policies that allow for the growth of incomes, particularly for the poorest. However, in the short term targeted food and nutrition programs, including income support programs, play a significant role in reducing food insecurity and preventing malnutrition.

What sort of programs work in the short term? In cases of severe and widespread food shortages and food distribution channel disruptions, such as in the case of the Asian tsunami in 2006 or in a conflict situation or refugee situation, direct food aid is effective in addressing hunger. The cost of direct food aid is increasing as world prices for grains and other foods common to food aid programs increase. Hence, advocates and aid administrators have focused attention on the design and implementation of food aid programs. In the United States, the U.S. Agency for International Development (USAID) is seeking flexibility to purchase some foods in regional markets (neighboring and nearby states) to avoid the costly shipping fees associated with U.S. sourced foods and to strengthen developing country agricultural markets (Barrett and Maxwell, 2005).

When the food distribution system is not broken but poverty makes some people vulnerable and food insecure, governments have turned to conditional cash transfers to provide small amounts of cash to support the poor and help facilitate desired behaviors. In developing countries, conditional cash transfer (CCT) programs give small amounts of cash to families meeting income and program targeting criteria (often, whether the household meets criteria for vulnerability, such as single mother, young children present, low asset level, low levels of parental education) if the family meets certain program requirements. The requirements usually include attendance at school for the children and regular attendance at a local primary health clinic for maternal and child health assessments and health education lessons. Conditional on these program requirements being met, the family then receives the cash payment, usually on the order of $10 to $30 per month.

Mexico's poverty reduction program PROGRESA (Programa de Educacion, Salud y Alimentacion) began in 1997. It is now called Oportunidades and is the centerpiece of Mexico's efforts to reduce poverty and improve the lives of its most vulnerable citizens. The program includes a cash transfer component linked to school enrollment and attendance as well as to clinic attendance. The program also provides health education lessons and in-kind benefits, such as nutritional supplements for a child up to the age of 5 and for pregnant and lactating mothers. In 2000, this program reached 2.6 million Mexican families, including approximately 40% of all families in rural areas. The amount of the cash transfers the households received were a function of the age and sex of the children and their compliance with the program requirements. The value of the transfer received equaled about 20% of total household consumption prior to the program. In an evaluation of the impact of PROGRESA on food consumption, Hoddinott and Skoufias (2004) concluded that the program induced a consumption of 6.4% more calories in communities where the program was delivered compared to similar but nonparticipating communities. Furthermore, they demonstrate that the program led to increases in the quality of the diet consumed, with greater portions of animal products and vegetables consumed in the participating communities. Additional evaluative research (Behrman and Hoddinott, 2005) shows the nutritional supplementation program as leading to a lower probability of child stunting and an increase of mean annual growth for participating children of approximately 15% for children 12 to 36 months. Thus, the CCT program leads to significant improvements in nutritional outcomes and food consumption.

While direct provision of food and cash (including food-for-work programs) work to ensure food security and improve nutrition in the short term, in the longer term countries can improve access to food by strengthening incomes, particularly for those people living on less than $1 or $2 per day. Broad-based economic growth, building on investments in human capital (education and primary health and public health services), public infrastructure (road, telecommunications networks, clean water, and adequate sanitation), as well as improved institutions and governance (legal frameworks, market access, equal rights under the law, equity in access to public goods), will serve to reduce food insecurity in the long run. Food programs also play a critical role in the long-run development strategies of countries, particularly in the area of human capital formation.

It has long been realized that child health and development depends critically on the education and skills of the mother (Glewwe, 1999). Yet in many countries, girls tend to attend fewer years of school than boys. For example, World Bank statistics show that in India in 2000, for females 15 years and older, average schooling was 4 years, whereas for males in the same age category it was 6 years (World Bank, 2008). To address this issue, as well as to provide a food supplement to children and to indirectly support low-income families through an in-kind transfer, many countries have turned to school lunch programs. School lunch programs increase school attendance for girls and thereby will help to promote human capital formation in the future (Vermeersch and Kremer, 2004). Afridi (2006) examined the Mid-Day Meal program (nationally mandated school meals) in India for its nutritional impact on the child program participants and found the program effectively increased the children's protein intake to 100% of daily requirements and reduced the overall caloric deficiency by roughly 30%. The program cost about 3 cents per child per day, so that small investments of funds can generate significant food security and nutrition benefits through a school lunch program.

To reduce childhood malnutrition, nutrition interventions targeted at pregnant women and young children have proven effective (Bhutta et al., 2008). Effective interventions include promotion of breastfeeding, supplementary feeding for the food insecure, reduction of the impact of malaria during pregnancy, therapeutic feeding for the acutely malnourished, and micronutrient interventions. These nutrition strategies help build human capital through higher levels of health, which lead to greater incomes and then higher food security in the long term.

CONCLUSIONS

Future food prices and affordability will depend crucially on developments in food production, marketing, and consumption; on how the global food system integrates these disparate activities; and on how both developed and developing country governments intervene to address short-term food shortages and long-term household food insecurity. Recent trends suggest that food availability and affordability are improving at the global level. However, a number of factors may prevent future increases in food production or divert crops or land from food to nonfood uses, and global-level calculations of food adequacy do not address the many millions of households that currently lack sufficient incomes to purchase adequate diets. Global food-marketing systems continue to evolve toward greater complexity and consolidation, which can result in both lower costs of food marketed to increasingly large numbers of urban consumers and increase the range of food product offerings. Population growth, urbanization, and income growth are key determinants of food consumption, which interacts with the food supply chain to affect affordability of food. Access to adequate food is increasingly viewed as a fundamental human right, and most governments have programs and policies in place to improve access to food for some, if not all, of their citizens. In the long term, ensuring access to food for all will require efforts to increase incomes and reduce poverty.

NOTES

1. These relationships are described as hypothesized not only because sometimes there is insufficient empirical evidence to support them but also because the process of diagramming a system is often iterative and assessments of causal relationships can change.
2. Often described with the words *ceteris paribus* in economics, this assumption implies that the hypothesized relationship is a partial one. In the example, an increase in household income would increase at-home food consumption relative to what it would have been in the absence of the increase in household income, but other factors can also change. Thus, if both the cost of food acquisition and household income increased, the net effect on at-home food consumption would depend on which of the effects is larger; at-home food consumption could in fact decrease even though household income increased.
3. Although diagrammatic tools can be helpful to understand systems, some authors (Sterman, 2000) have suggested that empirical models are necessary to address dynamically complex systems. A system as complex as the food system contains many feedback loops, and outcomes are best predicted by empirical models.
4. This equation is deceptively simple because it is essentially an accounting identity. That is, it tells us little about which factors influence the land area, crop yields, or crops per year. Each of these factors has dynamic and complex multiple causes at global, regional, and local levels.

5. Grain yields are often the focus of discussion, but roots and tubers play an important role in the diets of many of the world's poorer people, and increased yields have also been observed for them. Livestock, oilseed crops, and fisheries are other important food sources for which production and productivity have increased during the four decades.

6. One indicator for this "oversupply" was declining real grain prices during much of the past three decades.

REFERENCES

Afridi, F. 2006. *Child Welfare Programs and Child Nutrition: Evidence from a Mandated School Meal Program in India.* Syracuse, NY: Syracuse University, Department of Economics Working Paper.

Barrett, C.B. and D.G. Maxwell. 2005. *Food Aid After Fifty Years: Recasting Its Role.* New York: Routledge.

Behrman, J.R. and J. Hoddinott. 2005. Programme evaluation with unobserved heterogeneity and selective implementation: the Mexican PROGRESA impact on child nutrition. *Oxford Bulletin of Economics and Statistics* 67:547–569.

Bhutta, Z.A., T. Ahmed, R.E. Black, S. Cousens, K. Dewey, and E. Giugliani. 2008. What works? Interventions for maternal and child undernutrition and survival. *Lancet* 371:417–440.

Byrd, T.A. and N.W. Davidson. 2003. Examining possible antecedents of IT impact on the supply chain and its effect on firm performance. *Information and Management* 41:243–255.

Delgado, C., M. Rosegrant, H. Steinfeld, S. Ehui, and C. Courbois. 1999. *Livestock to 2020: The Next Food Revolution.* Washington, DC: International Food Policy Research Institute, Food and Agriculture Organization of the United Nations and the International Livestock Research Institute. Food, Agriculture and the Environment Discussion Paper 28.

Döös, B. R. 2002. The problem of predicting global food production. *Ambio* 31:417–424.

Elitzak, H. 2004. Behind the data. Calculating the food marketing bill. *Amber Waves* 2:43.

Food and Agriculture Organization of the United Nations (FAO). 2002. *World Agriculture Towards 2015/2030: Summary Report.* Rome: FAO.

Food and Agriculture Organization of the United Nations (FAO). 2004. Report on the FAO/AFMA/FAMA Regional Workshop on the Growth of Supermarkets as Retailers of Fresh Produce, October 4–7, Kuala Lumpur, Malaysia.

Gilland, B. 2002. World food population and food supply: can food production keep pace with population growth in the next half-century? *Food Policy* 27:47–63.

Glewwe, P. 1999. Why does mother's schooling raise child health in developing countries? Evidence from Morocco. *Journal of Human Resources* 34(1):124–159.

Haddad, L. and A. Orshaug. 1998. How does the human rights perspective help shape the food and nutrition policy research agenda? *Food Policy* 23:329–345.

Hazell, P. 2006, December. Developing bioenergy: a win-win approach that can serve the poor and the environment. In: *Bioenergy and Agriculture: Promises and Challenges.* Washington, DC: International Food Policy Research Institute (IFPRI). Brief 12 of 12, 2020 Vision for Food Agriculture and the Environment, Focus 14.

Hazell, P. and R.K. Pachuri. 2006, December. Bioenergy and agriculture: promises and challenges. Overview. In: *Bioenergy and Agriculture: Promises and Challenges.* Washington, DC: International Food Policy Research Institute (IFPRI). Brief 1 of 12, 2020 Vision for Food Agriculture and the Environment, Focus 14.

Hoddinott, J. and E. Skoufias. 2004. The Impact of PROGRESA on food consumption. *Economic Development and Cultural Change* 53:37–61.

Hu, D., T. Reardon, S. Rozelle, P. Timmer, and H. Wang. 2004. The emergence of supermarkets with Chinese characteristics: challenges and opportunities for China's agricultural development. *Development Policy Review* 22(5):557–586.

Hummel, D. 1999. *Have International Transportation Costs Declined?* Chicago: Working Paper, Graduate School of Business, University of Chicago.

Jayne, T.S., B. Zulu, and J.J. Nijhoff. 2006. Stabilizing food markets in eastern and southern Africa. *Food Policy* 31:328–341.

Kotler, P. and K.E. Keller. 2006. *A Framework for Marketing Management*, 3rd ed. Upper Saddle River, NJ: Prentice Hall.

Lewis, I. and A. Talayevsky. 2004. Improving the interorganizational supply chain through optimization of information flows. *Journal of Enterprise Information Management* 17:229–239.

McLaughlin, E.W. 2006. Grocery industry. Global trends and balance of power: is it shifting? Presentation at Cornell Food Executive program, July 11, Ithaca, NY.

Meadows, D., J. Randers, and D. Meadows. 2004. *Limits to Growth: The 30-Year Update*. White River Junction, VT: Chelsea Green.

Meadows, D.H. and J. Robinson. 1985. *The Electronic Oracle: Computer Models and Social Decisions*. Chichester, U.K.: Wiley.

Nicholson, C.F., R.W. Blake, R.S. Reid, and J. Schelhas. 2001. Environmental impacts of livestock in the developing world. *Environment*, March, pp. 7–17.

Pinstrup-Andersen, P., R. Pandya-Lorch, and M.W. Rosegrant. 1999. *World Food Prospects: Critical Issues for the Early Twenty-First Century*. Washington, DC: International Food Policy Research Institute. 2020 Vision Food Policy Report.

Prahalad, C.K. and A. Hammond. 2002. Serving the world's poor, profitably. *Harvard Business Review*, September, pp. 4–11.

Prahalad, C.K. and K. Lieberthal. 2003. Best of HBR 1998: the end of corporate imperialism. *Harvard Business Review*, August, pp. 1–12.

Population Reference Bureau. 2007. World population highlights. *Population Bulletin* 62(3) (September).

Randolph, T.F., E. Schelling, D. Grace, C.F. Nicholson, J.L. Leroy, D.C. Cole, M.W. Demment, A. Omore, J. Zinsstag, and M. Ruel. 2007. Role of livestock in human nutrition and health for poverty reduction in developing countries. *Journal of Animal Science* 85:2788–2800.

Reardon, T. and J.A. Berdegué. 2002. The rapid rise of supermarkets in Latin America: challenges and opportunities for development. *Development Policy Review* 20:317–334.

Regmi, A. and M. Gehlhar. 2005. Processed food trade pressured by evolving global supply chains. *Amber Waves* 1:12–19.

Rosegrant, M.W., S. Msangi, T. Sulser, and R. Valmonte-Santos. 2006, December. Biofuels and the global food balance. In: *Bioenergy and Agriculture: Promises and Challenges*. Washington, DC: International Food Policy Research Institute (IFPRI). Brief 3 of 12, 2020 Vision for Food Agriculture an the Environment, Focus 14.

Seale, J., Jr., A. Regmi, and J. Bernstein. 2003. *International Evidence on Food Consumption Patterns*. Washington, DC: Economic Research Service, USDA. Technical Bulletin 1904.

Sterman, J.D. 2000. *Business Dynamics: Systems Thinking and Modeling for a Complex World*. Boston: Irwin/McGraw Hill, pp. 37–38.

Vermeesch, C. and M. Kremer. 2005. *School Meals, Educational Achievement and School Competition: Evidence from a Randomized Evaluation*. World Bank Working Paper 3523, Washington, DC: World Bank.

von Braun, J., M.W. Rosegrant, R. Pandya-Lorch, M.J. Cohen, S.A. Cline, M.A. Brown, and M. Soledad B. 2005. *New Risks and Opportunities for Food Security: Scenario Analyses for 2015 and 2050*. Washington, DC: International Food Policy Research Institute. 2020 Discussion Paper 39.

von Braun, J. 2007. The world food situation: new driving forces and required actions. IFPRI Biannual Overview of the World Food Situation presented to the CGIAR Annual General Meeting, December 4, Beijing.

World Bank. 2008. Edstats data query. Available at: http://go.worldbank.org/47P3PLE940. Accessed March 4, 2008.

24 Food Availability and Quality:
Situations and Opportunities in Developing Countries

Dan L. Brown

CONTENTS

ABSTRACT

This chapter invites the reader to join international efforts to provide adequate high-quality food for all of the world's population. The life's work of a physician/educator, the addition of iodine to irrigation water, locally sourced ready-to-use therapeutic foods, and the assurance of food safety with limited resources are used to illustrate some of the features of successful efforts.

INTRODUCTION

The previous 23 chapters have described how the world feeds itself and presents the means by which all of humanity can be fed well, in perpetuity. We know how, and we have the resources required to provide adequate food for all. Never, in the history of the world, has malnutrition been so preventable, so unnecessary, its persistence so immoral as it is today. Fortunately, with the readers' help, marasmus, kwashiorkor, and micronutrient deficiencies may soon become a thing of the past, to follow small-pox into oblivion. But, without the readers' involvement and the involvement of those with the same knowledge, malnutrition will continue to shame our species.

There are probably macroeconomic policies, trade accords, and other large-scale interventions that will play a role in abolishing malnutrition, but before and after such measures come to pass, considerable progress will be made by thousands of more modest individual and collective efforts: region by region, village by village, child by child. A diverse range of projects, from short-term limited efforts to lifetime commitments, will all contribute, some because of, many more in spite of national and international scale policies.

This chapter surveys successful examples of bringing what we know about food production, affordability, availability, and distribution in rural and urban areas of developing countries to bear on wiping out undernutrition. More universal success will require interventions involving funding, educating, and developing community efforts in nutrient-rich plant and animal production for home consumption and for sale of products in the local market. Success of such efforts will benefit from financial services, often through governments, foundations, and private investments. But, while these obvious traditional routes to development have helped in the past, your new ideas and methods will be required to complete the job. Food systems need to be established for and by the people who need them in such a way that they are resistant to changes in government, economics, and whatever type of assistance is fashionable at any given time.

Although price rises in staple foods of plant origin have caused some recent problems around the world, the dominant type of malnutrition observed worldwide in the last 20 years has been due to the quality, not the quantity, of food. A lack of micronutrients (such as available iron, iodine, vitamin A, vitamin B_{12}) (Demment et al. 2003) and, in some cases, protein quality have become the leading concerns for nutritionists and poor families around the world. Some of these nutrient needs can be partially met by plant food sources, but the best predictor of healthy growth (Grillenberger et al. 2003), cognitive development (Whaley et al. 2003), and freedom from micronutrient and amino acid deficiency is the amount of animal source foods (ASFs) in the diets of children (see also Chapters 11 and 12).

Possibilities exist for improving nutrition, particularly in children, in low-income families in developing countries by increasing availability of ASFs through encouraging meat and milk production in rural communities. But, as valued as ASF is in the diet, many people cannot afford it and have found multiple barriers to getting it for their children. This chapter presents examples of some cases for which barriers have been identified and overcome with positive results for the nutrition of children around the world, as well as the most important case histories of all: your own projects that may not yet exist.

EXAMPLE 1: DR. STEPHEN WANJA

Stephen Wanja was a physician who spent most of his career in western Kenya at Fudumi. Dr. Wanja was born near the coast of Kenya into the Giriama tribe, was educated in both Kenya and in northern Europe, and performed his work among the Abaluhya people, principally the Tiriki. When Dr. Wanja came to Fudumi, he chose as his clinic site a few acres of steep, depleted, eroded, seemingly useless soil. As the early years progressed, he healed those people who came for medical assistance, and with strategic plantings of crops (nutrient-rich fruits and vegetables, along with some root, grain, and fodder crops), he healed the land as well.

Once the clinic and shamba (small farm) were running, Wanja was able to reduce return business among mothers with malnourished children by keeping them there during the treatment of the children and teaching them to increase the child's intake of essential nutrients. Lessons included growing nutritious crops they had not seen before; tending sheep, goats, swine and rabbits; and incorporating these animal source foods into their children's diets. These women in turn taught others, and this had a long-term effect on nutrition among the Tiriki. Innovative approaches to improving nutrition of rural populations like those of Dr. Wanja in Kenya are occurring in many developing countries through visionary cross-culture self help programs. [*]

European medical students also benefited from Fudumi's educational mission. While hiking among the fruit trees, I once encountered a sunburned Dutch medical student crouched down and cultivating some vegetables on the reclaimed hill farm at Fudumi. This young summer elective student used a short-handled jembe (a type of hoe) to weed and at the same time extend a shallow trench that carried the slurry from the home-built manure-powered methane digester at the top of the hill. [1] *"Fertigation" with the digester slurry provided both water and nutrients that would find their way through the food plant into otherwise-malnourished children. When asked what she was doing, the intern answered, "I am learning medicine," and returned to her work. Without a doubt, this young woman's experience taught her and her future patients that sound nutrition, appropriate seeds, and a hoe may be as important to health as pills and a hypodermic syringe.*

Another technique Stephen Wanja used to insert critical micronutrients into the young population was through his rabbit program. He noticed that children spent some time trying to catch wild rabbits and other small game to feed themselves, independent of their parents' provisions. This took a lot of time and energy with infrequent reward. To help the young children, Dr. Wanja brought some fast-growing domestic rabbits from Europe, taught the children to spend their time gathering foliage that could not run away and feeding it to the rabbits. In this way, they had an independent ASF supply that did not cost them so much in time and calories as hunting would. The children paid their gift forward by passing along rabbit rearing tips and a pregnant doe to the next child, and the rabbits spread across western Kenya, providing a segment of the population with its own independent supply of ASF.

Many of the aspects of Dr. Wanja's practice are instructive to those who would improve nutrition in settings such as his. Wanja not only knew how to contact experts in fields related to the health of his patients (I provided him with dual-purpose goats and the skills to deploy them); he had broad interdisciplinary knowledge. Too often, when extolled to take an interdisciplinary approach to a problem, we assemble an unwieldy team of experts. Frequently, it is more effective and efficient for an individual or a very few workers to learn several disciplines themselves. In many cases, that single person can carry out the interdisciplinary strategy, and where that person

---

[*] For example, Growing Hope Together, Inc. (www.growinghopetogether.org) is a non-profit organization established in a rural community in upstate New York. Its mission is to connect children, families, schools, and communities in different cultures. It is launching a coordinated cross-cultural school gardening initiative in Newfield, NY Central school and Simenya primary and secondary schools in Nyanza Province in western Kenya. The project is designed to improve experience and knowledge in agriculture while introducing nutritious foods into the diets of these children. This is only one of thousands of small non-government organizations (NGOs) now doing work once left to a few larger institutions and government agencies. The reader is encouraged to find one or start one in his/her own area.

may fall short, the breadth of acquired knowledge often helps the individual appreci-
ate and better utilize any outside expert advice.

An important aspect of Wanja's approach to farming, nutrition, and health is that it is
robust: It provided more nutrients to children without polluting the waters, depleting the
soils, or concentrating toxicant. And, it withstood colonial rule, independence, changes in
government, and at least one coup attempt. That is the very definition of sustainability.

EXAMPLE 2: IODINE IRRIGATION

In many regions of the world iodine (I) is too deficient in the environment to support
optimum crop, livestock, and human health. Much of the center of the Asian continent
is iodine deficient. When humans settle in regions particularly poor in iodine, developmental delays, cretinism, and increased infant mortality result, as well as problems
with agricultural production, particularly livestock production. Hotien county in the
southern part of Xinjiang province in China was just such a place. Attempts at curing the population with traditional salt and dietary supplements were ineffective, and
only reached a part of the food chain. After noticing that the area was served by a
limited number of irrigation water sources, a medical team rigged a potassium iodate
(5% solution) drip into the irrigation system sufficient to deliver water bearing iodine
at 10 to 80 µg/L to the entire supporting agroecosystem for 2 to 4 weeks (Cao et al.
1994). This application of iodine to the soils providing food for 37,000 people in three
townships resulted in a 50% reduction in infant mortality and a 43% increase in sheep
production (DeLong et al. 1997). These effects persisted for at least 4 years after the
initial treatment and were extremely cost-effective (less than U.S. 5 cents per person
per year) (Ren et al. 2008).

This incredible result is a rare example of an inexpensive, simple solution to a seri-
ous nutritional problem that will not work everywhere, but it illustrates the rewards
of paying attention to the specific properties of the ecology of the area. Failure to pay
attention to the environment and the efforts of multiple well-meaning organizations
could easily have resulted in oversupplementation with iodine in a number of loca-
tions. For example, enthusiastic salt supplementation, overlaid with imported food
and supplements from a variety of sources resulted in a very high incidence of iodine-
induced hyperthyroidism in Port Sudan on the Red Sea (Izzeldin et al. 2007). Although
iodine deficiency in much of Sudan has been noted and targeted by the medical com-
munity, this coastal site was overdosed.

EXAMPLE 3: READY-TO-USE THERAPEUTIC FOOD

The turn of the twenty-first century has brought an elegant solution to the problem of
treating childhood malnutrition in developing countries: ready-to-use therapeutic food
(RUTF). RUTFs are shelf stable at high tropical temperatures and deliver everything a
malnourished child needs to recover while at home with his or her family (Manary et
al. 2002). This home community-based therapeutic care prevents exposure to disease
in a hospital setting and frees the mother and rest of the family to continue most of
their normal activities while the child is getting better.

One class of RUTF is based on peanuts, but includes milk powder, oil, sugar, and
supplementary vitamins and minerals. In addition to having high-quality protein
compared with many plants, peanuts, unlike other nutrient-rich legumes, require very
little processing (light roasting at most) to release those nutrients for use by human
consumers and can be eaten raw. Other legumes require soaking, wet cooking, and
other expensive, time-consuming steps to become edible. And, virtually all alternative

plant protein sources usually suggested are also susceptible to aflatoxin infection. Unfortunately, importation from developed countries is expensive, creates dependency, and is no guarantee of safety: maize and peanuts exported from the U.S. and Europe can also be a source of this toxin (Flett, 2001). For these reasons, populations who can grow them prefer local peanuts as a protein source.

Peanut-based RUTF products have been effective in treating child malnutrition throughout the world (Briend et al. 1999, Maleta et al. 2004, Sandige et al. 2005, Kuusipalo et al. 2006), such as in Ethiopia, Sudan, Malawi, and Haiti. Available under various names, this particular type of RUTF is manufactured both in developed countries for sale and distribution by NGOs and by local organizations in the very countries where they are needed. Local manufacture and sourcing of ingredients should be the goal wherever possible: The short life spans of aid programs, the short attention span of world leaders with regard to food aid, the unpredictability of world food trade policies, and common sense dictate that a successful, long-term sustainable program focused on the elimination of malnutrition be based on locally available materials as much as possible. This is especially true of countries similar to Haiti, an island nation with little ability to generate the foreign exchange needed to purchase expensive food ingredients from overseas.

The greatest challenge to the manufacture of specialty food such as RUTF in a third-world setting is to ensure the identity, safety, and consistency of such a product without access to the equipment, reagents, personnel, and laboratory infrastructure that is standard in developed countries. To do so requires using local resources and the availability of critical information.

EXAMPLE 4: FOOD SAFETY IN LIMITED-RESOURCE SETTINGS

A good example of effecting quality assurance in a limited-resource environment is the reduction of aflatoxin contamination of RUTF in a Haitian malnutrition program: In 2006, in Haiti, my laboratory discovered high levels (380–1556 parts per billion [ppb]) of aflatoxins in peanuts used for the preparation of a ready-to-eat therapeutic food and in the RUTF itself. (As a point of reference, anything more than 20 ppb is illegal to feed to U.S. cows.) These are extremely high levels, potentially harmful, quite variable, and completely unacceptable. The general population was probably exposed to these high levels on a regular basis. The local RUTF manufacturers became more selective in the farmers they chose to buy peanuts from and added an enteroabsorbent clay (bentonite) to the formulation. The process helped some, but there was never much evidence that the bentonite was binding the aflatoxins.

Why were we so concerned about this compound in the peanuts? Aflatoxin is a potent mycotoxin produced by Aspergillus flavus *and* Aspergillus parasiticus. *This toxin is produced in a variety of cereals, root crops, spices, oil seeds, and pulses but most commonly reported to reach the human population directly through the ingestion of maize and peanut products (Williams et al. 2004). Aflatoxin suppresses the immune system (Turner et al. 2003), interferes with the uptake and absorption of nutrients (Shane 1993), stops protein synthesis, stunts children (Gong et al. 2002, 2003, 2004), destroys livers (Gorelick et al. 1993), and at lower chronic intakes, leads to liver cancer (Egner et al. 2001). In Haiti, there was little awareness of aflatoxin in the health care community or the general population.*

Since Haitian manufacturers of peanut butter and peanut-based RUTF source local peanuts from farmers unaware of aflatoxin's threat, there was a special need to be sure this special product for vulnerable, malnourished children is as free of aflatoxins as possible. Once informed of aflatoxin threat, local producers began sorting the peanuts visually and removing kernels that floated in water. They began to make steady progress. In September 2006, manufacturers found market peanuts (rejected and never included in product) that analyzed at 412.5 ± 32.1 ppb. In November 2006, after stringent bulk selection on a farm in Port Margot, they found peanuts with 26.8 ± 7.0 ppb, still above U.S. standards but much better. In January 2007, controlled experimentation with sorting and floating procedures in Cap Haitien resulted in a peanut supply that tested at 0.20 ± 0.10 ppb, showing that Haitians can produce a peanut supply that is virtually free of aflatoxin by applying some low tech, but effective methods. Developing countries do not have to settle for a food supply contaminated with this toxin, despite statements to the contrary from workers supporting the use of drugs, herbs, and other approaches.

This success created another problem: What happens to highly contaminated rejected peanuts? All of the contamination became concentrated in very few peanuts, and because of widespread hunger and poverty, these rejected peanuts would find their way back into the food chain and be even more poisonous to whoever ate them than the lot from which they were sorted. If some way could not be found to destroy the peanuts and shells while retaining some of their economic value, then the sorting procedure would protect children at the expense of others in the community. Fortunately, by March 2007 we successfully tested cooking fuel briquettes made from the shells, rejected peanuts, and cassava flour glue so the toxin can be burned usefully. These are not charcoal; they are direct burn resources to exploit the heat value of the peanut oil. In addition, a local aluminum recycler used the oil-rich rejected peanuts to help fire a small foundry for making cooking pots.

Thus childhood malnutrition was alleviated by an innovative local product that served as a critically needed source of cooking fuel. But, if the last step had not been taken, the entire scheme threatened to increase rather than decrease the health of the population. Burning the bad peanuts became the linchpin of the system that permitted better nutrition and health. (The reader can imagine the reaction of traditional health agencies such as NIH to requests for funds to create briquettes to help cure kwashiorkor.)

YOUR PROGRAM

How will you serve the cause of providing adequate good-quality food to those who currently lack it? Will you join a government agency, found a micro-NGO, or conduct lab bench research to produce a better food preservative, supplement system, or cure for a devastating plant pest? Will you become a commercial producer of crops and livestock or a manufacturer of food products destined for third-world countries? One thing that is almost certain is that you will not do exactly the same thing as in the four examples of successful projects just discussed.

What we hope the reader takes from those examples includes

1. If serendipity and your powers of observation indicate a simple solution to the problem you wish to solve, find out if the people will try it. Iodine is not for every watershed, but you may find something that is.
2. It may take a lifetime to build the experience, infrastructure, and supportive following needed to be effective in this field. Start early.
3. Be a master at a few things, but do not limit your journeyman-level knowledge of many other fields. Your 10 years of education as a microbiologist

may be at the heart of your contribution to helping a country improve dairy foods, but your skill at installing a rainwater collection system, fixing a tractor, and raising chickens along the way may be the extra bits that make an integrated program work.

FINAL OBSERVATIONS

This text has presented both barriers to a universal adequate food supply and a wide range of strategies to overcome them. There are two more barriers, perhaps the most important, now that we have the technical knowledge to feed everyone well. Those two barriers are cynicism and dogmatism.

Peter, Paul and Mary once sang of killers and cynics as dancing partners (Yarrow 1972). That poetic image still informs us, as we work to eliminate deadly food shortages once and for all. One unfortunate aspect of the culture and practice of too many engaged in international development work is that a cynical view of efforts to improve food systems around the world is frequently seen to be more wise and knowing than an optimistic outlook. The roots of this bias are beyond the scope of this chapter and my capabilities, but the damage to motivation and morale brought on by the derision of the enthusiastic among us is beyond dispute. Certainly, there have been failures along the way in the attempt to make sure everyone has a sustainable, accessible food supply, but there are successes, too, many of them. Poverty and hunger have not been abolished, but they are in retreat. Even in Africa, where there have been recent reverses, a growing sense of purpose among the people, experience from Asia and Latin America, together with spectacular improvements in communication at all levels show great promise of helping to prevent the repetition of mistakes and promote the replication of successes.

Dogma can also be a barrier. The world is a complex diverse place, and the probability that a narrow set of solutions is going to solve malnutrition at a given location is slim. Free trade might be just the ticket for some countries, nurturing protectionism best elsewhere. Some ecosystems will support great improvements in rangeland management; others are far too constrained by climate or soils. Indigenous knowledge may be a rich source of ideas to restore a country's food system torn by war or famine or a shallow pool of superstition preventing it. Do you send money or send food? It depends. Fortunately, the specter of hungry children and starvation tends to favor pragmatism over ideology, if allowed. Be well advised to align yourself with organizations that permit innovation and flexibility over dogma, even if they are not as well funded. To do so will prevent cynicism.

STUDY QUESTIONS

1. How do you find out if malnutrition exists and what causes it in a given community?
2. Give an example of how a nutrient deficiency can be reversed without addressing the original cause?

3. List at least three places in the world where you could, by personal action, help reverse malnutrition. Include the problem, the cause, the apparent solution, and barriers to that solution.
4. If you find that the solutions may fall outside your area of expertise, what are your options?

NOTE

1. Wanja had one of the few methane digesters that actually worked well in the country in 1982. He had discovered that adding hog manure to cattle manure was needed to keep the fermentation going. The feces from coarse forage-fed zebu cattle lacked some of the substrate needed to sustain productive fermentation. Fudumi clinic lights, kitchen, and lab ran on the low-pressure methane resulting from the proper mixture.

REFERENCES

Briend, A., R. LaSalle, C. Prudhon, B. Mounier, Y. Grellety, and M.H.N. Golden. 1999. Ready-to-use therapeutic food for treatment of marasmus. *Lancet* 353:1767–1768.

Cao, X.Y., X.M. Jiang, A. Kareem, Z.H. Dou, M. Abdul Rakeman, M.L. Zhang, T. Ma, K. O'Donnell, N. DeLong, and G.R. DeLong. 1994. Iodination of irrigation water as a method of supplying iodine to a severely iodine-deficient population in Xinjiang, China. *Lancet* 344(8915):107–110.

DeLong, G.R., P.W. Leslie, S.H. Wang, X.M. Jiang, M.L. Zhang, M. Rakeman, J.Y. Jiang, T. Ma, and X.Y. Cao. 1997. Effect on infant mortality of iodination of irrigation water in a severely iodine-deficient area of China. *Lancet* 350(9080):771–773.

Demment, M.W., M.M. Young, and R. Sensenig. 2003. Providing micronutrients through food-based solutions: a key to human and national development. *J. Nutrition* 133:3879S–3885S.

Egner, P.A., J. B. Wang, Y. R. Zhu, B. C. Zhang, Y. Wu, Q. N. Zhang, G. S. Qian, S.Y. Kuang, S.J. Gange, L.P. Jacobson, K.J. Helzlsouer, G. S. Bailey, J. D. Groopman, and T.W. Kensler. 2001. Chlorophyllin intervention reduces aflatoxin-DNA adducts in individuals at high risk for liver cancer. *Proc. Natl. Acad. Sci. U. S. A.* 98:14601–14606.

Flett, B. 2001. Maize our staple food. Should toxins concern us? Science in Africa. November 2001. Available at http://www.scienceinafrica.co.za/2001/november/maize.htm

Gong, Y.Y., K. Cardwell, A. Hounsa, S. Egal, P.C. Turner, A.J. Hall, and C.P. Wild. 2002. Dietary aflatoxin exposure and impaired growth in young children from Benin and Togo: cross sectional study. *BMJ* 325:20–21.

Gong, Y.Y., S. Egal, A. Hounsa, P.C. Turner, A.J. Hall, K.F. Cardwell, and C.P. Wild 2003. Determinants of aflatoxin exposure in young children from Benin and Togo, West Africa: the critical role of weaning. *Int. J. Epidemiol.* 32:556–562.

Gong, Y., A. Hounsa, S. Egal, P.C. Turner, A.E. Sutcliffe, A.J. Hall, K. Cardwell, and C.P. Wild. 2004. Postweaning exposure to aflatoxin results in impaired child growth: a longitudinal study in Benin, West Africa. *Environ. Health Perspect.* 112:1334–1338.

Gorelick, N.J., R.D. Bruce, and M.S. Hoseyni. 1993. Human risk assessment based on animal data: inconsistencies and alternatives. In: *The Toxicology of Aflatoxins: Human Health, Veterinary and Agricultural Significance,* D. Eaton and J. Groopman, Eds. London: Academic Press, pp. 508–511.

Grillenberger, M., C.G. Neumann, S.P. Murphy, N.O. Bwibo, P. van't Veer, J.G.A.J. Hautvast, and C.E. West. 2003. Food supplements have a positive impact on weight gain and the addition of animal source foods increases lean body mass of Kenyan schoolchildren. *J. Nutr.* 133:3957S–3964S.

Izzeldin, H.S., M.A. Crawford, and P.L. Jooste. 2007. Population living in the Red Sea State of Sudan may need urgent intervention to correct the excess dietary iodine intake. *Nutr. Health* 18(4):333–341.

Kuusipalo, H., K. Maleta, A. Briend, M. Manary, and P. Ashorn. 2006. Growth and change in blood haemoglobin concentration among underweight Malawian infants receiving fortified spreads for 12 weeks: a preliminary trial. *J. Pediatr. Gastroenterol. Nutr.*43:525–532.

Maleta, K., J. Kuittinen, M.B. Duggan, A. Briend, M.J. Manary, J. Wales, T. Kulmala, and P. Ashorn. 2004. Supplementary feeding of underweight, stunted Malawian children with ready-to-use food. *J. Pediatr. Gastroenterol. Nutr.* 38:152–158.

Manary, M.J., M. Ndekha, A. Briend, and P. Ashorn. 2002. Home-based Therapy using ready-to-use food in the treatment of severe childhood malnutrition. Poster presentation, 2002 Pediatric Academic Societies' meeting, Baltimore, MD, May.

Ren, Q., J. Fan, Z. Zhang, X. Zheng, and G.R. DeLong. 2008. An environmental approach to correcting iodine deficiency: supplementing iodine in soil by iodination of irrigation water in remote areas. *J. Trace Elem. Med. Biol.* 22(1):1–8.

Sandige, H., M.J. Ndekha, A. Briend, P. Ashorn, and M.J. Manary. 2005. Home-based treatment of malnourished Malawian children with locally produced or imported ready-to-use food. *J. Pediatr. Gastroenterol. Nutr.* 39:141–146.

Shane, S.M. 1993. Economic issues associated with aflatoxins. In: *The Toxicology of Aflatoxins: Human Health, Veterinary and Agricultural Significance,* D. Eaton and J. Groopman, Eds. London. Academic Press, pp. 513–517.

Turner, P.C., S.E. Moore, A.J. Hall, A.M. Prentice, and C.P. Wild. 2003. Modification of immune function through exposure to dietary aflatoxin in Gambian children. *Environ. Health. Perspect.* 111:217–220.

Whaley, S.E., M. Sigman, C. Neumann, N. Bwibo, D. Guthrie, R.E. Weiss, S. Alber, and S.P. Murphy. 2003. The impact of dietary intervention on the cognitive development of Kenyan school children. *J. Nutr.* 133: 3965S–3917S.

Williams, J.H., T.D. Phillips, P.E. Jolly, J.K. Stiles, C.M. Jolly, and D. Aggarwal. 2004. Human aflatoxicosis in developing countries: a review of toxicology, exposure, potential health consequences and interventions. *Am. J. Clin. Nutr.* 80:1106–1122.

Yarrow, P. 1972. Greenwood. In *Peter* (album). Burbank, CA: Warner Brothers Music.

Index